Biotechnology in Horticulture

Biotechnology in Horticulture
Methods and Applications

Edited by
Prof. K.V. Peter, Ph D FNASS.FNASc. FNABS.FHSI.
Director
World Noni Research Foundations
No. 12, Srinivasa Nagar, 2nd Street
World Wellness Open University Building
Rajiv Gandhi Road, Old Mahabalipuram Road
Perungudi, Chennai-600096.
Tamil Nadu, India

NEW INDIA PUBLISHING AGENCY
New Delhi – 110 034

NEW INDIA PUBLISHING AGENCY
101, Vikas Surya Plaza, CU Block, LSC Market
Pitam Pura, New Delhi 110 034, India
Phone: + 91 (11)27 34 17 17 Fax: + 91(11) 27 34 16 16
Email: info@nipabooks.com
Web: www.nipabooks.com

Feedback at feedbacks@nipabooks.com

ISBN : 978-93-81450-91-8

Composed and Designed by NIPA

Devotion

I devote the book "Biotechnology in Horticulture : Methods and Applications" to my late Sister Miss. Elizabeth Varkey Kuruppacharil Kumbalanghy who left us from the laps of my brother Prof. K V Thomas and myself during sleep 60 years ago. I also devote this series to the memory of Late Rev. Sr. Mary Jacob (Sr Alice), my niece who led a divine life and is now with the Almighty.

Dr. K.L. CHADHA
Ph.D. (Hort.). D.Sc. (Honoris causa), FNAAS, FHSI
Formerly
National Professor (Hort.) ICAR
Deputy Director General (Hort.) ICAR
Horticulture Commissioner, Ministry of Agri., Govt. of India
Executive Director, National Horticulture Board, Ministry of Agri., Govt. of India
Director, Indian Institute of Horticultural Research (ICAR), Bangalore
Project Coordinator (Fruits), ICAR, Lucknow

Residence: House No. 7281, Block-B, Pocket-10
Vasant Kunj, New Delhi-110 070
Phone: (011) 26896529
Telefax: (011) 26124702
Mob.: 9312005646
E-mail: klchaha@yahoo.com
klchadha@gmail.com
Office: The Horticultural Society of India
F-1, Societies Block, NASC Complex
New Delhi-110 012
Tel.: (011) 25842127
E-mail: hsi42@rediffmail.com

President

The Horticultural Society of India, New Delhi (1990-till date)

Orchid Society of India, Chandigarh (1990-1996)

Indian Society of Ornamental Horticulture, New Delhi (1990-1998)

Mushroom Society of India, Solan, H.P. (1991-1998)

Bougainvillea Society of India, New Delhi (1993-1996)

Indian Soc of Agril. Sciences, New Delhi (1994-1996)

Indian Society of Oil Palm, Pedavegi, A.P. (1996-1998)

Association for the Improvement in Production & Utilization of Banana, Trichy, Tamil Nadu (1996-2007)

Vice-President

National Academy of Agricultural Sciences, New Delhi (2000-2002)

Indian Society of Veg. Sciences, Varanasi, U.P. (1986-till date)

Rose Society of India, New Delhi (1989-1996)

Delhi Agri-Horticultural Society, New Delhi (1987-1996)

Delhi Dahlia Society, New Delhi (1993-1996)

Chief Patron

All India Kitchen Garden Association, New Delhi (2000-till date)

Country/Member

International Society for Horticultural Sciences (ISHS), Belgium (1982-2002)

Foreword

As per International Society of Horticulture Science, horticultural crops include tree, bush and perennial vine fruits; perennial bush and tree nuts; vegetables, aromatic and medicinal plants; cut flowers, potted ornamental plants and bedding plants, trees, shrubs turfs and ornamental grasses; honey and edible fungi. The Planning Commission of India expanded the coverage of horticulture as "the science of growing and management of fruits, vegetables including tubers, ornamentals, aromatic and medicinal crops, spices, plantation crops and their processing, value addition and marketing". Right to food is incomplete with out a nutritious food supplemented with fruits, vegetables, spices, medicinal and aromatic herbs. Sixty five percent of the world's anaemic among women and children are in India and 50% of the World's anaemic are Indians, a matter of national shame despite the possibility of growing fruits, vegetables, spices and medicinal and aromatic through out the country. Horticulture accounts for 30% of India's Agricultural GDP from 13.08% of cropped area. It also provides about 37% of the total exports of agricultural commodities. A growth rate of 6% was targeted for the IX plan onwards, but the achievement was less than 4% even in XIth plan. This is a matter of concern despite increased financial allotment of Rs 24.7 crores in IVth Five year plan to Rs 16,000 in XIth plan besides allocation for tea, coffee and rubber. There has been significant strengthening in horticultural R and D infrastructure. At present there are 10 central institutes, 13 project directorates, 13 AICRPs besides State Horticulture Universities in HP, AP, TN and Karnataka and separate colleges for horticulture in private and government sectors. National Horticulture Board(NHB), APEDA, NCDC, NAFED, NHRDF, SFAC, Bee Board, Coffee Board, Tea Board, Rubber Board, Coconut Development Board, Coir Board and Spices Board are a few organizations established to strengthen the respective sectors. Central sector schemes like National Horticultural Mission, Technology Mission for North Eastern and Himalayan States, Bamboo Mission etc were established to fill up critical gaps in development.

There has been explosion in generation and flow of scientific information pertaining to horticulture as science and technology. Horticulture Science includes science of fruits, vegetables, ornamentals, spices and plantation crops, aromatic and medicinal plants, mushrooms, honey and bamboos. Good number of scientific societies in horticulture started promoting horticulture by publishing scientific journals and also holding National and International Symposia/ Seminars. The Horticultural Society of India established in January, 1942, which I have the privilege to preside over is one of the oldest scientific societies. With advancement in science and technology, there has been significant demand for books and serials on Horticulture Science especially molecular biology, breeding for resilieance, bio-fortification; ideotypes for mechanization; amenability for long term storage; novelty, uniformity, distinctiveness and stability etc.

Dr. K.V. Peter Director, World Noni Research Foundation and Former Professor of Horticulture Kerala Agriculture University envisioned to edit and author a Horticulture Science Series published by New India Publishing Agency, New Delhi. The present book "Biotechnology in Horticulture : Methods and Applications" covers 16 chapters authored by 24 scientists from 13 institutions from India, Japan and Taiwan. I congratulate Dr K.V. Peter and NIPA New Delhi for this important initiative.

K L Chadha

(K L Chadha)

Centre for Natural Biological Resources and Community Development

CNBRCD

Dr. D Joseph Bagyaraj

M.Sc. (Ag.), Ph.D., FNA., F.N.A.Sc., F.N.A.A.S., F.N.A.B.S., F.P.S.I., F.A.M.I., F.M.S.I., F.S.A.B.

NASI Sr. Scientist & Chairman, CNBRCD, Bangalore

Preface

The title Biotechnology In Horticulture : Methods And Applications is interesting to breeders as well as microbiologists. Microbes were considered causing diseases both to plants, animals and of course human beings. Discovery of endophytic bacteria, several antibiotics, pollutants scavenging bacteria, fermentation facilitating microorganisms especially yeasts and now biotic food manufacturing lactic acid bacteria are revolutionizing food and feed industry. Once the food base is horticultural crops-vegetables, fruits, plantation crops, spices, mushrooms, products of microbial actions are as fascinating as yogurts, cheeses and several biotic foods to meet the palate of people of all ages, especially the sick and bed ridden. Cabbage preserves, capsicum, chilly and vegetable pickles and fermented fruit juices-mango, pineapple, apple, pear, grapes, gooseberry, black plum (jamun) and a plethora of products are enriching the nutritional value of daily diets. The present book has two exclusive chapters on biotic foods.

With GM technology, plant breeding has taken a new turn -lesser time and tailored to add/delete genes. Food safety of GM plant based diets is a matter of debate in India.

I congratulate Dr K.V. Peter, a long time colleague of mine, for the efforts made to collect 16 chapters from 24 eminent scientists from 14 Research Institutions and transform to an outstanding book.

41, RBI Colony, Anand Nagar, Bangalore - 560 024, India Ph : +91 080 23335368 Fax : +91 080 23335368 Email : djbagyaraj@vsnl.com / jeravi@bgl.vsnl.net.in

I congratulate New India Publishing Agency, New Delhi for the excellent printing.

Dr. Joseph Bagyaraj
NASI Sr. Scientist & Chairman
CNBRCD
41 RBI Colony
Anand Nagar
Bangalore - 560 024

Acknowledgement

I acknowledge with gratitude Prof.P I Peter Chairman NoniBiotech, Chennai for the encouragement and facilities provided for my ventures in science journalism. He gave me free time and provided academic environment to edit the several series. I express my gratitude to Dr Kirti Singh Chairperson, World Noni Research Foundation for all the academic encouragement. I profusely thank my wife Vimala and sons Anvar and Ajay and daughter in law Cynara for the encouragement. I also acknowledge the scientists who contributed the 16 chapters.

My special gratitude to Padma Shri Dr K L Chadha, President Horticultural Society of India for contributing Foreword to the book. I express gratidue to Dr. D.J. Bagyaraj, NASI Senior Scientist and micro biologist for writing Preface. I also express gratitude to the publisher New India Publishing Agency, New Delhi -for meticulous printing and publishing.

List of Contributors

D.J. Bagyaraj: NASI Senior Scientist and Chairman, CNBRCD, 41,RBI Colony Anand Nagar, Bangalore-560 024, Karnataka, India

Debashis Chakraborthy: Division of Agricultural Physics, Indian Agricultural Research Institute, New Delhi-110 012, India

Deepu Mathew: Centre for Plant Biotechnology and Molecular Biology,Kerala Agricultural University, P O KAU, Thrissur Kerala-680656, India

Dinesh Kumar: Directorate of Wheat Research, Karnal-132 001, Haryana India

E V Soniya: Plant Molecular Biology Division, Rajiv Gandhi Centre for Biotechnology, Thycaud Post, Poojappura, Thiruvananthapuram-695 014 Kerala, India

K. Nirmal Babu: Indian Institute of Spices Research, P O Marikunnu Calicut-673 012, Kerala, India

K.V. Peter: Director, World Noni Research Foundation, P O Perungudi Chennai-400 096, Tamilnadu, India

L. Pugalendhi: Department of Vegetable Crops, Horticulture College and Research Institute, TNAU, Coimbatore-641 003, Tamilnadu India

M.K.Naik: Department of Plant Pathology, University of Agricultural Sciences Raichur-584 102, Karnataka, India

Neelima Garg: Central Institute for Subtropical Horticulture Rehmankhera Lucknow-226 101, Uttar Pradesh, India

Parveen Kumar: Central Soil Salinity Research Institute, Karnal-132 001 Haryana India

R. Srinivasan: Entomologist, AVRDC – The World Vegetable Center Shanhua Tainan, Taiwan, (ROC)

R. Sujatha: College of Horticulture, Kerala Agricultural University, P O KAU Thrissur, Kerala-680 656, India

Renu Agrawal: Department of Food Microbiology, Central Food Technological Research Institute, Mysore-570 020, Karnataka, India

S. Natarajan: Department of Vegetable Crops, Horticulture College and Research Institute, TNAU, Coimbatore-641 003, Tamilnadu, India

S. Praneetha: Department of Vegetable Crops, Horticulture College and Research Institute, TNAU, Coimbatore-641 003, Tamilnadu, India

S. Sudha: Department of Plant Pathology, University of Agricultural Sciences, Raichur-584 102, Karnataka, India

S.K.Pandey: Central Potato Research Institute, Shimla, Himachal Pradesh India

S.V. Singh: Central Potato Research Institute, Shimla, Himachal Pradesh India

Shigeru Satoh: Graduate School of Life and Environmental Sciences Kyoto Prefectual University, Sakyo-ku, Kyoto 606-8522, Japan

Sourav Saha: Division of Agricultural Physics, Indian Agricultural Research Institute, New Delhi-110 012, India

Sunil Pareek: Department of Horticulture, Rajasthan College of Agriculture MPUAT, Udaipur-313 001, Rajasthan, India

Sweda Sreekumar: Plant Molecular Biology Division, Rajiv Gandhi Centre for Biotechnology, Thycaud Post Poojappura, Thiruvananthapuram - 695 014 Kerala, India

T.Vanaja: College of Agriculture, Kerala Agricultural University Padannakad Kasargod-671328, Kerala, India

Takako Narumi: Department of Horticulture, Faculty of Agriculture Kagawa University, Miki-cho, Kagawa, Perfecture 761-0795, Japan

V. Rajashree: Department of Vegetable Crops, Horticulture College and Research Institute, TNAU, Coimbatore-641 003, Tamilnadu, India

V.A. Parthasarathy: Indian Institute of Spices Research, P O Marikunnu Calicut-673 012, Kerala, India

V.K. Singh: Central Institute for Subtropical Horticulture, Rehmankhera Lucknow-226 101, Uttar Pradesh, India

Yusuke Kosugi: Department of Horticulture, Faculty of Agriculture Kagawa University, Miki-cho, Kagawa, Perfecture 761-0795 Japan

Introduction

Horticulture Science includes science of fruits, vegetables, ornamentals, spices and plantation crops, aromatic and medicinal plants and mushrooms. The science has contributed to improving nutritional value to basic foods, aesthetic values to a clean human habitat, generating accelerated values to human endeavours and above all wellness and health. Attempts are made continuously to upgrade productivity and quality of produces through conventional and now using the most cutting edge technologies like transgenics and genetically modified plant system. Conventional breeding methods have stood the test of time and many varieties are coming out of the laboratories to farmers fields. Mode and mechanism of pollination are vital knowledge inputs for breeders and innovative farmers to evolve varieties for a scenario of space saving, water saving and labour saving horticulture. Novelty, distinctiveness, uniformity and stability are the keys to ownership of evolved lines to breeders and farmer -innovators. Vanaja (2012) elaborates on the mode and mechanism of pollination in horticultural crops.

Transgenic technology is made use of to develop novel plants with uniformity, distinctiveness and uniformity in horticultural crops especially ornamental plants. Barriers to gene transfer are broken through gene bombarding using gene guns, use of carriers, recombinant DNA technology and lately by GM technology. Nanotechnology is getting application in development of transgenic plants especially in ornamentals-foliages, flowers, flower buds, greens, palms, aquatic plants, orchids, ferns, trees- and vegetables-tomato, capsicum, eggplant, cucumbers, watermelons, plantation crops-oil palm, rubber-, tubers-cassava, sweet potato-, fruits-strawberry, cherries, apple, plum, pear -and spices-pepper, ginger-. Shiguru *et al.*(2012) elaborates on transgenic ornamentals with reduced ethylene production and perception. Ethylene is a primary plant hormone involved in the senescence of plant tissues including flowers and leaves. In carnation flowers, inhibition of the synthesis or action of ethylene delays the onset of senescence symptoms and lengthens the vase life of the flowers. Leaves of cut chrysanthemum of some cultivars frequently become yellow, sometimes prior to the onset of flower senescence, in response to exogenous ethylene. Generation of transgenic plants with suppressed production or action of ethylene is an excellent way to lengthen the longevity of these ornamental plants. Carnations are used as ornamental plants in the form of potted plants as well as in the form of cut flowers. Preservatives are administered to cut flowers through vascular transport by immersing the stem end in solutions . This administration mars the beauty and look of cut flowers. Generation of transgenic carnations with long shelf life addresses the weakness of physical insertion through cut ends.

Biopiracy and misuse of names of popular varieties/hybrids/ synthetic hybrids are matters of concerns. Convention on Biodiversity, later ratified by 148 nations and subsequent legislations like Intellectual Property Right and Geographycal Indications Act(IPR and Gi), National Biological Diversity Act (NBA), Protection of Plant Variety and Farmers Right Act(PPV and FR act) and establishment of authorities to implement above legislations demand characterization -descriptors, molecular characterization, DNA finger printing, DNA barcoding-. DNA bar coding has come in handy for very detailed and authentic description of flora, fauna-terrestrial, aquatic- and micro organisms. Deepu Mathew(2012) reviewed DNA bar coding and its application in horticultural crops. DNA barcoding is a technique in which species identification is performed by determining and comparing the sequences of several hundreds of base pairs of nucleotides from a small fragment of an identified gene, with the aim of contributing to a wide range of ecological and conservation studies in which traditional taxonomic identification is not practical. Two main objectives of DNA barcoding are to assaign unknown specimens to species and to enhance the discovery of new species and facilitate identification, particularly in cryptic, microscopic and other organisms with complex or inaccessible morphology. In addition, identification of species previously defined by other criteria also be an objective of barcoding. DNA barcoding may also be used for forensic purposes by linking the specimens with already identified data bases. In three situations, species identification must necessarily be molecular based. Determination of taxonomic identification of damaged specimens or fragments, molecular based identification to match adults with immature specimens and in cases where molecular traits do not clearly discriminate species especially if species have polymorphic life cycles and /or exhibit pronounced phenotypic plasticity. 'A consortium for the 'Bar code of Life' based in University of Guelph in Ontario, Canada operates a project 'Barcode of life' which aims to create a universal system for a eukaryotic species inventory based on a standard molecular approach. The DNA Barcode of life data system has progressively been developed since 2004 which enables acquisition, storage, analysis and publication of DNA barcode records. As for horticulture crops, DNA bar coding is an emerging field.

Genome mapping and gene tagging for marker assisted selection accomplish study of genetic organization, knowing the gene/genes influencing a trait, locating a gene on a chromosome, aiding selection of traits and studying the evolutionary relation among the species. Genome mapping is the determination of the relative positions and distances of the genes and other DNA sequences (molecular markers) of a linkage group on chromosomes. The entire linkage groups in an organism showing the specific order of linear arrangement of genes and the distance between them is known as the genome map of that organism. The techniques of genome mapping and linkage analysis were developed in 1911 by D.H.Morgan and his graduate student Alfred H Sturtevant and are used till today in much the same way but with far more advanced technique. Types of genome maps-Genetic/ linkage map, physical-, markers used in genome mapping, statistical concepts behind mapping, how mapping is done, frame work map and saturated map, genetic and physical distances, marker assisted selection and associated problems are discussed by Sujatha and Peter (2012).

Climate change is manifesting in many forms and expressions of living organisms including terrestrial,aquatic,polar and mountainous flora and fauna. Global warming and associated melt down of polar glaciers and ice, inundation of flood water to submerged agrarian lands, sea water rise and unexpected and unpredictable earthquakes and hurricanes, dwindling biodiversity and slow continental shifts are resultant phenomina necessitating intensive research in mission mode. Green house gas(GHG) emission, change in nitrogen:oxygen: carbon dioxide ratio, emission of refrigeration gases and depletion of ozone layer, emittance of infra red radiation, environmental pollution from industries, deforestation (<33% of land area), rift and dislocation of coral around islands, overexploitation of natural resources-coal, petroleum, gas, ores-, overuse of pesicides and weedicides, residues of radio active particles, wars and man made catastrophies are a few events leading to climate change. The earth we live is the 5th largest planet in the solar system and third from the sun. The earth's atmosphere is made of nitrogen(77%) and oxygen(21%). Carbon dioxide consists of a very small fraction(0.03%) but is more vital for the life cycle. This extremely small quantity of CO_2 keeps life on earth by maintaining earth's temperature at a comfortable 14 degree C(without CO_2, earth's temperature will be -21degree C). This effect is called "Green house effect". Green house gases other than CO_2 are nitrous oxide, ozone, methane and fluorinated gases-hydrofluorocarbons, perfluorocarbons and sulfur hexafluorides-. Contributions of these non-CO_2 gases in radiative forcing is commendable. Debashis Chakraborthy and Sourava Saha(2012) review climate change research facilities with special emphasis on CO_2 fertilization effect.

Understanding biology and genetics at molecular level has become very important for better understanding and manipulation of genome architecture and this creates new opportunities to streamline plant breeding. Molecular markers have contributed significantly and are used in plant science in a number of ways, including genetic finger printing, diagnostics, identification of duplicates and selecting core collections, determination of genetic distances, genome analysis, identification of markers associated with desirable breeding traits usefull in marker assisted breeding,genomics and development of transgenics. Nirmal Babu *et al.* (2012) elaborate advances in molecular breeding of perennial spice crops-black pepper, small cardamom, ginger, vanilla and cinnamon-. There are 109 spices as per document ISO 676. These spices impart colour, flavor, fragrance and odour to main food. A few-black pepper, ginger, turmeric, fenugreek- have medicinal properties and are used in Grandammas prescriptions. Biotic stresses like Phytophthora foot rot in black pepper, rhizome rot in ginger and wilts in seed spices-fenugreek, cumin, coriander, fennel, dill- cause considerable losses. Drought is a major abiotic stress where productivity is affected adversely. It is apparent that genetic modifications are necessary to enable these plants to resist stress, thereby paving the way for crop improvement.

Crop productivity is limited by stress factors-biotic and abiotic-. Drought, salinity, high or low temperature, light, deficient or excess nutrients, heavy metals, pollutants etc either or in combination causes abiotic stresses. Biotic stress refers to any stress provided by living organisms, be it unicellular or multicellular. Plants facing a variety of biotic and abiotic stresses are able to protect themselves by means of tolerance mechanisms which result from changes at the molecular level. They are often confronted by a multitude of stresses in their natural environment. To overcome this, plants have evolved various defence strategies, which involve comprehensive

The Technical Advisory Committee(TAC) of the Consultative Group of International Agricultural Research(CGIAR) defined sustainability as "successful management of resources for agriculture to satisfy changing human needs while maintaining or enhancing natural resources". A sustainable system must be sustainable both ecologically and economically. Production of horticultural crops has undergone tremendous change in both developed and developing countries due to use of fertilizers, pesticides, mechanization, soil fumigation, container production in soil-less media, breeding, micro propagation, pathogen elimination and a number of other innovations for attaining future stability and sustainability. Solutions for many of the illeffects of over fertilizer use, pesticides and crop residues lie in new technology with reduced use of fertilizers and pesticides besides maintaining a sustainable level of productivity. One such effort was put forth in the management of microbes which live in the rhizosphere soil around the roots of nearly all terrestrial plants. Mycorrhizal fungi play a vital role in fertility enhancement and disease management. Ectomycorrhizae are common among temperate forest trees and can be cultured in laboratory media. Endomycorrhizae include arbutoid, ericoid, orchid and arbuscular mycorrhizal (AM) forms, the most common and widely occurring of all the mycorrhizal associations and having great economic significance.This cannot be cultured in laboratory media. Bagyaraj(2012) reviews mycorrhizal symbiosis for sustainable horticulture.

Vegetables and fruits are living perishable biological entities performing a number of metabolic functions and have a short life span(Neelima Garg and Singh, 2012). They review carbohydrate metabolism during fruit spoilage. The gradual changes causing deterioration are due to cetain organisms and chemicals in the fruit and outside the fruit. Carbohydrates play a major role in the fruit spoilage process. Major classes of cell wall polysaccharides which undergo modifications during ripening are pectins, cellulose, starch and hemicelluloses. Fruit cell wall is a complex structure composed of polysaccharides imposing cell shape and rigidity. These are carbohydrate polymers consisting of tens to hundreds to several thousand of monosaccharide units. Primary cell walls of plants are networks of cellulose microfibrils embedded in a matrix of hemicellulose and pectin. Cell walls of fruits do not have secondary walls and contain very low amounts of lignins. Cellulose fibres consist of over 500,000 cellulose molecules.

Kumar *et al.*(2012) take a view on potatoes for processing in sub-tropical climates. Potato is one of the four most important food crops of the world, the other three being rice, wheat and maize. It is a nutritive crop producing the highest dry matter, carbohydrates, edible proteins, minerals and vitamins C and B per unit area and time. Potato has high potential to solve the problems of hunger and malnutrition in the country. Being a processable crop to French fries(chips), starches and starch based products, potatoes for processing are special characteristics. Shape, size and depth of eyes; tuber dry matter content and reducing sugar content are parameters determining the processing fitness of a potato variety. Being temperature sensitive, potato is a short duration crop in tropicasub-tropical India while it is a long duration crop in temperate Europe and USA. The Central Potato Research Institute, Shimla(H.P.) has developed processing varieties like Kufri Chipsona-1,2,3; Kufri Himsona and Kufri Frysona. CPRI has also worked out storage technologies for processing potatoes for off season availability of quality potatoes to the industry.

Arid fruits-pome granate, aonla, dates, lasoda, ber, bael-are attracting attention for their higher nutritional value coupled with medicinal uses.Being limited water requiring and tolerant to drought, these fruit trees demand special research and development in the context of desertification and bioprotection and soil conservation.Pome granate is both a fruit and a spice. Sunil Pareek(2012) provides an overview of nutritional composition, health benefits, post harvest physiology and technology of pome granates in relation to maintaining their quality between harvest and fresh consumption or processing.The pome granate fruit is used fresh, processed and put to non-food usages. Processed usages are for tanning, dyeing, ornamentals and essences. Processed forms are juices, syrup, jams or wine. The seeds from wild grown pome granates are used as spices.

Adding value to popular foods is a recent trend. Awareness to innovative probiotic fermented food products containing friendly bacteria is resulting in novel foods with higher nutritive value, higher absorbtive properties and pro-nature functionalities. Probiotic foods are called Foods for Specialized Health Uses(FOSHU). According to

American Dietic Association(ADA), these foods are potential health foods or ingredients which provide health benefits beyond the traditional nutrients. Probiotics(for life)are fermented foods cultured by beneficial micro organisms. Probiotic foods like yogurt and sauerkraut are very familiar. Probiotic functional foods contain significant levels of biologically active components which provide health benefits beyond basic nutrition. Lactic acid bacteria enhance gastrointestinal functions. Renu Agrawal(2012) reports probiotic functional foods in recent times. She also elaborates on recent trends in food biotechnology for value addition with special reference to probiotics. People are moving towards newer trends in food diversification for value addition. Beneficial effects could be either maintenance, promotion or wellbeing of health or reduction of the risk of a pathogenic process or a disease. Genetically modified (GM) crops/foods, algae, mushrooms and probiotics are among the important functional foods which apart from improving health also protect against infections and/or stimulate immune systems. The interactive research between physiology, microbiology, food technology and molecular biology may produce many healthful and diversified value based food products for human consumption.

Aflatoxins are toxins produced by moulds which occur in plants and their edible parts.The word "aflatoxins" was coined in early 1960s when the death of thousands of turkeys and other domestic animals was attributed to presence of *Aspergillus flavus* toxins in ground meal imported from South America. Aflatoxin contamination of agricultural commodities gained global significance as a result of their deleterious effects on human and animal health and its adverse impact on international trade. Aflatoxins are the most potent carcinogens belonging to mycotoxins, most commonly found in chillies, reported as a potent carcinogenic,mutagenic and immunosuppressive agent produced as a secondary metabolite by the fungus *Aspergillus flavus* and *A. parasiticus*. Sudha and Naik(2012) review the aflatoxin contamination in chilli and its management.

Seed as an input contributes to 20-30% of crop productivity. Heterozygote advantage was exploited over years to realize higher performance in cross pollinated crops like maize, cotton, sorghum and pearlmillet. In vegetable crops, heterosis was utilized in brinjal, okra, cucurbits, chillies, cole crops and root vegetables. In fruits- mango, pineapple, guava- heterotic hybrids are reported. In ornamentals, heterosis was reported in carnations, chrysanthemums,impatiens etc. In spices, blackpepper and cardamom have hybrids released. In plantation crops hybrid coconut, arecanut and oil palm are available. Heterosis along with resistance to biotic and abiotic stresses are breeding objectives realized in many crops. Exploitation of hybrid vigour in self pollinated crops like tomato, cowpea,peas and beans is an evolutionary step of genetic significance. Pugalendhi *et al.* (2012) reviews practical application of hybrid vigour taking ashgourd as a typical vegetable.

Srinivasan(2012) reviews safe management of borers in vegetables.Borer pests cause significant yield losses and are one of the most serious production constraints in tropical vegetable production. Eggplant fruit and shhot borer reduces yield as much as 70%. Tomato borer causes about 5 to 55% loss in almost all the tomato producing areas in the world. Up to 80% yield losses are reported in various vegetables and grain legumes due to legume pod borer. Due to higher yield losses, growers tend to apply extremely high quantities of chemical pesticides to control borer pests in vegetables. A survey of pesticide use in Bangladesh indicated that farmers spray up to 180 times with chemical insecticides during a year to protect their eggplant crop against eggplant fruit and shoot borer. Pesticide application often exceeds 50 sprays per tomato crop season in South India. In Bengladesh, lab lab bean is being sprayed at weekly or biweekly intervels -sometimes every day- to control Lab lab Pod Borer. On average, farmers used 10-24 kg/ha of pesticide in cabbage and cauliflower in India to manage diamond back moth and other pests. Safe management of borers including biological control, use of pheromones and attractants are elaborated with case studies.

The book "Biotechnology in Horticulture : Methods and Applications" carries 16 chapters contributed by 24 scientists from 13 Institutions in India, Japan and Taiwan.

Contents

1

Transgenic Ornamentals with Reduced Ethylene Production and Perception

Shigeru Satoh, Yusuke Kosugi and Takako Narumi

Ethylene is a primary plant hormone involved in the senescence of plant tissues including flowers and leaves. In carnation, inhibition of the synthesis or action of ethylene delays the onset of senescence symptoms and lengthens the vase life of flowers. Leaves of cut chrysanthemum of some cultivars frequently become yellow, sometimes prior to onset of flower senescence, in response to exogenous ethylene. Generation of transgenic plants with suppressed production or action of ethylene is an excellent way to lengthen the longevity of these ornamental plants. This paper reviews our studies on generation and characterization of carnations with reduced ethylene production and chrysanthemum with reduced sensitivity to ethylene.

Ethylene is a primary plant hormone involved in the senescence of cut flowers of many ornamentals such as carnation, sweetpea and delphinium. It is produced on a large amount at a later stage of flower senescence. The increased ethylene production accelerates withering and dropping of petals. Inhibition of synthesis or action of ethylene delays the onset of senescence symptoms and extends the vase life of flowers.

In cut flowers susceptible to ethylene, preservatives containing silver thiosulphate anionic complex ion (STS) (Veen 1979) are currently being used to lengthen their vase life. However, STS is harmful to the environment, and may be banned in the near future. Thus, transgenic ornamentals with suppressed production or action of ethylene are excellent alternatives for the preservation of flower longevity. So far, transformation with transgenes related to the biosynthesis and action of ethylene was successfully used to down-regulate ethylene production or responsiveness to ethylene in petunia (Wilkinson *et al.*, 1997), torenia (Aida *et al.*, 1998) and carnation (Savin *et al.*, 1995, Bovy *et al.*, 1999). Cut flowers of these transgenic plants have a prolonged vase life compared with those of non-transgenic plants.

Carnations are used as ornamental plants in the forms of potted plants and of cut flowers. The preservatives described above cannot be used for potted carnation plants, since the chemicals are usually administered to cut flowers through vascular transport by immersing the cut stem end in solutions containing the preservative. Therefore, the generation of transgenic plants is a promising way to both lengthen the display time of potted carnation plants and the vase life of cut flowers of ordinary carnation plants.

Cut chrysanthemum is one of the most popular cut flowers and is of economic importance in the floriculture industry. Leaves of cut chrysanthemum frequently become yellow spontaneously, sometimes prior to onset of flower senescence (Doi *et al.*, 2003, 2004, Ferrante *et al.*, 2003). Moreover, exposure to ethylene during storage, transportation or display of the cut flowers accelerates leaf yellowing although the sensitivity to ethylene differs with the cultivar (Reyes-Arribas *et al.*, 2001, Doi *et al.*, 2003). Leaf yellowing is a characteristic of leaf senescence; it makes cut flowers unsightly, lowers their quality, value and shortens vase life. Therefore, transgenic chrysanthemum with reduced ethylene sensitivity is expected to have very green leaves, and a greatly extended vase life.

1. Transgenic Carnations with Reduced Ethylene Production

1.1. Senescence and ethylene production of carnation flowers

Cut carnation flowers produce a large amount of ethylene several days after full opening of the flower (Borochov and Woodson 1989, Abeles *et al.*, 1992, Reid and Wu 1992). The produced ethylene causes in-rolling of petals and wilting of the whole flower. Therefore the prevention of synthesis or action of ethylene inhibits the onset of senescence symptoms and extends vase life of flower.

Ethylene is synthesized in plant tissues through the following pathway: L-methionine → *S*-adenosyl-L-methionine → 1-aminocyclopropane-1-carboxylate (ACC) → ethylene. ACC synthase and ACC oxidase catalyze the last two reactions (Yang and Hoffman 1984, Kende 1993). Previously, three genes encoding ACC synthase (*DC-ACS1, DC-ACS2* and *DC-ACS3*) and one gene encoding ACC oxidase (*DC-ACO1*) were identified in carnation (Wang and Woodson 1991, Park *et al.*, 1992, Henskens *et al.*, 1994, Jones and Woodson 1999). Recently, Harada *et al.* (2009) identified another ACC synthase gene, *DcACS1b*, which is very similar to *DcACS1a* (renamed from *DC-ACS1*), and Norikoshi *et al.* (2008) that for another ACC oxidase [tentatively *DcACO2(DC-ACO2)*]. The expression of these genes is regulated in a tissue-specific manner in relation to flower senescence. *DC-ACS1* and *DC-ACO1* genes play a crucial role in autocatalytic and massive ethylene production in the petals of carnation flowers undergoing senescence.

So far, transgenic carnation plants with suppressed production or action of ethylene have been generated; e.g., the line transformed with ACC oxidase (*DC-ACO1*) cDNA in antisense orientation (Savin *et al.*, 1995) and the line transformed with an *Arabidopsis thaliana etr1-1* allele capable of rendering ethylene insensitivity (Bovy *et al.*, 1999). Cut flowers of these transgenic lines have a prolonged vase life compared with those of non-transgenic plants.

In a series of our studies, several lines of transgenic carnation were generated by transforming with *DC-ACO1* cDNA in sense orientation (Kosugi *et al.*, 2002) and *DC-ACS1* cDNA in sense orientation (Iwazaki *et al.*, 2004). In addition, carnation plants, which are used as potted plants, were transformed with *DC-ACO1* cDNA in sense and antisense orientation (Kinouchi *et al.*, 2006).

1.2. Transformation vectors

The transformation vectors harboring carnation ACC oxidase cDNA (*DC-ACO1, s/a ACO* transgenes) in sense or antisense orientation were constructed using pMLH2113-GUS (Mitsuhara *et al.*, 1996) and pIG121-Hm (Ohta *et al.*, 1990) (Table. 1). The transformation vector pMLH2113/*sACO* was prepared by replacing a *GUS* gene with the coding sequence of *DC-ACO1* cDNA in sense orientation in pMLH2113-GUS (Kosugi *et al.*, 2000). The transformation vectors, pIG121/*aACS* and pIG121/*sACS*, were prepared by replacing the *GUS* gene with *DC-ACS1* cDNA in sense (*sACS* transgene) or antisense (*aACS* transgene) orientation in pIG121-Hm (Iwazaki *et al.*, 2004). Similary, the transformation vectors, pIG121/*aACO* and pIG121/*sACO*, were prepared using *DC-ACO1* cDNA instead of *DC-ACS1* cDNA (Fig. 1). The transformation vectors have an *NPT II* gene and an *HPT* gene, which enable selection of the transformants in the presence of Kanamycin or Hygromycin, respectively.

1.3 Generation and cultivation of transgenic carnation plants

Kosugi *et al.* (2000, 2002) transformed a carnation cultivar 'Nora' with *DC-ACO1* cDNA in sense orientation (*sACO* transgene) and generated a transgenic carnation line (sACO-1).

Transformation by *Agrobacterium*-mediated gene transfer and regeneration of carnation plants were performed by the method of Firoozabady *et al.* (1995) with modifications. Briefly, leaf explants from shoot culture were mixed with *Agrobacterium* in a nutrient solution [BD medium: MS medium supplemented with B5 vitamins, 3% sucrose, 0.059% MES, 0.5 mg l^{-1} BA and 0.5 mg l^{-1} 2,4-D] for 15 min, and cocultivated on another medium supplemented with 100 mM acetosyringone for a week at 24 °C in the dark. For selection of transformants, Hygromycin at 20 mg l^{-1} was used as a selection agent, since a preliminary observation showed that Kanamycin up to 100 mg l^{-1} allowed frequent escape formation. Survived explants were subcultured repeatedly on solid IT medium [MS medium supplemented with B5 vitamins, 3% sucrose, 0.059% MES, 0.5 mg l^{-1} indolebutyric acid, 0.5 mg l^{-1} thidiazuron and 0.25% Gerlite] until they formed normal shoots with distinct internodes. The shoots were rooted and transplanted to vermiculite in pots, then to soil. The plants were grown in a containment greenhouse until they flowered. Similarly, Iwazaki *et al.* (2004) generated 2 lines of 'Nora' carnation transformed with *DC-ACS1* cDNA in sense orientation (*sACS* transgene) and 3 lines transformed with *DC-ACS1* cDNA in antisense orientation (*aACS* transgene), respectively.

'Lillipot' carnation, which is used as potted plants, were transformed with *aACO and ACO* transgenes (Fig. 1) by the method described above, but the following modifications were done to improve transformation efficiency. Namely, acetosyringone at 100 mM was added to the medium for shoot culture prior to leaf explant preparation and pre-culture of *Agrobacterium* in addition to media for infection and coculture, and water was used instead of BD medium for infection and coculture.

Table 1 : Transformation of 'Lillipot' carnation with constructs harboring *DC-ACO1* cDNA in antisense or sense orientation (*a/sACO* transgenes)

Transformation vector	No. of explants	No. of lines containing *HPT* gene	No. of lines containing *a/sACO* transgenes	*a/sACO* transgene-containing plants/ explants (%)
pIG121/*sACO*	200	4	3	1.5
pIG121/*aACO*	200	10	10	5.0
pMLH2113/*sACO*	200	16	14	7.0
Total	600	30	27	4.5

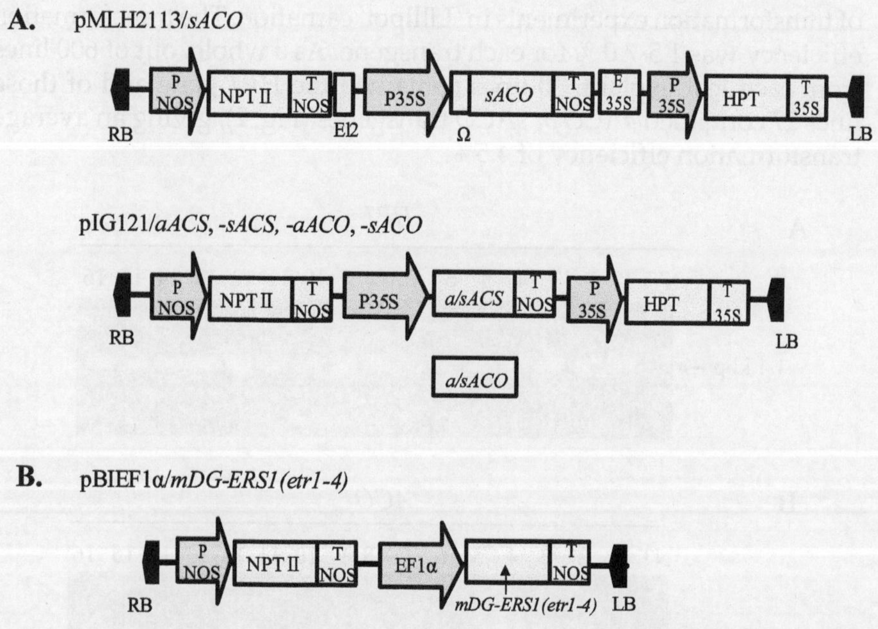

Fig. 1 : Schematic representation of transformation vectors. A. pMLH2113/*sACO*, pIG121/*aACO, -sACO, -aACS* and *-sACS* which harbor carnation *DC-ACO1* cDNA or *DC-ACS1* cDNA in sense or antisense orientation. B. pBIEF1α/*mDG-ERS1(etr1-4)* which harbor a chrysanthemum ethylene receptor *DG-ERS1* cDNA with one-nucleotide substitution corresponding to that present in *Arabidopsis etr1-4*. RB and LB, right and left border of T-DNA, respectively; NPT II, neomycin phosphotransferase II gene; P35S, cauliflower mosaic virus 35S (CaMV 35S) promoter; T35S, CaMV 35S terminator; PNOS, the promoter of nopalin synthase gene; TNOS, the terminator of nopalin synthase gene; E35S,

CaMV 35S enhancer; El2, the upstream sequence (–419~–90) of CaMV 35S promoter x 2; Ω, 5'-untranslated sequence of tobacco mosaic virus; HPT, Hygromycin phosphotransferase gene; *sACS* and *αACS*, *DC-ACS1* cDNA in sense and antisense orientation, respectively; EF1α, the promoter of tobacco elongation factor 1α gene.

1.4 Selection of transformants and transformation efficiency

An example is shown below with the results of transformation of 'Lillipot' carnation. The leaf explants (shoot explants) of 'Lillipot' carnation were transformed with transformation vectors harboring *aACO or sACO* transgenes and subjected to selection with Hygromycin for one month, and subsequent re-differentiation of leaflets for one month. The leaflets were grown further for 3 months to form shoot clusters. The integration of *HPT* gene and *aACO or sACO* transgenes was confirmed by PCR-analysis (Fig. 2) using genomic DNAs extracted from leaflet segments of shoot clusters. Table 1 summarizes the results of transformation experiments in 'Lillipot' carnation. The transformation efficiency was 1.5-7.0 % for each transgene. As a whole, out of 600 lines from each leaf explant, 30 lines contained the *HPT* gene and of those lines 27 contained *aACO or sACO* transgenes (Fig. 2), giving an average transformation efficiency of 4.5%.

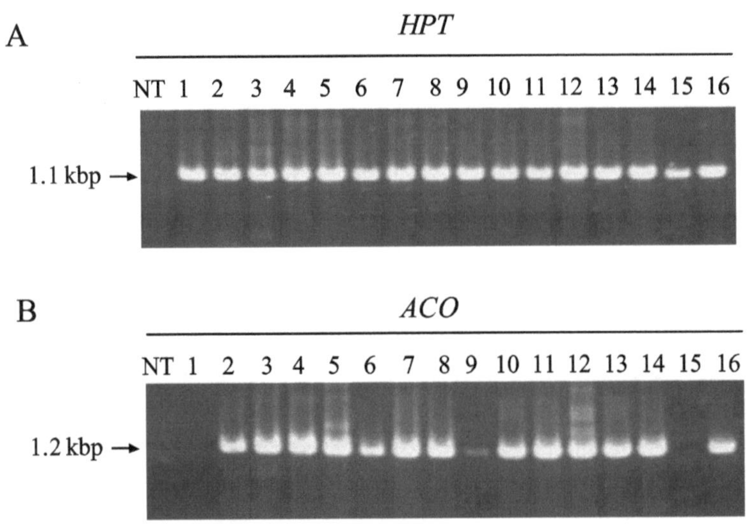

Fig. 2 : PCR analysis of integrated *HPT* gene and *sACO* transgene in 'Lillipot' transgenic lines. Genomic DNAs extracted from lines transformed with pMLH2113/*sACO* were analyzed by PCR with combinations of appropriate upstream and downstream primers for *HPT* gene, and those for *sACO* transgene. The arrows show 1.1 kbp PCR amplificates for *HPT* gene and 1.2 kbp PCR amplificates for *sACO* transgene. NT, non-transformed 'Lillipot'; Nos 1-16, respective lines transformed with pMLH2113/*sACO*.

1.5 Confirmation of integration of transgenes by southern blot analysis

Integration of respective transgenes was confirmed by Southern blot analysis. Preliminary experiments showed that the sACO-1 line harbored at least 5 copies of the *sACO* transgene (Satoh *et al.*, 2000). Among 5 transgenic lines transformed with *sACS* and *a ACS* transgenes (Iwazaki *et al.*, 2004), the sACS-1 line harboring *sACS* transgene had the severest reduction in ethylene production in flowers (see below). Southern blot analysis revealed that the sACS-1 line had two copies of the *sACS* transgene (Fig. 3). Fig. 4 shows Southern blot analysis of four lines of 'Lillipot' carnation transformed with pMLH2113/*sACO*. This analysis indicated the integration of multiple copies of *sACO* transgenes into the lines.

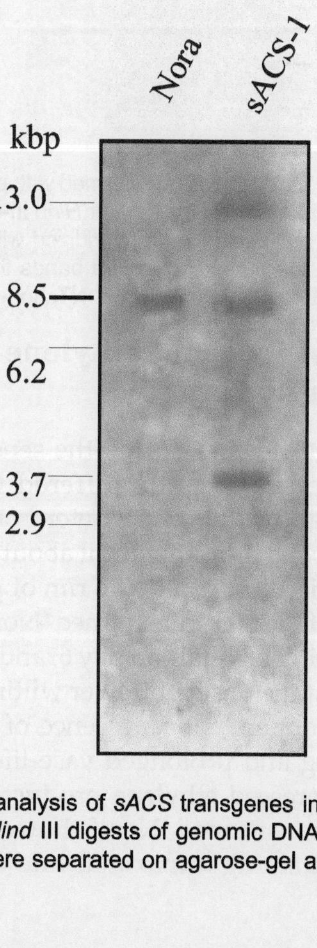

Fig. 3 : Southern blot analysis of *sACS* transgenes in the sACS-1 line derived from 'Nora' carnation. The *Hind* III digests of genomic DNAs from non-transformed 'Nora' and the sACS-1 line were separated on agarose-gel and hybridized with a DC-ACS1 probe (560 bp).

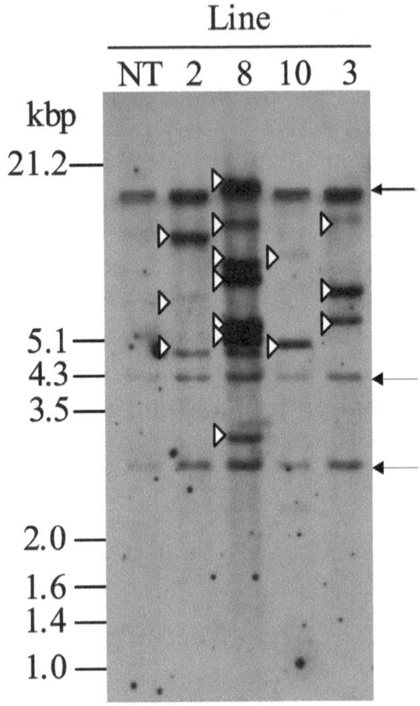

Fig. 4 : Southern blot analysis of 'Lillipot' lines transformed with pMLH2113/*sACO*. Four lines were randomly selected from 14 lines (Table 1) for *Hin*d III-digestion, respectively. The digests were separated on agarose-gel and hybridized with a 986-bp DC-ACO1 OR probe. Open arrowheads in the panel show extra bands that were absent in the digests of the non-transformed 'Lillipot' genomic DNA. NT, non-transformed 'Lillipot'.

1.6 Suppressed senescence and ethylene production of transgenic carnation flowers

The appearance of cut flowers during the senescence period (10 days) in the sACO-1 line significantly differed from that in non-transformed 'Nora' carnation (Fig. 5). Flowers of the sACO-1 line remained turgid without petal in-rolling until about day 10, but started to show desiccation and discoloration in the rim of petals on day 11 or later. By contrast, flowers of the non-transformed 'Nora' remained turgid until day 5, showed in-rolling of petals on day 6, and completely wilted on day 9. Petal in-rolling at the onset of flower wilting is a well-known characteristic of ethylene-dependent senescence of carnation flowers. The lack of petal in-rolling and prolonged vase-life in flowers of the sACO-1 line suggest suppressed ethylene production in the flowers. Actually, flowers of non-transformed 'Nora' showed a climacteric rise in ethylene production attaining a maximal rate on day 5, whereas flowers of the sACO-1 line produced a negligible amount of ethylene during the senescence period (Fig. 6).

Fig. 5 : Senescence of carnation flowers of the non-transformed 'Nora' (top) and the sACO-1 transgenic line (bottom). Flowers of both lines were harvested at the full-opening stage (day 0), and then left in a glass vessel with their cut stem end in distilled water. Photographs were taken 0, 5, 6, 7, 8, 9 and 10 days after the start of the experiment.

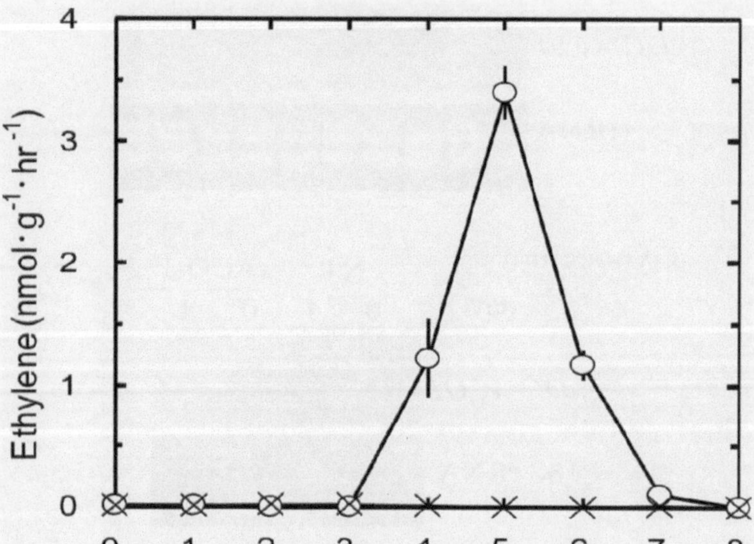

Fig. 6 : Ethylene production from cut carnation flowers during senescence in the non-transformed 'Nora' (control) and the sACO-1 lines. Five flowers each of both lines were harvested at the full-opening stage (day 0) and their ethylene production measured daily. Data are shown by the mean ± SE of 5 flowers. 0, 'Nora' control; ×, sACO-1.

The suppressed ethylene production in the sACO-1 line was accompanied by no accumulation of mRNA for the *DC-ACO1* gene in the gynoecia and mRNAs for the *DC-ACO1* and *DC-ACS1* genes in the petals (Fig. 7). Previously, it was shown that in naturally senescing carnation flowers, ethylene is first produced in the gynoecium; and the evolved ethylene acts on petals, as a diffusible signal, to induce the expression of *DC-ACO1* and *DC-ACS1* genes resulting in autocatalytic ethylene production in the petals (Shibuya *et al.,* 2000). The results indicate that the *sACO* transgene inhibits the expression of *DC-ACO1,* probably by cosuppression, in the gynoecium, which then results in no ethylene production in the whole flower (gynoecium plus petals) of the sACO-1 line.

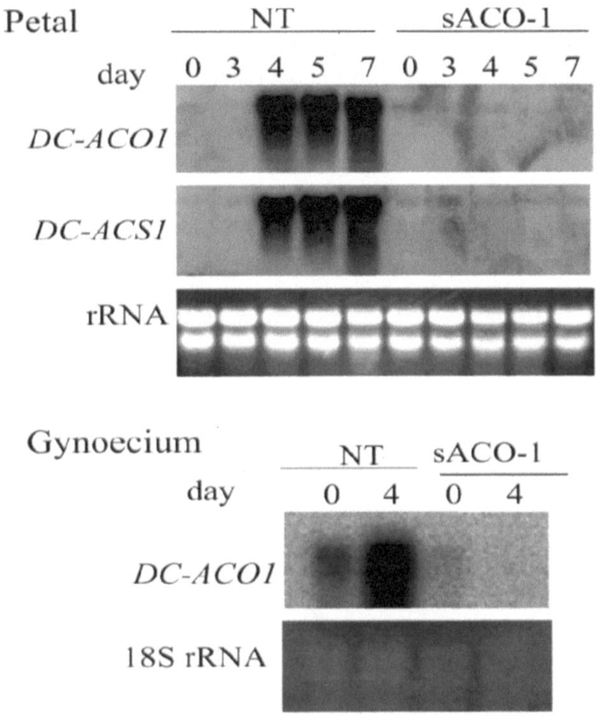

Fig. 7 : RNA gel blot analysis of mRNAs for *DC-ACO1* and *DC-ACS1* in petals and gynoecium of the non-transformed 'Nora' control (NT) and the sACO-1 lines during natural senescence. Petals were isolated from cut flowers 0, 3, 4, 5 and 7 days, but gynoecia 0 and 4 days after full opening of flowers. Total RNAs isolated from the petals were separated on an agarose gel and hybridized to DIG-labeled *DC-ACO1* and *DC-ACS1* probes, but in the case of total RNAs isolated from the gynoecium a ^{32}P-labeled DC-ACO1 probe was used .

Furthermore, cut flowers of 2 lines transformed with *sACS* transgene and 3 lines with *aACS* transgene (Iwazaki *et al.*, 2004) showed suppressed ethylene production during natural senescence as compared with flowers of non-transformed 'Nora' carnation. Among 5 transgenic lines, the sACS-1 line harboring *sACS* transgene had the severest reduction in ethylene production in flowers, and the flowers lost their vase-life by drying and discoloring in the rim of petals, which are typical to ethylene-independent senescence of carnation flowers.

In 5 'Lillipot' lines transformed with *sACO* and *aACO* transgenes, cut flowers of the pMLH-sACO-2 line and -12 lines had vase-lives of 7.6 ± 0.4 and 6.3 ± 0.5 days, respectively, whereas that of the NT control line had 2.8 ± 0.2 days. Other transgenic lines had vase-lives insignificant from that of the non-transformed (NT) control. Flowers of the pMLH-sACO-2 and -12 lines eventually deteriorated with desiccation and discoloration in the rim of petals. Flowers of the NT control showed a climacteric rise in ethylene production, attaining a maximal rate (2.90 $nmol^{-1}$ g^{-1} h^{-1}) on day 3. Flowers of the pMLH-sACO-2 and -12 lines produced very small amounts of ethylene during senescence period of 8 days and their maximum ethylene production rates were around 10% that of the NT control. With pMLH-sACO-2 and -12 lines, *DC-ACS1* and *DC-ACO1* transcripts were not detected in their petals during senescence of 6 days, but they were accumulated abundantly in the petals of the NT control flowers.

1.7 Analysis of ethylene production and wilting in carnation petals using the transgenic sACO-1 line

Ethylene, evolved from the gynoecium, induces autocatalytic ethylene production and wilting in carnation petals. The autocatalytic ethylene production is due to expression of *DC-ACS1* and *DC-ACO1*. Petal wilting in flower senescence is caused by decomposition of cell constituents by hydrolytic enzymes such as proteinase, lipase and nuclease. Cysteine proteinase (CP ase) is probably one of the enzymes responsible for hydrolytic degradation of cell components leading to cell death during senescence of petals (Jones *et al.*, 1995, Panavas *et al.*, 1998, 1999). A CPase gene (*DC-CP1*) is up-regulated during natural and exogenous ethylene-induced senescence of carnation petals. Previously, it was thought that the ethylene production and wilting were regulated in concert and could not be separated since the two events occur simultaneously.

Kosugi *et al.* (2000) examined the response of petals of transgenic carnation line sACO-1 to exogenous ethylene with regard to the induction of petal in-rolling, ethylene production and some genes involved in these events. As described above, cut flower of the sACO-1 line produced only a trace amount of ethylene during the senescence period, and had a vase life about twice longer than flower of the control 'Nora' carnation. Interestingly, the treatment with ethylene of petals detached from the sACO-1 flowers caused accumulation of transcripts for *DC-CP1* and petal wilting, but it did not induce accumulation of transcripts for *DC-ACO1* and *DC-ACS1* and ethylene production (Fig. 8). These findings indicated that the exogenous ethylene-induced expression of *DC-ACS1* and *DC-ACO1* genes, leading to autocatalytic ethylene production, and that of *DC-CP1* genes, leading to petal wilting, were regulated separately in the petals of the sACO-1 flowers. Based on these findings, Satoh *et al.* (2000) suggested the presence of two subset of ethylene responses leading to the expression of respective genes on carnation petals (Fig. 9).

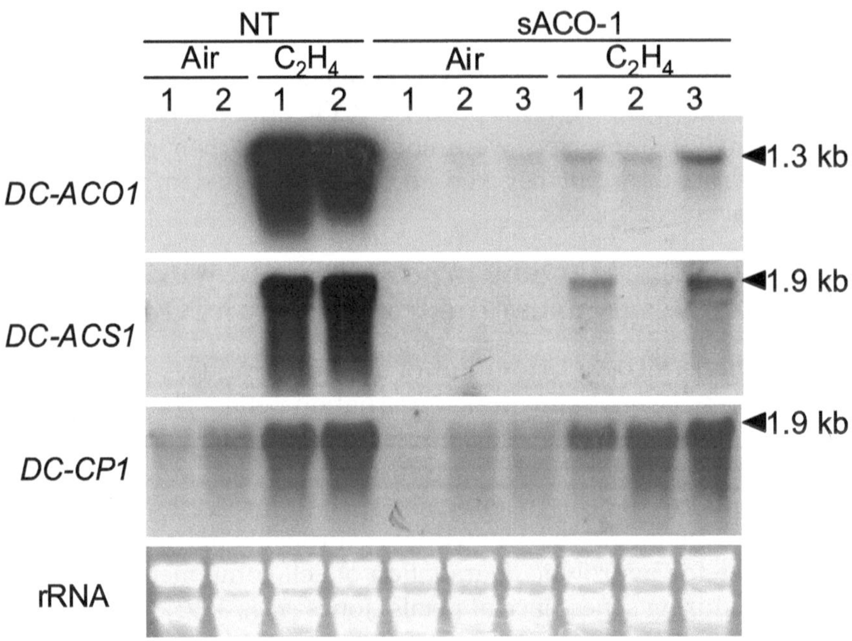

Fig. 8 : Accumulation of transcripts for *DC-ACO1*, *DC-ACS1* and *DC-CP1* in petals of control (NT) and sACO-1 transgenic lines after treatment with ethylene at 10 μl l^{-1} for 18 h. Analysis was performed with petals obtained from two individual flowers (1 and 2) for the control line and three individual flowers (1, 2 and 3) for the transgenic line.

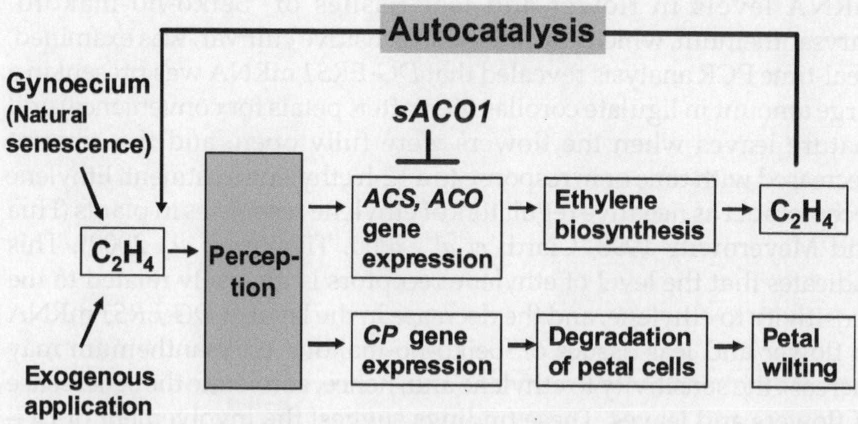

Fig. 9 : Two subsets of ethylene responses in carnation petals. The figure depicts two putative ethylene signaling pathways responsible for autocatalytic ethylene production and wilting during petal senescence in carnation flowers.

2. Transgenic Chrysanthemum with reduced Ethylene Sensitivity

2.1 Leaf senescence and an ethylene receptor gene of chrysanthemum

As described in Introduction, it is possible that leaf yellowing is prevented and the vase life greatly extended in transgenic chrysanthemum plants with reduced ethylene sensitivity. Previously, a mutated ethylene receptor gene (*Arabidopsis etr1-1* gene) was successfully used to reduce sensitivity to ethylene in petunia (Wilkinson *et al.*, 1997) and carnation (Bovy *et al.*, 1999). However, in many cases transgenes of heterologous origin do not work well in chrysanthemum plants, probably because of gene silencing, raising doubt on the use of heterologous transgenes like the *etr1-1* gene in chrysanthemum. To solve this problem, it may be helpful to use ethylene receptor genes obtained from chrysanthemum itself to down-regulate ethylene responsiveness in transgenic chrysanthemum plants, because the native gene might be less affected by the gene-silencing mechanism. Narumi *et al.* (2005a) obtained a cDNA encoding ethylene receptor (DG-ERS1) from chrysanthemum [*Chrysanthemum morifolium*, formerly *Dendranthema grandiflorum* (Ramat.) Kitamura]. The cDNA (2427 bp) contained an open reading frame of 1920 bp coding for 640 amino acids. The predicted DG-ERS1 protein had an amino-terminal ethylene sensor domain and a histidine kinase domain, but lacked a receiver domain. Then, *DG-ERS1* mRNA levels in flower and leaf tissues of 'Seiko-no-makoto' chrysanthemum, which is an ethylene-sensitive cultivar, was examined. Real-time PCR analysis revealed that *DG-ERS1* mRNA was present in a large amount in ligulate corollas (hereafter, petals for convenience) and mature leaves when the flowers were fully open, and the amount decreased with time or in response to a 12-h ethylene treatment. Ethylene receptors act as negative regulators of ethylene responses in plants (Hua and Meyerowitz 1998, Ciardi *et al.,* 2000, Tieman *et al.,* 2000). This indicates that the level of ethylene receptors is inversely related to the sensitivity to ethylene, and the decrease in the level of *DG-ERS1* mRNA in flower and leaf tissues of 'Seiko-no-makoto' chrysanthemum may increase the sensitivity to ethylene and, hence, accelerate the senescence of flowers and leaves. These findings suggest the involvement of DG-ERS1 in the perception of ethylene in cut chrysanthemum plants, especially in those of 'Seiko-no-makoto' cultivar.

2.2 Transformation vectors

Narumi *et al.* (2005b) produced mutated ethylene receptor genes (*mDG-ERS1*s) from the *DG-ERS1* cDNA by introducing one-nucleotide substitutions corresponding to those present in *Arabidopsis etr1-1, etr1-2, etr1-3, etr1-4,* and tomato *Nr,* and introduced each of the *mDG-ERS1*s into chrysanthemum plants. However, only the results with the mutated ethylene receptor gene *mDG-ERS1(etr1-4)* are shown hereafter, since it was revealed that this construct tended to be more effective in conferring reduced ethylene sensitivity in chrysanthemum than the others.

A one-nucleotide substitution was introduced into *DG-ERS1* cDNA with site-directed mutagenesis by PCR using appropriate primers; the mutation replaced A with T at the position 187 bp (when +1 is A of the ATG starting codon of *DG-ERS1* cDNA), resulting in the conversion of Ile-63 in DG-ERS1 to Phe-63 in the resultant mutated protein. This one-nucleotide substitution was the same as that present in the nucleotide sequence of mutated *Arabidopsis etr1-4* gene (Chang *et al.,* 1993).

A transformation vector pBIEF1a/*mDG-ERS1(etr1-4)* (Fig. 1B) was constructed from pBIEF1a by replacing the *GUS* gene with the *mDG-ERS1(etr1-4)* cDNA. In turn, the pBIEF1a vector was constructed from pBI101 (Clonetech) by inserting the promoter of a tobacco elongation factor 1a (EF1a) gene. The construct in the transformation vector was named *EF1a/mDG-ERS1(etr1-4)*. The transformation vector contained the *NPT II* gene for the selection of transformants.

2.3 Generation and cultivation of transgenic chrysanthemum plants

The transformation vector was introduced into *Agrobacterium tumefaciens* EHA105 (Hood *et al.*, 1993), and to 'Sei-Marine' chrysanthemum plants according to the procedure described previously (Aida *et al.*, 2004). In brief, about one month after *Agrobacterium* inoculation, putative transformants (calli) were formed on the infected leaf explants. About two months after inoculation, adventitious buds emerged from the calli. The adventitious buds were detached, transplanted to the elongation medium and grown to normal plantlets. As a whole, a total of 62 independent regenerated lines were obtained from 1600 explants transformed with the transformation vector pBIEF1a/*mDG-ERS1(etr1-4)*. Table 2 summarizes the results of transformation experiments.

Table 2 : Transformation of 'Sei-Marine' chrysanthemum with pBIEF1a/*mDG-ERS1 (etr1-4)*

No. of explants	No. of regenerants	No. of Paromomy-cin* -resistant lines	No. of *EF1a/mDG-ERS1 (etr1-4)* containing lines	*EF1a/mDG-ERS1(etr1-4)* containinglines/ explants (%)
1600	62	45	42	2.6

* Paromomycin is an analog of the antibiotic Kanamycin.

2.4 Selection of transformants and transformation efficiency

Forty-five regenerated lines out of 62 were Paromomycin (an antibiotic related to Kanamycin)- resistant, i.e. putative transgenic lines. Finally, it was shown that 42 out of 45 lines contained the transgene *EF1a/mDG-ERS1(etr1-4)*, with a 2.6% transformation efficiency (Fig. 10).

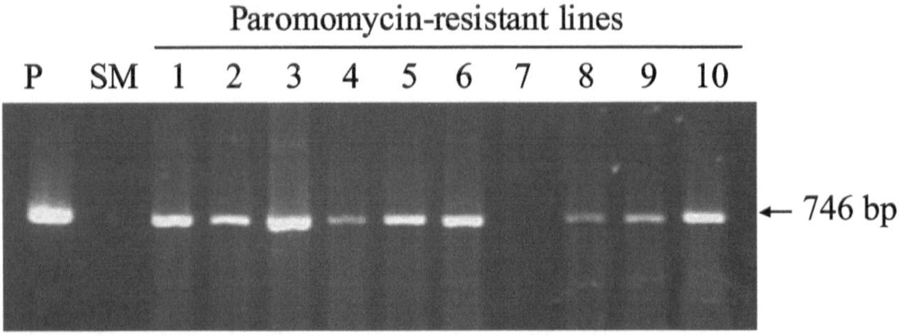

Fig. 10 : PCR analysis of integrated *FF1a/mDG-ERS1 (etr1-4)* transgene in non-transformed and transformed lines of 'Sei-Marine' chrysanthemum. Genomic DNAs extracted from Paromomycin-resistant lines of FF1α/mDG-ERS1(etr1-4) transformant were analyzed by PCR with a combination of appropriate upstream and downstream primers, whose complementary sequences located in the EF1α promoter and *mDG-ERS1(etr1-4)*, respectively. PCR analyses were conducted on all of the transformed lines, but only the results for ten randomly-selected lines are shown. In the figure, lines were randomly numbered for convenience. The arrow shows the PCR amplificate of 746 bp. P, pBIEF1α::*mDG-ERS1(etr1-4)* plasmid. SM, non-transformed 'Sei-Marine'.

2.5 Confirmation of integration of the transgene by Southern blot analysis

The integration of the transgene was confirmed by PCR and Southern blot analyses. PCR analysis gave amplification corresponding to a partial sequence of the *EF1a/mDG-ERS1(etr1-4)* transgene (Fig. 10). Southern blot analysis was conducted with *Hin*d III- and *Xba* I-digests of genomic DNAs obtained from the lines and the 300-bp long DG-ERS1 DNA probe. In Fig. 11, the results with 'Sei-Marine' and representative three transformed lines are shown. Ornamental chrysanthemums are usually hexaploids, and are known to be generated from diploid wild chrysanthemums by cross breeding. In the previous study, Southern blot analysis of genomic DNAs extracted from several cultivars of hexaploid chrysanthemums revealed presence of multiple *DG-ERS1* genes and its homologous genes in the hexaploid chrysanthemum (Narumi *et al.*, 2005a). Six distinct bands hybridized to the DG-ERS1 DNA probe were detected in the *Hin*d III- and *Xba* I-digests of the genomic DNA of 'Sei-Marine' (Fig. 11). Irrespective of sensitivity to ethylene (as described in the next section), all the lines had one or two (and three in the line not shown in Fig. 11) extra bands in the *Hin*d III-digests of genomic DNAs, which were absent in 'Sei-Marine' genomic DNA (Fig. 11A). Moreover, with *Xba* I-digests of genomic DNAs, the DNA fragment of 1514-bp long, which corresponded to the partial sequence of the *EF1a/mDG-ERS1(etr1-4)* transgene, was detected in all of the lines of the mDG-ERS1(etr1-4) transformant (Fig. 11B), but not in non-transformed 'Sei-Marine'.

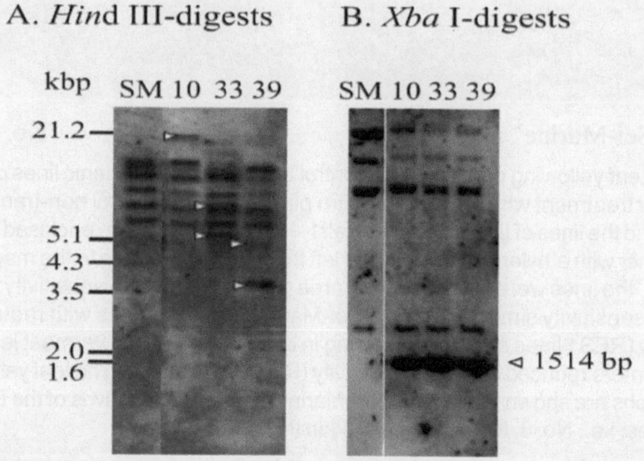

Fig. 11 : Southern blot analysis of lines of the mDG-ERS1(etr1-4) transformant of 'Sei-Marine' chrysanthemum. Three lines were randomly selected from the mDG-ERS1(etr1-4) transformant. The *Hind* III- or *Xba* I-digests of genomic DNAs from the three lines were separated on agarose-gel and hybridized with a 300-bp DG-ERS1 DNA probe for *DG-ERS1* cDNA. Open arrowheads in the panel of *Hind* III-digests (A) show extra bands that were not obtained with the non-transformed 'Sei-Marine' genomic DNA. The 1514-bp *Xba* I-fragment is indicated in the panel of *Xba* I-digest (B). SM, non-transformed 'Sei-Marine'.

2.6 Reduced ethylene sensitivity in *in vitro* plantlets of the transgenic lines

Sensitivity to ethylene was examined using the Paromomycin-resistant lines of the putative transformant. Four *in vitro* plantlets, 7-9 cm high with 8-10 leaves, from each of 40 independent transformed lines were treated with ethylene at 10 ml l^{-1} for 6 days. The plantlets of respective lines showed various degrees of leaf yellowing in response to *in-vitro* exposure to ethylene, and were classified into three groups according to the magnitude of leaf yellowing (Fig. 12). Out of 40 lines, 25 lines showed ethylene sensitivity similar to that of the non-transformed 'Sei-Marine' (severe leaf yellowing), 8 lines reduced ethylene sensitivity (yellowing in one or two of the lowermost leaves of the plants), and 7 lines markedly reduced ethylene sensitivity (no leaf yellowing).

'Sei-Marine' No. 8 No. 18 No. 41

Fig. 12 : Leaf yellowing profiles of the control and putative transgenic lines of chrysanthemum after treatment with ethylene. In vitro plantlets of the control non-transformed 'Sei-Marine' and the lines of the mDG-ERS1(etr1-4) transformant were enclosed in a tight plastic container with ethylene at 10 μl l^{-1} and left for 6 days. According to the magnitude of leaf yellowing, the lines were classified into three groups of ethylene sensitivity; the lines with ethylene sensitivity similar to that of 'Sei-Marine' (ES), the lines with reduced ethylene sensitivity (RES1 lines showing yellowing in one or two of the lowermost leaves) and the lines with more reduced ethylene sensitivity (RES2 lines showing no leaf yellowing). In the photographs are shown the control 'Sei-Marine' and representatives of the ES, RES1 and RES2 lines, i.e., No. 8, No. 18 and No. 41 in this order.

Narumi (unpublished results) detected the transcript of the mutated ethylene receptor gene *mDG-ERS1(etr1-4)* in leaves of all the transgenic lines, but did not find any relationship between the amount of *mDG-ERS1(etr1-4)* transcript and the magnitude of reduction in sensitivity to ethylene. The expression of the integrated *mDG-ERS1(etr1-4)* gene might confer reduced ethylene sensitivity through a mechanism similar to that responsible for ethylene insensitivity of the *Arabidopsis etr1-1* mutant (Chang *et al.,* 1993).

2.7 Reduced leaf senescence (yellowing) in soil-grown mature plants after exposure to exogenous ethylene

In the following experiments, the non-transformed (NT) control 'Sei-Marine' (SM) and seven transgenic lines (No 10, 14, 17, 19, 33, 39 and 41) were used. These transformants were preliminaril selected from *in vitro* plantlets by observing reduced sensitivity to exogenous ethylene as compared with SM. When *in vitro* plantlets, which had been grown aseptically in a plastic culture vessel, were transplanted to soil in a plastic container, all plants of SM and transgenic lines grew vigorously and flowered almost normally.

Four or 5 shoots, 20 cm high with 8-10 leaves, were cut from soil-grown plants maintained as perennial plants in plastic containers, and treated with ethylene at 10 ml l^{-1} for 9 days. The shoots of some transgenic lines showed reduced leaf yellowing as compared with those of Sei-Marine (SM). A typical profile of leaf yellowing is shown in Fig. 13, in which a shoot of SM showed severe yellowing (almost all the leaves showed yellowing), but a shoot of No. 33 line did not. Fig. 14 shows the degree of leaf yellowing caused by exposure to exogenous ethylene in SM and seven transgenic lines. The leaf yellowing was significantly suppressed in shoots of transgenic lines, No. 10, 17, 19 and 33, as compared with SM, whereas, the reduction in leaf yellowing was not significant in shoots of the transgenic lines No. 14, 39 and 41. These results suggested that the sensitivity to ethylene, shown by leaf yellowing was reduced in the transgenic lines No. 10, 17, 19 and 33.

Fig. 13 : Leaf yellowing profiles of the non-transformed 'Sei-Marine' control and a transgenic line No. 33 after treatment with ethylene. Shoots of 20-cm long were detached from soil-grown plants, enclosed in a glass container with 10 μl l^{-1} ethylene, and left for 9 days at 25°C under continuous white fluorescent light.

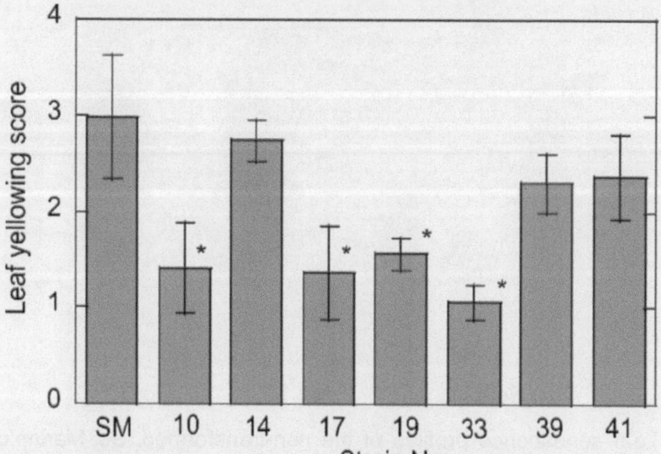

Fig. 14 : Leaf yellowing after treatment with ethylene in the 'Sei-Marine' and transgenic lines. Shoots of 20-cm long were detached from soil-grown plants of the 'Sei-Marine'

and transgenic lines, enclosed in a glass container with 10 μl l^{-1} ethylene, and left for 9 days at 25°C under continuous white fluorescent light. Leaf yellowing score was determined as 0, 1 , 2, 3 and 4 when leaf yellowing covered 0, 1-10, 11-40, 41-70 and 71-100% from the bottom of the shoot, respectively. Leaf yellowing score was shown as the average of 4 or 5 shoots. Data were shown by the mean ± SE of 4 experiments. *shows significant difference from the value in 'Sei-Marine' at p=0.05 by Dunnett's multiple test.

2.8 Reduced leaf deterioration during storage in darkness without ethylene treatment

We examined the leaf deterioration (yellowing, browning and rotting) of cut shoots kept in darkness without ethylene treatment, in SM and the transgenic lines. Shoots (20-cm long with 8-12 leaves) cut from the soil-grown plants (about 40 cm in height) were enclosed in a plastic bag with minimum air space and left in the dark for one week at 25°C. This treatment caused leaf deterioration characterized by yellowing, browning and rotting in the shoots of SM. In some transgenic lines, however, leaf deterioration was alleviated showing less yellowing, deterioration and rotting of the leaves than in SM. Fig. 15 shows typical profiles of leaf deterioration. The shoot of SM showed severe leaf deterioration, yellowing, browning and rotting, but the shoots of transgenic lines No. 17 and 33 showed little or no leaf deterioration. Reduced rotting probably resulted from a significant reduction in susceptibility to bacteria or fungi, because some pathogens often infect senescent tissue.

Sei-Marine No. 17 No. 33

Fig. 15 : Leaf senescence profiles of the non-transformed 'Sei-Marine'control and transgenic lines No.17 and No. 33 of chrysanthemum after natural senescence. Shoots of 20-cm long were detached from soil-grown plants, enclosed in a plastic bag and left for 1 week in the dark at room temperature.

These results indicated that the integration of the transgene *mDG-ERS1(etr1-4)* suppressed the senescence of leaves in the transgenic chrysanthemum plants, irrespective of ethylene-induced or dark-induced senescence. Chrysanthemum plants belong to the family Compositae. Compositae plants include leafy vegetables such as lettuce, garland chrysanthemum, chicory and butterbur. These vegetables are vulnerable to senescence, resulting in yellowing, browning and rotting of leaves. Usually, to prevent senescence, harvested vegetables are handled and stored under cool temperature or controlled atmosphere conditions. However, the Compositae vegetables cannot be stored for a long time in cool temperature because of cold injury, which is possibly accelerated by the action of ethylene. Thus, the senescence of these vegetables could be preventable by the generation of transgenic plants with reduced leaf senescence. The present study suggests that *mDG-ERS1(etr1-4)* could be used to generate transgenic Compositae leafy vegetables with retarded leaf senescence resulting in healthy green leaves and a greatly extended shelf life.

3. Conclusion

In our studies, carnation plants, irrespective of their use for cut flowers and potted plants, were successfully transformed with *sACO* or *aACO* transgenes. Cut flowers of the resultant transformants had reduced ethylene production and showed extended vase life. The results suggest that these transgenic techniques could be used to lengthen the vase life of carnation flowers instead of currently-used chemical preservatives of the flowers. In addition, chrysanthemum plant was transformed with a mutated ethylene receptor cDNA, *mDG-ERS1(etr1-4)*. The resultant transformants had reduced sensitivity to exogenous ethylene, exhibiting less yellowing in their leaves after ethylene treatment. Moreover, the transformants showed reduced deterioration during storage without ethylene. This result suggests that this transgenic technique is useful to retard leaf deterioration and improve postharvest performance in Compositae leafy vegetables. Since ornamentals like carnations and chrysanthemums are exclusively cultivated by vegetative propagation, To plants which have gained useful traits will be practically useful without the establishment of homozygous descendent lines.

References

Abeles FB, Morgan PW and Saltveit ME, Jr 1992 Ethylene in Plant Biology. 2nd ed. Academic Press, San Diego

Aida R, Yoshida T, Ichimura K, Goto R and Shibata M 1998 Extension of flower longevity in transgenic torenia plants incorporating ACC oxidase transgene. *Plant Science* 138 : 91-101

Aida R, Ohira K, TanakaY, Yoshida K, Kishimoto S, Shibata M and Ohmiya A 2004 Efficient transgene expression in chrysanthemum, *Dendranthema grandiflorum* (Ramat.) Kitamura, by using the promoter of a gene for chrysanthemum chlorophyll-a/b-binding protein. *Breeding Science* 54 : 51-58

Borochov A and Woodson WR 1989 Physiology and biochemistry of flower petal senescence. *Horticultural Review* 11 : 15-43

Bovy AG, Angenent GC, Dons HJM and van Altvorst A-C 1999 Heterologous expression of the *Arabidopsis etr1-1* allele inhibits the senescence of carnation flowers. *Molecular Breeding* 5 : 301-308

Ciardi JA, Tieman DM, Lund ST, Jones JB, Stall RE and Klee HJ 2000 Response to *Xanthomonas campestris* pv. *vesicatoria* in tomato involves regulation of ethylene receptor gene expression. *Plant Physiology* 123 : 81-92

Chang C, Kwok SF, Bleecker AB and Meyerowitz EM 1993 *Arabidopsis* ethylene-response gene *ETR1*: similarity of product to two-component regulators. *Science* 262 : 539-544

Doi M, Nakagawa Y, Watanabe S, Aoe K, Inamoto K and Imanishi H 2003 Ethylene-induced leaf yellowing in cut chrysanthemum (*Dendranthema grandiflora* Kitamura). *Journal of the Japanese Society for Horticultural Science* 72 : 533-535

Doi M, Aoe K, Watabe S, Inamoto K and Imanishi H 2004 Leaf yellowing of cut standard chrysanthemum (*Dendranthema grandiflora* Kitamura) 'Shuho-no-chikara' induced by ethylene and postharvest increase in ethylene sensitivity. *Journal of the Japanese Society for Horticultural Science* 73 : 229-234

Ferrante A, Tognoni A, Mensuali-Sodi A and Serra G 2003 Treatment with thidiazuron for preventing leaf yellowing in cut tulips and chrysanthemum. *Acta Horticulturae* 624 : 357-363

Firoozabady E, Moy Y, Tucker W, Robinson K and Gutterson N 1995 Efficient transfomation and regeneration of carnation cultivars using *Agrobacterium. Molecular Breeding* 1 : 283-293.

Harada T, Murakoshi Y, Tanase K, Onozaki T and Satoh S 2009 Structure of genomic DNA of *DcACS1*, a 1-aminocyclopropane-1-carboxylate synthase gene expressed in senescing carnation (*Dianthus caryophyllus*) petals. *Horticultural Research (Japan)* 8 (Suppl.1) : 257

Henskens JAM, Rouwendal GJA, ten Have A and Woltering EJ 1994 Molecular cloning of two different ACC synthase PCR fragments in carnation flowers and organ-specific expression of the corresponding genes. *Plant Molecular Biology* 26 : 453-458

Hood EE, Gelvin SB, Melchers LS and Hoekemma A 1993 New *Agrobacterium* helper plasmids for gene transfer to plants. *Transgenic Research* 2 : 208-218

Hua J and Meyerowitz EM 1998 Ethylene responses are negatively regulated by a receptor gene family in *Arabidopsis thaliana. Cell* 94 : 261-271

Iwazaki Y, Kosugi Y, Waki K, Yoshioka T and Satoh S 2004 Generation and ethylene production of transgenic carnations harboring ACC synthase cDNA in sense or antisense orientation. *Journal of Applied Horticulture* 6 : 67-71

Jones ML, Larsen PB and Woodson,WR 1995 Ethylene-regulated expression of a carnation cysteine proteinase during flower petal senescence. *Plant Molecular Biology* 28: 505-512

Jones ML and Woodson WR 1999 Differential expression of three members of the 1-aminocyclopropane-1-carboxylate synthase gene family in carnation. *Plant Physiology* 119 : 755-764

Kende H 1993 Ethylene biosynthesis. *Annual Review of Plant Physiology and Plant Molecular Biology* 44 : 283-307

Kinouchi T, Endo R, Yamashita A and Satoh S 2006 Transformation of carnation with genes related to ethylene production and perception: Towards generation of potted carnations with longer display time. *Plant Cell, Tissue and Organ Culture* 85 : 27-35

Kosugi Y, Shibuya K, Tsuruno N, Iwazaki Y, Mochizuki A, Yoshioka T, Hashiba T and Satoh S 2000 Expression of genes responsible for ethylene production and wilting are differently regulated in carnation (*Dianthus caryophyllus* L.) petals. *Plant Science* 158 : 139-145

Kosugi Y, Waki K, Iwazaki Y, Mochizuki A, Tsuruno N, Mochizuki A, Yoshioka T, Hashiba T and Satoh S 2002 Senescence and gene expression of transgenic non-ethylene-producing carnation flowers. *Journal of the Japanese Society for Horticultural Science* 71 : 638-642

Mitsuhara I, Ugaki M, Hirochika H, Ohshima M, Murakami T, Gotoh Y, Katayose Y, Nakamura S, Honkura R, Nishiyama S, Ueno K, Mochizuki A, Tanimoto H, Tsugawa H, Otsuki Y and Ohashi Y 1996 Efficient promoter cassettes for enhanced expression of foreign genes in dicotyledonous and monocotyledonous plants. *Plant and Cell Physiology* 37 : 49-59

Narumi T, KannoY, Suzuki M, Kishimoto S, Ohmiya A and Satoh S 2005a Cloning of a cDNA encoding an ethylene receptor (DG-ERS1) from chrysanthemum and comparison of its mRNA level in ethylene-sensitive and -insensitive cultivars. *Postharvest Biology and Technology* 36 : 21-30

Narumi T, Aida R, Ohmiya A and Satoh S 2005b Transformation of chrysanthemum with mutated ethylene receptor genes: *mDG-ERS1* transgenes conferring reduced ethylene sensitivity and characterization of the transformants. *Postharvest Biology and Technology* 37 : 101-110

Norikoshi R, Niki T, and Ichimura K 2008 A new ACO gene associated with ACO activity in carnation gynoecium. *Horticultural Research (Japan)* 7(Suppl 1) : 233

Ohta S, Mita S, Hattori T and Nakamura K 1990 Construction and expression in tobacco of b-glucronidase (GUS) reporter gene containing an intron within the coding sequence. *Plant and Cell Physiology* 31 : 805-813

Panavas T, Pikula A, Reid PD, Rubinstein B, and Walker EL 1998 Programmed cell death of daylily petals: activities of wall-based enzymes and effects of heat shock. *Plant Physiology and Biochemistry* 36 : 379-388

Panavas T, Pikula A, Reid PD, Rubinstein B and Walker EL 1999 Identification of senescence-associated genes from daylily petals. *Plant Molecular Biology* 40 : 237-248

Park KY, Drory A and Woodson WR 1992 Molecular cloning of an 1-aminocyclopropane-1- carboxylate synthase from senescing carnation flower petals. *Plant Molecular Biology* 18 : 377-386

Reid MS and Wu M-J 1992 Ethylene and flower senescence. *Plant Growth Regulation* 11 : 37-43

Reyes-Arribas T, Barrett JE, Huber DJ, Nell TA and Clark DG 2001 Leaf senescence in a non-yellowing cultivar of chrysanthemum (*Dendranthema grandiflora*). *Physiologia Plantarum* 111 : 540-544

Satoh S, Kosugi Y, Iwazaki Y, Shibuya K and Waki K 2000 Two ethylene signaling pathway in senescing carnation petals: exogenous ethylene-induced expression of genes for 1-aminocyclopropane-1-carboxylate (ACC) synthase and ACC oxidase is different from that of the gene for cysteine proteinase. *Journal of Plant Biotechnology* 2 : 83-87

Savin KW, Baudinette SC, Graham MW, Michael MZ, Nugent GD, Lu C-Y, Chandler SF and Cornish D (1995) Antisense ACC oxidase RNA delays carnation petal senescence. *HortScience* 30 : 970-972

Shibuya K, Yoshioka T, Hashiba T and Satoh S 2000 Role of the gynoecium in natural senescence of carnation (*Dianthus caryophyllus* L.) flowers. *Journal of Experimental Botany* 51 : 2067-2073.

Tieman DM, Taylor MG, Ciardi JA and Klee HJ 2000 The tomato ethylene receptors NR and LeETR4 are negative regulators of ethylene response and exhibit functional compensation within a multigene family. *Proceedings of National Academy of Science USA* 97 : 5663-5668

Veen H 1979 Effects of silver on ethylene synthesis and action in cut carnations. *Planta* 145 : 467-470

Wilkinson JQ, Lanahan MB, Clark DG, Bleecker A, Chang C, Meyerowitz EM and Klee HJ 1997 A dominant mutant receptor from *Arabidopsis* confers ethylene insensitivity in heterologous plants. *Nature Biotechnology* 15 : 444-447

Wang H and Woodson WR 1991 A flower senescence-related mRNA from carnation shares sequence similarity with fruit ripening-related mRNAs involved in ethylene biosynthesis. *Plant Physiology* 96 : 1000-1001

Yang SF and Hoffman NE 1984 Ethylene biosynthesis and its regulation in higher plants. *Annual Review of Plant Physiology* 35 : 155-189

2

DNA Barcoding and its Applications in Horticultural Crops

Deepu Mathew

Delineating species boundaries correctly and also identifying species are crucial to the discovery of life's diversity because it determines whether different individual organisms are members of the same entity or not (Dayrat, 2005). The identification of species depends on the knowledge held by taxonomists whose work cannot cover all taxon identification requested by non-specialists. To deal with these difficulties, the 'DNA Barcode of Life' project aims to develop a standardized, rapid and inexpensive species identification method accessible to non-taxonomists.

DNA barcoding is a technique in which species identification is performed by determining and comparing the sequences of several hundred base pairs of nucleotides from a small fragment of an identified gene, with the aim of contributing to a wide range of ecological and conservation studies in which traditional taxonomic identification is not practical (Krishnamurthy and Francis, 2012). The two main ambitions of DNA barcoding are to assign unknown specimens to species and to enhance the discovery of new species and facilitate identification, particularly in cryptic, microscopic and other organisms with complex or inaccessible morphology (Hebert *et al.*, 2003a). Apart from these, identification of species previously defined by other criteria to also is an objective of barcoding. Another major objective could be the definition of operational units for ecological studies. This may also be used for forensic purposes by linking the specimens with already identified databases.

In three important situations, relevant Species identification must necessarily be molecular-based. First is in the determination of the taxonomic identity of damaged specimens or fragments. The DNA barcoding tool is thus potentially useful in the food industry, diet analyses and in preventing illegal trade and poaching of endangered species. Second, molecular-based identification is necessary when there are no obvious means to match adults with immature specimens. The third case is when morphological traits do not clearly discriminate species especially if species have polymorphic life cycles and/or exhibit pronounced phenotypic plasticity.

The idea of a standardized molecular identification system emerged progressively during the 1990s with the development of PCR-based approaches for species identification. Several universal systems for molecular systematic analyses were used for lower taxa but were not successfully implemented for broader scopes. The Barcode of Life project aims to create a universal system for a eukaryotic species inventory based on a standard molecular approach. It was initiated in 2003 by researchers at the University of Guelph in Ontario, Canada (http://www.barcoding.si.edu) and promoted in 2004 by the international initiative 'Consortium for the Barcode of Life' (CBOL). By then, it had more than 150 member organizations from 45 countries including natural history museums, zoos, herbaria, botanical gardens, university departments as well as private companies and governmental organizations. The DNA barcode project does not have the ambition to build the tree of life nor to perform molecular taxonomy (Erbach and Holdrege, 2005; Gregory, 2005), but rather to produce a simple diagnostic tool based on strong taxonomic knowledge that is collated in the DNA barcode reference library (Schindel and Miller, 2005). The DNA Barcode of Life Data System (BOLD, http://www.boldsystems.org) has progressively been developed since 2004 and was officially established in 2007 (Ratnasingham and Hebert, 2007). This data system enables the acquisition, storage, analysis and publication of DNA barcode records.

Using the mitochondrial Cytochrome C Oxidase 1 (*CoI*) gene, DNA barcoding is well established in animals (Hebert *et al.*, 2003a; Smith *et al.*, 2005; Vences *et al.*, 2005; Ward *et al.*, 2005) and by March 2004, the Consortium for the Barcode of Life started to promote the use of a standardized DNA barcoding approach, consisting of identifying a specimen as belonging to a certain animal species based on the DNA barcode sequence as a single universal marker. The DNA barcode in animals consists of a 648 bp region 58-705 from the 5′-end of the *Co*I gene using the mouse mitochondrial genome as a reference. It is based on the postulate that every species will most likely have a unique DNA barcode (there may be 4650 possible ATGC-combinations compared to an estimated 10 million species remaining to be discovered, Wilson, 2004) and that genetic variation between species exceeds variation within species (Hebert *et al.*, 2003b; Hebert *et al.*, 2004). The mitochondrial DNA is proposed to be advantageous because it is present in several copies per mitochondrion that may be present in hundreds in every eukaryotic cell. This high copy number makes it far simpler to recover mitochondrial genes than their nuclear counterparts from small amounts of tissue or when DNA preservation is poor. In contrary, Rubinoff *et al.* (2006a) have registered their strong reservation against the use of mitochondrial DNA barcodes, proving that this strategy is associated with technical and practical shortcomings and may degrade the longstanding synthesis of genetic and organism based research.

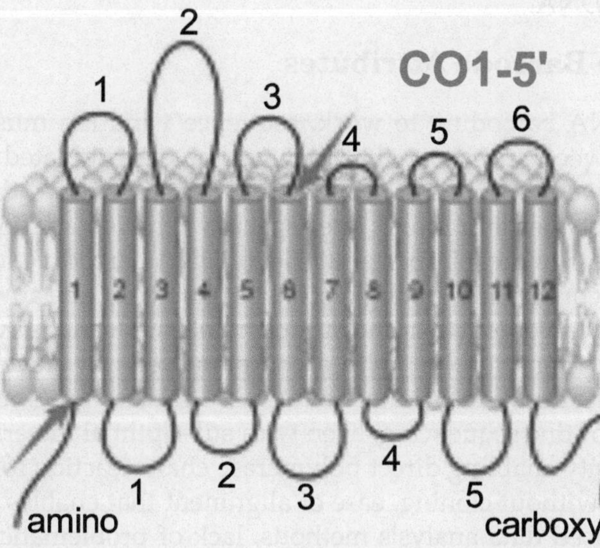

Horticulture represents a wide range of plants suited for garden land conditions and identified for their economic potential. Every crop holds the enormous potential of improvement especially towards the identified character. Crop husbandry and in particular, the heterosis breeding solely depends on the hybrid vigour that could be generated by combining the genetically diverse parents. Even though the morphological strategies for genetic distance estimation are well established, the precision in many occasions may be inferior to that offered by molecular techniques. In this situation, barcoding will be useful to establish the identity of the materials at inter and intra-specific levels. Many a times, it is observed that the cross pollination among the parents fail due to the wrong morphological species allocation. The exact identification could save such efforts since the alternates such as embryo rescue may be advocated based on the results. Further, the wild and newly identified species that are recently identified may harbour precious genes. Under such situation also, the DNA barcoding will be essential for identifying and allocating it to a certain species and estimating its distance from other members of the species, if at all already identified. Even though the *CoI* mitochondrial gene is identified as a universal candidate for barcoding the members of animal kingdom, the situation is more complicated in plant kingdom. Detection of suitable barcoding locus remains to be a major challenge in each new genus. As included in our forthcoming discussions, many candidate genes are reported from the nuclear coding and non-coding regions and chloroplast DNA.

Desirable Barcode Attributes

For DNA barcoding to work, sequence variation must be high enough between species so that they can be discriminated from one another; however, it must be low enough within species that a clear threshold between intra- and inter-specific genetic variations can be defined. The two most important traits of DNA barcoding loci are the presence of conserved flanking regions to enable routine amplification across highly divergent taxa and sufficient internal variability to enable species discrimination but with a comparatively low level of intra-specific variation. Additional favourable factors are short length facilitated routine sequencing, even with sub-optimal material, lack of heterozygosity enabling direct polymerase chain reaction followed by sequencing without cloning, ease of alignment that enables the use of character-based data analysis methods, lack of problematic sequence composition, such as regions with several microsatellites, that reduces sequence quality, universal capability to get amplified/sequenced with standardized primers, easy alignability and capability to get recovered easily from herbarium samples and other degraded DNA samples (Hollingsworth *et al.*, 2009).

However, it may be noted that DNA barcoding is not intended to supplant or invalidate existing taxonomic practice. Also, it is not DNA-taxonomy ie., it does not equate species identity, formally or informally, with a particular DNA sequence. Further, barcoding does not aim to duplicate or compete with efforts to resolve deep phylogeny.

Proposed Barcodes in Plants

Though there are a few reports that some of the barcodes are universally useful for plants, it still remains mandatory to screen out the suited barcode for any new species (Rubinoff *et al.*, 2006b; Pennisi, 2007; Ledford, 2008). Tests of potential DNA barcodes have been based on a taxonomic coverage approach, necessarily encompassing just a few representatives from a wide range of distantly related groups of land plants (Kress *et al.*, 2005; Chase *et al.*, 2007; Kress and Erickson 2007; Newmaster *et al.*, 2006). In general, the genes used in angiosperms are *mat*K, *rpo*C1, *rpo*B, *acc*D, *YCF5* and *ndh*J whereas in non-angiosperms *mat*K, *rpo*C1, *rpo*B, *acc*D, and *ndh*J are used (http://www.rbgkew.org.uk/barcoding/index.html). With higher potential to identify the variation, easy amplification and alignment, a portion of the plastid *mat*K gene was proposed as a universal DNA barcode for flowering plants (Lahaye *et al.*, 2008). Studies by Newmaster *et al.* (2008) in Myristicaceae and Seberg and Petersen (2009) in *Crocus* have confirmed *mat*K and the intergenic spacer *trn*H-*psb*A as suitable land plant barcodes.

Internal transcribed spacer regions of nuclear ribosomal DNA (*ITS*) is commonly recommended based on the facts that these are often highly variable in angiosperms at the generic and species level and divergent copies are often present within single individuals (Chase *et al.*, 2005; Kress *et al.*, 2005). Although *ITS* works well in many plant groups and may be a useful supplementary locus, numerous cases of incomplete concerted evolution and intra-individual variation make it unsuitable as a universal plant barcode.

Kress *et al.* (2005) also proposed that the *trn*H-*psb*A plastid intergenic spacer region would be a suitable universal barcode for land plants. Concurrently, the Plant Working Group of Consortium for the Barcoding of Life (CBOL-PWG) tested a series of other genomic regions disregarding *trn*H-*psb*A because of its complex molecular evolution (Chang *et al.*, 2006). Taberlet *et al.* (2007) proposed the chloroplast *trn*L intron as a potential barcoding candidate gene in angiosperms. Further, three regions viz. *atp*F-*atp*H, *mat*K, and *psb*K-*psb*I were proposed (Pennisi, 2007). Devey *et al.* (2009) argued that non-coding regions *atp*F-*atp*H and *trn*H-*psb*A should be considered as suboptimal barcodes due to prevalence of microsatellites.

It was also proposed that, because the plastid genome is evolving so slowly relative to other genomes and shows intra-molecular recombination (Mower *et al.,* 2007), more than one barcode may be necessary to provide enough variation for this technique to work Newmaster *et al.,* 2006 (Chase *et al.,* 2007; Kress and Erickson 2007;). Kress and Erickson (2007) proposed to combine the original *trn*H-*psb*A barcode from Kress *et al.,* (2005) with *rbc*L (which is not variable enough at species level for many plant groups), following analyses from Newmaster *et al.* (2006). This combination is also potential enough to be used as a universal barcode.

Hollingsworth *et al.* (2009) evaluated the seven main candidate plastid regions *rpo*C1, *rpo*B, *rbc*L, *mat*K, *trn*H-*psb*A, *atp*F-*atp*H and *psb*K-*psb*I in three divergent groups of land plants. *rpo*C1 was the most universal locus and amplified well across all three groups. *trn*H-*psb*A showed greatest universality of the noncoding regions. In individual taxonomic groups, for species discrimination, the best performing locus was in each case a coding locus. Across these groups, no single locus showed high levels of universality and resolvability. Inter-specific sharing of sequences from individual loci was common. However, when multiple loci were combined, fewer barcodes were shared among species. Minor improvements on these were obtained by various new three-locus combinations involving *rpo*C1, *rbc*L, *mat*K and *trn*H-*psb*A, but no single combination clearly outperformed all others. This report further points to the necessity to carry out enormous work towards standardizing the candidate genes in each plant species.

Chase *et al.* (2007) proposed to make the universal barcodes with the combination *rpoc*1+*rpo*B+*mat*K and *rpoc*1+*mat*K+*trn*H-*psb*A. Of late, the two-locus combination of *rbc*L+*mat*K has been recommended as the core barcode for land plants (CBOL Plant Working Group, 2009).

In bryophytic plants, five loci, viz. *rbc*L, *rpo*C1, *rps*4, *trn*H-*psb*A and *trn*L-*trn*F, were easy to amplify and sequence and showed significant inter-specific genetic variability, making them potentially useful DNA barcodes. The best performing single loci were *rbc*L and *rpo*C1 coding regions (Liu *et al.,* 2010b).

Table 1 : Summary of studies comparing DNA barcoding regions in plants (Hollingsworth *et al.*, 2009)

Study	Regions compared	Sampling strategy	Universality (% success)	Sequence divergence/variation	Barcode recommendation
Kress *et al.*, 2005	*atpB-rbcL*, ITS, *psbM-trnD*, *trnC-ycj6*, *tmH-psbA*, *trnL-F*, *trnk-rps 16*, *IntV-atyE*, *rpl36-rps8*, *ycf6-psbM*	19 individuals/19 species from 7 angiosperm families	*tmH-psbA*, *rpl136-rpf8*, *trnL-F = 100%* *truC-ycfb*, *ycj6-psbM* = 90% Other regions = 73-80%	% sequence divergences: ITS (2.81%) ***tmH-psbA*** (1.24%) *trnH-psbA* had = 2 x sequence divergence of other plastid regions	ITS and tmH-*psbA*
	ITS, rbcL′, tmH-psbA	R3 individuals/83 species from 50 families	tmH-psbA - ***100%*** rbcL * **95%** ltSS88%	*tmH-psbA* >> *rbcL*	
Kress & Erickson 2007	accD, ITSI, *ndhj*, *matk*, *tmH-psbA*, *rixL rpoB*, *rpoCl*, *ycfB*	96 individuals/96 species from 43 families of land plants	*trnH-psbA*, *irbcL* - 95% *trpoCl* = 90% *accD*, *rpoB* = 80% *ndhj* = 70%, *ITSI* = 60% *ycf5* = 50% *matK*- 40%	ITS (5.7%) *trnH-psbA* (2.69%) *rpoB* (2.05%) *ycfS* (1.55%) *rpoCl* (1.38%) *rbcL* (1.29%) *accD* (1.2%) *matK* (1.13%) *ndhj* (0.2%)	*rbcL* and *trnH-psbA*
Sass *et al.*, 2007	*accD*, ITS, *ndhj*, *matk*, *trnH-psbA*, *rpoB*, *rpoCl*,*ycf5*	21 Individuals/21 species from 10 genera of cycads. more individuals for some regions (up to 96)	*rpoCl*, *ycf5* = 100% ITS= 100%? *accD* = 96% *ndhj*== 57% rpoB = 33% matK* 24% *trnH-psbA* double-banded in most samples	ITS most variable; quantitative figures on relative sequence divergence of other regions not given. other thane. 10%of bases variable in each region	ITS (considered most promising)

Contd.

Newmaster *et al.*, 2008	*accD, matK, traH-psbA, rbcL, rpoB, rpoCl*, UPA	40 individuals/8 species in Myristicaceae	*trnH-psM, rpoCl,* UPA = 100% rpoB, accD, ≥ *95%* *rbcL* a 90% *niatK* required primer redesign, then-98%	Uncorrected interspecific *p*-distances: *tmH-psbA* (0.060) *matK* (0.042) *accD* (0.003) *rpoCl* (0.002) *rbcL* (0.002) *rpoB* (0.001) UPA (0.001)	*matK and tmH-psbA*
Lahaye *et al.*, 2008a	*accD, iulhl, nwlK IrnH-psbA, rbcL, rpoB, rpoCl,ycf5*	172 individuals/86 species total (consisting of 71 individuals/48 orchid species +101 individuals/38 species from 13 angiosperm families)	In the orchids,*ycf5* did not amplify and *ndhj* amplification was patchy. All other regions .95-100%	K2P interspecific sequence divergence: *tmH-psbA* (0.(1216) *matK* (0.0125) *ycf5* (0.01) *rbcL* (0.0079) *rpoB* (0.0061) *ndh]* (0.0046) *accD* (0.0038) *rpoCl* (0.0019)	*matK* (or *mat*K and trnH-psbA)
Fazekas *et al.*, 2008	*coxl*,23SrDNA,*rpoB, rpoCl, rbcL, matK*, tmH-psbA, atpFntpH, psbK-psbl	251 individuals/92 species from 32 genera of land plants	23S rDN A = 100% *rbcL* = 100% (2 primer pairs used) tmH-psbA = *99%* *rpoCl* = 95% (3 primer pairs used) *rpoB* - 92% (5 primer pairs used) ***matK*** -88% (10 primer pairs used) psbK-psbl = ***85%*** ***eoxl***» 72% atpF-atpH * ***65%***	No. of parsimony-informative characters: *matK* (386) *trnH-psbA* (350) *atpF-atpH* (308) *psbK-psbl* (263) ***rbcl*** (242) ***rpoB*** (179) ***Coxl*** (146) ***rpoCl*** (134) 23SrDNA(19)	Broadly equivalent performance from various combinations of loci; suggested selecting 3-4 regions front: *rbcL, rpoB, matK. trnH-psbA, atpF-atpH*

The Barcoding Primers for Plants

In every species, primer information is the most vital in initiating the screening of various reported candidate genes towards their suitability. The forward and reverse sequences should be carefully combined, the PCR program including the specific annealing temperature has to be standardized and the efficacy should be tested for each gene. Sometimes, none the reported sequences for any of the candidate genes may not work and under such situation, we have to design our own primers for each gene, after retrieving the respective gene sequences from the public domains such as NCBI. For this purpose, the on-line open access tools such as Primer3 may be used. These softwares along with all the possible forward and reverse primer combinations, provide the annealing temperature for PCR programme.

Table 2 : Primer sequences for the candidate genes for barcoding in plants

Plastid Gene	Direction	Sequence 5'-3'
*mat*K	F	CCTATCCATCTGGAAATCTTAG
	F	ATCCATCTGGAAATCTTAGTTC
	R	GTTCTAGCACAAGAAAGTCG
	R	CTTCCTCTGTAAAGAATTC
*mat*K primers	F	CGA TCT ATT CAT TCA ATA TTT C
(Jarret, 2008)	R	TCT AGC ACA CGA AAG TCG AAG T
*mat*K primers	F	(T)AATTTACGATCAATTCATTC
(Dunning and	F	CGATCTATTCATTCAATATTTC
Savolainen (2010)	F	ACCCAGTCCATCTGGAAATCTTGGTTC
	F	TCAGAATTTACGATCTATTC
	F	TCATAATTTACGATCAATYC
	F	TCATAATTTACGATCAATTCA
	F	TAATTTACGATCAATTCAITC
	F	TAATTTACGATCAATTCAKTC
	F	TAATTTACGATCAATTCATKC
	F	TAATTTACGATCAATTCATTCA
	F	CGATCTATTCATTCAATATT
	F	GATCTATTCATTCAATATTTYC
	F	AATATACTAATACCTTAYC
	F	CCTTACCCAGCTCATCTGGAAAT
	F	ACCCAGTCCATCTGGAAATCTTGGTIC

Contd.

	F	TTCAAACCCTTCGTTACTGGATVAA
	R	TCTAGCACACGAAAGTCGAAGT
	R	GTTCTAGCACAAGAAAGTCG
	R	CGTACAGTACTTTTGTGTTTACGAG
	R	TAATGAGAAAGATTTCTGCATAT
	R	AACCTGTTCTTTGAGGABCC
	R	CACGAAAGTCGAAGTATATAY
	R	GCACAAGAAAGTCGAAGTATATA
	R	ACAAGAAAGTCGAAGTAT
	R	GCACAAGAAAGTCGAAGTAT
	R	CGAGACAAAGTTCTAGCCCAVG
	R	CTTTTGTGTTTACGAGCCAADG
	R	CGTACTTTTATGTTTACAGGCYAA
	R	GCTTTTGTGTTTACGAGCTA
	R	CGTACAGTACTTTTGTGTTTACGIG
	R	CAAATAATATCCAAATACCAA
	R	ATACAAATAATATCCAAATAC
*mat*K primers	F	CGATCTATTCATTCAATATTTC
(Lahaye *et al.*, 2008;	R	TCTAGCACACGAAAGTCGAAGT
Yu *et al.*, 2011)		
*rpo*C1	F	GTGGATACACTTCTTGATAATGG
	F	GGCAAAGAGGGAAGATTTCG
	R	TGAGAAAACATAAGTAAACGGGC
	R	CCATAAGCATATCTTGAGTTGG
*rpo*C1 primers	F	GGCAAAGAGGGAAGATTTCG
(Chase *et al.*, 2007;	R	CCATAAGCATATCTTGAGTTGG
Yu *et al.*, 2011)		
rpoB	F	AAGTGCATTGTTGGAACTGG
	F	ATGCAACGTCAAGCAGTTCC
	R	CCGTATGTGAAAAGAAGTATA
	R	GATCCCAGCATCACAATTCC
*rpo*B primers	F	AAGTGCATTGTTGGAACTGG
(Chase *et al.*, 2007;	R	GATCCCAGCATCACAATTCC
Yu *et al.*, 2011)		
*acc*D	F	AGTATGGGATCCGTAGTAGG
	F	GGRGCACGTATGCAAGAAGG
	R	TTTAAAGGATTACGTGGTAC
	R	TCTTTTACCCGCAAATGCAAT

Contd.

YCF5	F	GGATTATTAGTCACTCGTTGG
	F	ACTTTAGAGCATATATTAACTC
	R	ACTTACGTGCATCATTAACCA
	R	CCCAATACCATCATACTTAC
*ndh*J	F	CATAGATCTTTGGGCTTYGA
	F	TTGGGCTTCGATTACCAAGG
	R	ATAATCCTTACGTAAGGGCC
	R	TCAATGAGCATCTTGTATTTC
ITS2	F	GCGATACTTGGTGTGAAT
	R	GACGCTTCTCCAGACTACAAT
ITS (Li *et al.*, 2011a)	F	GGAAGTAAAAGTCGTAACAAGG
	R	TCCTCCGCTTATTGATATGC
*rbc*L (Kress and Erickson, 2012; Yu *et al.*, 2011)	F	ATGTCACCACAAACAGAAAC
	R	TCGCATGTACCTGCAGTAGC
*rbc*L (Stoeckle *et al.*, 2011)	F	ATGTCACCACAAACAGAGACTAAAGC
	R	GAAACGGTCTCTCCAACGCAT
*trn*S-*trnf*M (Jarret, 2008)	F	GAG AGA GAG GGA TTC GAA CC
	R	CAT AAC CTT GAG GTC ACG GG
trnL-trnT (Jarret, 2008)	F	CAAATGCGATGCTCTAACCT
	R	TCTACCGATTTCGCCATATC
trnF-trnL (Jarret, 2008)	F	CAAAATCGTGTAGACGCTACG
	R	ATTTGAGGTGACACTGAG
trnD-trnT (Jarret, 2008)	F	ACCAATTGAACTACAATCCC
	R	CTACCACTGAGTTAAAAGGG
trnC-rpoB (Jarret, 2008)	F	CACCCRGATTYGAACTGGGG
	R	CKACAAAAYCCYYCRAATTG
*rps*16 (Jarret, 2008)	F	AAACGATGTGGTARAAAGCAAC
	R	AACATCWATTGCAASGATTCG ATA
trnH-psbA (Jarret, 2008)	F	CGCGCATGGTGGATTCACAATCC
	R	GTTATGCATGAACGTAATGCT C
waxy (Jarret, 2008)	F	TGCCTGGGATACAAGCAT TA
	R	AACTGGAGGCAATGTGAAAT

Barcoding Initiatives in Horticultural Plants

As far as the Science of Horticulture is concerned, DNA barcoding is an emerging field. The large quantum of work to be done in every species in standardization of candidate gene as well as the primer sequences remains to be a challenge. Recently, reports on very many horticultural plants are arriving.

Table 3 : Horticultural plants in which DNA barcoding is successfully implemented

Sl. No.	Plants	Candidate gene	Reference
1	*Compsoneura* (Myristicaceae)	Coding loci Universal Plastid Amplicon (UPA), *rpoB, rpoC1, accD, rbcL, matK*Non-coding chloroplast locus*trnH-psbA*	Newmaster *et al.* (2008)
2	Medicinal plants	*psbA-trnH, matK, rbcL, rpoC1, YCF5,* ITS2, and ITS	Chen *et al.* (2010)
3	322 species of *Astragalus* (Fabaceae), 234 species of *Indigofera* (Fabaceae), 223 species of *Trifolium* (Fabaceae), 206 species of *Miconia* (Melastomataceae), 199 species of *Draba* (Brassicaceae), 185 species of *Centaurea* (Asteraceae), 178 species of *Veronica* (Plantaginaceae), 176 species of *Oxalis* (Oxalidaceae), 174 species of *Ficus* (Moraceae), 162 species of *Solanum* (Solanaceae), 161 species *Senecio* (Asteraceae), 138 of species of *Aspalathus* (Fabaceae), 127 species of *Acacia* (Fabaceae), 124 species of *Rubus* (Rosaceae), 124 species of *Begonia* (Begoniaceae), 123 species of *Polygala* (Polygalaceae), 118 species of *Artemisia* (Asteraceae), 118 species of *Cliffortia* (Rosaceae), 117 species of *Ruellia* (Acanthaceae), 117 species of *Euphorbia* (Euphorbiaceae), 117	ITS2	Yao *et al.* (2010)

Contd.

	species of *Impatiens* (Balsaminaceae), 113 species of *Eryngium* (Apiaceae), 106 species of *Eucalyptus* (Myrtaceae), 104 species of *Croton* (Euphorbiaceae), 99 species of *Calceolaria* (Calceolariaceae), 98 species of *Cuscuta* (Convolvulaceae), 97 species of *Dianthus* (Caryophyllaceae), 96 species of *Salvia* (Lamiaceae), 94 species of *Berberis* (Berberidaceae), 86 species of *Rhododendron* (Ericaceae), 84 species of *Macaranga* (Euphorbiaceae), 83 species of *Acer* (Sapindaceae), 82 species of *Prunus* (Rosaceae), 81 species of *Pilea* (Urticaceae), 81 species of *Coffea* (Rubiaceae), 273 species of *Allium* (Alliaceae), 43 species of *Cyrtanthus* (Amaryllidaceae), 34 species of *Crinum* (Amaryllidaceae), 49 species of *Pinanga* (Arecaceae), 52 species of *Kniphofia* (Asphodelaceae), 50 species of *Costus* (Costaceae), 318 species of *Carex* (Cyperaceae), 52 species of *Eleocharis* (Cyperaceae), 31 species of *Lachenalia* (Hyacinthaceae), 45 species of *Luzula* (Juncaceae), 42 species of *Juncus* (Juncaceae), 79 species of *Gagea* (Liliaceae), 78 species of *Lilium* (Liliaceae), 49 species of *Fritillaria* (Liliaceae), 37 species of *Musa* (Musaceae), 227 species of *Maxillaria* (Orchidaceae), 139 species of *Oncidium* (Orchidaceae),	

Contd.

	121 species of *Dendrobium* (Orchidaceae), 120 species of *Disa* (Orchidaceae), 100 species of *Ophrys* (Orchidaceae), 85 species of *Paphiopedilum* (Orchidaceae), 56 species of *Phalaenopsis* (Orchidaceae), 48 species of *Masdevallia* (Orchidaceae), 46 species of *Gomesa* (Orchidaceae), 42 species of *Satyrium* (Orchidaceae), 42 species of *Dendrochilum* (Orchidaceae), 41 species of *Cyrtochilum* (Orchidaceae), 38 species of *Telipogon* (Orchidaceae), 36 species of *Dichaea* (Orchidaceae), 33 species of *Diuris* (Orchidaceae), 33 species of *Scaphyglottis* (Orchidaceae), 30 species of *Cymbidium* (Orchidaceae), 115 species of *Poa* (Poaceae), 46 species of *Alpinia* (Zingiberaceae), 37 species of *Amomum* (Zingiberaceae)		
4.	Fern *(Cheilanthes wrightii)*	Plastids *rbcL, atpA, trnG-R*	Pryer *et al.* (2010)
5	Medicinal plant – *Paris* spp.	ITS2	Zhu *et al.* (2010)
6	*Tree Peony*	Chloroplast gene fragments *ndh*F, *rps*16-*trn*Q, *trn*L-F and *trn*S-G	Zhang *et al.* (2009)
7	Many genus in the family Rosaceae	*ITS2*	Pang *et al.* (2010)
8	Medicinal plants in the family Fabaceae	*ITS2*	Gao *et al.* (2010a)
9	Members of Asteraceae family	*ITS2*	Gao *et al.* (2010b)
10	Chinese medicinal plant - *Cistanches herba*	nr*ITS2*	Sun *et al.* (2012)
11	*Dendrobium*	**Chloroplast** *psbA-trnH* **intergenic region**	Yao *et al.* (2009)

Contd.

12	Many genus belonging to Oleaceae, Fagaceae and Aceraceae	Non-coding intergenicspacer *trn*H-*psb*A	de Nicola *et al.* (2008)
13	*Fragaria* spp.	Nuclear ribosomal internal transcribed spacer (nr*ITS*), chloroplast spacer *psb*A-*trn*H	Njuguna and Bassil, (2011)
14	19 *Citrus* spp., *Microcitrus* spp., *Eremocitrus glauca*, 3 *Poncirus* spp., *Clymenia polyandra*, 5 *Fortunella* spp.,	*mat*K	Yu *et al.* (2011)
15	Ferns	*rbc*L and *mat*K	Li *et al.* (2011b)
16	Mango (*Mangifera* spp.)	*mat*K	Hidayat *et al.* (2012)
17	7 *Capsicum* spp.	*trn*H-*psb*A or *trn*L-*trn*T, and *waxy* nuclear introns	Jarret, (2008)
18	Tea	*rbc*L, *mat*K	Stoeckle *et al.* (2011)
19	Many cryptic orchid species	*mat*K	Lahaye *et al.* (2008)
20	Many cryptic *Conocephalum* (Bryophyte) species	rbcL	Miwa *et al.* (2009)
21	Many cryptic *Taxus* species	nr*ITS*, *trn*L-F	Liu *et al.* (2010a)
22	*Cinnamomum* spp.	*trnL–trnF* chloroplast DNA	Kojoma *et al.* (2002)

Table 4 : DNA barcoding applications in plants other than species identification

Sl. No.	Application	Candidate gene/s	Reported by
1	Community phylogenetics of a tropical forest plot	*rbc*L+*trn*H-*psb*A +*mat*K	Kress *et al.* (2009, 2010)
2	Identifying the composition of herbal medicines (reference to *Phyllanthus* spp.)	*trn*H-*psb*A	Srirama *et al.* (2010)
3	Assessing the plant components of honey	*trn*L P6 loop	Valentini *et al.* (2010)
4	Identification of invasive species	nr*ITS*, *trn*L, *trn*L-F	Van DeWiel *et al.*, (2009)
5	Identification of CITES listed ramin timber and products	*mat*K (SNP assay of barcode sequence)	Ogden *et al.* (2008)

Contd.

6	Proof-of-concept for application to forensics	*trn*H-*psb*A, *trn*L-F	Ferri *et al.* (2009)
7	Identification of plant components of herbivore diet	*trn*L P6 loop	Valentini *et al.* (2009)
8	Identification of adulteration in Chinese herbal drug Baiying (substitution of *Solanum lyratum with Aristolochia mollissimal*)	ITS, *matK, rbcL, trnH-psbA* and *trnL-trnF*	Li *et al.* (2012)

Major Agencies and Projects

The major agencies involved in the large scale DNA barcoding projects are listed in Table 5.

Table 5 : The major agencies involved in barcoding projects (Hollingsworth *et al.*, 2011)

Sl. No.	Project	Further information	Lead institute
1	Consortium for the Barcode of Life (CBOL)	http://barcoding.si.edu/DNABarCoding.htm)	
2	International Barcode of Life Project (iBOL)	http://www.ibolproject.org, Vernooy *et al.*, 2010); http://www.dnabarcoding.ca/research.php	Canadian Centre for DNA barcoding
3	Barcoding the world's tree species (TreeBOL)	http://www.bgci.org/worldwide/news/0463/	The New York Botanical Garden
4	Barcoding grasses and grass-like plants (GrassBOL)		Adelaide University and University of British Columbia
5	Flora of the Kruger National Park		University of Johannesburg
6	Flora of the Area de Conservacion Guanacaste Costa Rica		University of Pennsylvania
7	Flora of Korea		Korea University
8	Plant Barcoding China: DNA barcoding of 5000 Chinese plant species		Kunming Institute of Botany

Contd.

9	All-genera: DNA barcoding of representatives of all angiosperm genera		The New York Botanic Garden
10	DNA barcoding of Centre for Tropical Forestry Plots		Smithsonian Institute
11	DNA barcoding Chinese medicinal plants		Institute of Medicinal Plant Development Beijing
12	DNA barcoding the flora of Wales		National Botanic Garden of Wales
13	DNA barcoding British bryophytes		Royal Botanic Garden Edinburgh

Procedure for DNA Barcoding

General procedure followed in any barcoding process includes total DNA extraction, PCR amplification of candidate genes at barcoding regions using specific forward and reverse primers, cleaning the PCR products (illustra DNA and Gel Band purification kits may be used), elution in buffer, cycle sequencing (2–5μL PCR product and 2μL DTCS may be optimum in a 10μL reaction, cleaning by ethanol precipitation), sequence analysis (Beckman Coulter CEQ 8800 or 8000 Genetic Analysis System are generally used), sequence editing (CEQ Genetic Analysis System software v. 8.0 or CLUSTAL X v. 1.81 are generally used) and sequence assembling (with Sequencher 4.6, GeneCodes Corporation).

PCR Mix

Buffer X 1, Mg^{2+} 1.5mM, dNTPs 0.2mM, FW test primer 1μM, RE test primer 1μM, *Taq* DNA polymerase 2 units, BSA 0.1mg/ml, Template (variable) and make the volume to 20 μl with water

Thermal Profile

Initial denaturation - 94°C	1min	1 cycle
Denaturation - 94°C	30sec	
Annealing - 53°C	40sec (to be standardized for each genus) Up to	40 cycles
Extension - 72°C	40sec	
Final extension - 72°C	5min	1 cycle

Protocol issued by CBOL (Ivanova *et al.*, 2006)

Follow any of the fast and cheap extraction protocols and quantify the DNA. Critically design the primers. Primers with 2-4 degenerate positions will often rescue barcodes from recalcitrant specimens. Single bp mismatches at the 3′-end of a primer usually prevent PCR amplification. This problem can often be solved by the use of degenerate or inosine containing primers. Standardize the PCR protocols and complete the reactions. For projects that are examining compliant specimens, it is possible to proceed directly from the barcode PCR to a sequencing reaction. However, it is often critical to screen PCR reactions for product when working with older specimens or with a new taxonomic group. The traditional method of agarose gel screening of the PCR product is very laborious. The alternates suggested are microfluidic technology and pre-cast agarose gels. When the PCR product is sequenced in just a single direction, PCR products should be cleaned up to remove unincorporated nucleotides and residual primers. The cleaning and ethanol precipitation may be done with any commercial kits such as Milllipore MultiScreen Filter Technology or QIAGEN QIAquick Robo96 PCR Clean-up Kit or Promega's Wizard® MagneSil TM. Dispense the sequencing reaction cocktail into either 96-well plates or into larger volume tubes and then freeze it for up to 3 months. Trehalose is added to ensure stability of the enzyme during freeze-thaw cycles. Sequencing primer concentrations of 10 and 3 pM are optimal for analyzing PCR products without and with clean up, respectively. Sequencing reaction clean up is generally done by Sephadex clean up protocol. For high-throughput labs, machines such as Applied Biosystems 3730 or 3730XL DNA Analyzer with 48-96 capillaries and automated polymer delivery may be used. Employ the Barcode of Life database to organize sequence records and keep them in alignment. This software also allows results from different projects to be merged into a virtual project enabling sequence comparisons and tree generation for specimens from different projects.

Barcode of Life Data System (BOLD) and the Sequence Database

An essential criterion for depositing the species sequences in BOLD is a minimum length of 500 bp for *Cox*I, *rbc*L or *mat*K sequences and 100bp for ITS. The sequence of more than 3 individuals per species should be submitted. Under International Barcode of Life project (iBOL) by the Canadian International Consortium Initiative (ICI), researchers from 25 countries are involved in building a comprehensive DNA barcode registry for eukaryotic life. The program plans to acquire DNA barcode records for 5 million specimens representing 500,000 species (out of more than an estimated 10 million species to be discovered) within 2015. The BOLD barcode Submission tool associates the 'Barcode submissions' with a further submission to NCBI using the 'My NCBI' user name.

Query Sequence Assignment at BOLD

This assigns one query sequence to a set of referenced tagged-specimen sequences extracted from BOLD. The method currently used in BOLD combines similarity methods with distance tree reconstruction in the following way:

First, the query sequence is aligned to the global alignment through a Hidden Markov Model (HMM) profile of the candidate gene specific protein followed by a linear search of the reference library. The 100 best hits are selected as a pre-set of "closely related tagged-specimens. Second, a Neighbor-Joining tree is reconstructed on this preset plus the query sequence to assess the relationship between the query sequence and its neighboring referenced sequences. The query sequence is then assigned to the species name of its nearest-neighboring referenced sequence, whatever the distance between the two sequences.

Various alternative methods have been proposed to analyze DNA barcode data amongst which we can distinguish four main categories of approaches:

(i) similarity approaches, based solely on the similarity between the total DNA barcode sequences or small parts of them

(ii) classical phylogenetic approaches, using either genetic distances or maximum likelihood/ Bayesian algorithms and assuming different mutational models. Online softwares such as Taxl are launched exclusively for analyzing the barcoding data (Steinke *et al.*, 2005).

(iii) multiple-character based analysis

(vi) pure statistical approaches based on classification algorithms without any biological models or assumptions

(v) genealogical methods based on the coalescent theory using demo-genetic models and maximum likelihood /Bayesian algorithms

Table 6 : The bioinformatics tools available for DNA barcode sequence analysis (Casiraghi *et al.*, 2010).

Tipology	Method (s)	Software/ tool(s)	Resources
Threshold (distance)	Similarity	Blastall-BLASTn	ftp:/ftp.rKbi.nih.gov/blast/
	Similarity	BLAT .edu/~kent/exe/	htt p//ge nom e-test.cse.ucsc
	Similarity	Blastall-megaBLAST	ftp://ftp.ncbi.nih.gov/ blast/
	Pair wise distance	Taxl	axel meyer@unl-konscanz.de
	Pair wise distance	TaxonDNA	http:www.taxondna.sf.net/
	K2P distance	MUSCLE, MEGA	maurizio.casiraghi@unlimib.it
	K2P distance	BOLD-IDS	http://www.barcodinglife.org/ views/idrequest.php
	Patristic distance	Mr Bayer, PAUP, APE, Perl scripts	lefebure@univlyonl.fr
Phylogenetic	Neighbour Joining	MUSCLE, MEGA	marianne.elias@ed.ac.uk
	Parsimony	MUSCLE, TNT	dittle@nybg.org
	Maximum likelihood	MUSCLE, SPRI, PHYML2	http://atgc.lirmm.fr/spr/
	Bayesian inference	SAP	http://fisher.berkeley.edu/cteg/ software/munch
	Coalescent based		rasmus@binf.ku.dk
	Coalescent based		rasmus@binf.ku.dk
	Coalescent based	COALESCENCE, FLUCTUATE, PAUP. Seq.Gen	golding@mcmaster.ca
	Coalescent based	COAL, MESQUITE	knowlest@mumich.edu
	Coalescent based	general mixed Yule-coalescent (GMYC) model	monaghan@g.b.-berlin.de
Character based	Diagnostic	CAOS	http://www.genomecurator.org/ CAOS/CAOSindex.html
	Diagnostic	MATLAB, local perl scripts	drichardson@rsmas.miami.edu
	Diagnostic	DNA-BAR (degentar)	http://dna.engr.uconn.edu/ ~software/DNA-BAR/
	Diagnostic	DOME ID (local perl scripts)	dittle@nybg.org
Combined	Yule model/coalescence	TCS, MEGA, Arlequin, PAUP. PAUPRat script, Phylip, rBs, R	http://www.imedea.uib.es/ ~jpons/jPWPhome.htm
	BLAST/parsimony ratchet	BLAST, MUSCLE, TNT	dittle@nybg.org
	BLAST/SPR	BLAST, MUSCLE, SPR	dittle@nybg.org
	BLAST/Neighbour Joining	BLAST, MUSCLE, neighbour	dittle@nybg.org
Aligment-free	Tree-based	ATIM:TNT, local scripts	dittle@nybg.org
	Component vector	CV Tree alpha 1.0	http://cvtree.cbi.pku.edu.cn
	Spectrum kernel method	Spectrum	vladimira@cs.rutgers.edu
Web tool		Web browser	http://www.ibarcode.org
		Web browser	http://www.dnabarcodeliker.com/
		Web browser	http://www.asianbarcode.org
Pther		ConFind, Phthon	http://www.colorado.edu/ chemistry/RGHP/software/

Challenges Ahead

Ways to extract DNA from formalin-conserved specimens are not available at present. This procedure should be standardized and the DNA barcodes should be harvested from them.

The sampling shortage across taxa can sometimes lead to 'barcoding gaps'. The individuals chosen to represent each taxon in the reference database should cover the major part of the existing diversity. Indeed, in the interrogation of BOLD, identification difficulties arise when the unknown specimens come from a currently under-described part of biodiversity.

The diversity of mitochondrial DNA (*mt*DNA) is strongly linked to the female genetic structure due to maternal inheritance. The use of mitochondrial loci can thus lead to overestimate sample divergence and render conclusions on species status unclear.

Nuclear mitochondrial DNAs (NUMTs) are nuclear copies of mitochondrial DNA sequences that have been translocated into the nuclear genome. In eukaryotes, the number and size of NUMTs are variable, ranging. Disturbance due to NUMTs must be seriously considered, in both DNA barcode library construction and further specimen identification.

The rate of genome evolution (mitochondrial or nuclear) is not equal for all living species. In higher plants, the mitochondrial genome evolves much more slowly than in animals. Considering mitochondria solely would not solve problems of differential evolutionary rates among close genera with inheritance discrepancy of *mt*DNA introgression processes and of the intron-size variations that prevent the sequence alignment. Thus CBOL-PWG agrees that plant barcoding will be multi-locus, with one 'anchor' (universal across the plant kingdom) and 'identifiers' to distinguish closely related species. Preferably, the combination will certainly contain non-coding intergenic spacers (e.g. *trn*H-*psb*A) and plastidial coding sequences (e.g. *mat*K).

In case of lack of resolving power of any specific gene sequence reported for any taxa, CBOL envisages the transition from the primary single-gene method to a multi-region barcoding system.

Geographical structure, if ignored, can blur and distort species delineation. In clades where speciation has been very recent, or rates of mutation are very slow, barcode sequences may be shared among related species. Polyploid speciation can also lead to incongruence between barcode sequences and taxon concepts (Fazekas *et al.*, 2009). In species where dispersal is poor, populations are relatively isolated from one another and hence the efficiency of barcoding may have an inverse correlation between intra- and inter-specific gene flow (Petit and Excoffier, 2009). Actually, high rates of intra-specific divergence can derive from geographically isolated populations and hence should be considered standardizing the DNA barcode reference database.

Conclusions

The purpose of the DNA barcoding is to rapidly assemble a precise and representative reference library. Thus it is based on conventional and inexpensive protocols for DNA extraction, amplification and sequencing. This technique is proven useful in international trade of horticultural crops and it had already exposed a case of mistaken identity in the fern export (Pryer *et al.,* 2010). Further, barcoding was proven useful to identify the cortex herbs of Chinese pharmacopoeia (Song *et al.,* 2009; Li *et al.,* 2011a; Sun and Chen, 2012). Even though the barcoding can not substitute morphological identification, especially in few complicated species, with time, the reference library will become increasingly useful, enabling the rapid identification of low taxonomic level taxa with specific short-DNA sequences (Will and Rubinoff, 2004).

References

Casiraghi, M., Labra, M., Ferri, E., Galimberti A. and De Mattia F. 2010, DNA barcoding: a six-question tour to improve users' awareness about the method. *Briefings in Bioinformatics,* 11: 440-453.

CBOL Plant Working Group. 2009. A DNA barcode for land plants. *Proc. Natl. Acad. Sci. USA,* 106: 12794–12797.

Chang, C.C., Lin, H.C., Lin, I.P.,Chow, T.Y., Chen, H.H., Chen, W.H., Cheng, C.H., Lin, C.Y., Liu, S.M., Chang, C.C., *et al.,* 2006, The chloroplast genome of *Phalaenopsis aphrodite* (Orchidaceae): comparative analysis of evolutionary rate with that of grasses and its phylogenetic implications. *Mol. Biol. Evol.,* 23: 279–291.

Chase, M.W., Cowan, R.S., Hollingsworth, P.M., van den Berg, C., Madrinan, S., Petersen, G., Seberg, O., Jorgsensen, T., Cameron, K.M., Carine, M., *et al.,* 2007, A proposal for a standardised protocol to barcode all land plants.*Taxon,* 56: 295–299.

Chase, M.W., Salamin, N., Wilkinson, M., *et al.,* 2005, Land plants and DNA barcodes: short-term and long-term goals. *Philosophical Transactions of the Royal Society B: Biological Sciences,* 360: 1889–1895.

Chen, S., Yao, H., Han, J., Liu, C., Song, J., Shi, L., Zhu, Y., Ma, X., Gao, T., Pang, X., Luo, K., Li, Y., Li, X., Jia, X., Lin, Y. and Leon, C. 2010, Validation of the ITS2 region as a novel DNA barcode for identifying medicinal plant species. *PLoS One,* 5: e8613.

Dayrat, B. 2005, Towards integrative taxonomy. *Biol. J. Linn. Soc.,* 85: 407-415.

de Nicola, S., Ferreira, R., D'Ambrosio, C., Piredda, R., Puddu, G., Simeone, M.C., Attimonelli, M. and Schirone, B. 2008, DNA barcoding of wild italian dendroflora: first evidences of application in *Oleaceae, Fagaceae* and *Aceraceae.* In: *Proceedings of the 52nd Italian Society of Agricultural Genetics Annual Congress,* Padova, Italy – 14/ 17 September, 2008 (ISBN 978-88-900622-8-5), p. F01.

Devey, D.S., Chase, M.W. and Clarkson, J.J. 2009, A stuttering start to plant DNA barcoding: microsatellites present a previously overlooked problem in non-coding plastid regions. *Taxon*, 58: 7–15.

Dunning, L.T. and Savolainen, V. 2010, Broad-scale amplification of *matK* for DNA barcoding plants, a technical note. *Botanical Journal of the Linnean Society*, 164: 1–9.

Erbach, M.C. and Holdrege, C. 2005, DNA barcoding is no substitute for taxonomy. *Nature*, 434: 697.

Fazekas, A.J., Burgess, K.S., Kesanakurti, P.R., Graham, S.W., Newmaster, S.G., *et al.*, 2008, Multiple multilocus DNA barcodes from the plastid genome discriminate plant species equally well. *PLoS One*, 3: e2802.

Fazekas, A.J., Kesanakurti, P.R., Burgess, K.S., Percy, D.M., Graham, S.W., *et al.*, 2009, Are plant species inherently harder to discriminate than animal species using DNA barcoding markers? *Molecular Ecology Resources*, 9: 130–139.

Ferri, G., Alu,' M., Corradini, B. and Beduschi, G. 2009, Forensic botany: species identification of botanical trace evidence using a multigene barcoding approach. *International Journal of Legal Medicine*, 123: 395–401.

Gao, T., Yao, H., Song, J.Y., Liu, C., Zhu, Y.J., Ma, X.Y., Pang, X.H., Xu, H.X. and Chen, S.L. 2010a, Identification of medicinal plants in the family Fabaceae using a potential DNA barcode ITS2. *J. Ethnopharmacol.*, 130: 116-121.

Gao, T., Yao, H., Song, J.Y., Zhu, Y.J., Liu, C. and Chen, S.L. 2010b, Evaluating the feasibility of using candidate DNA barcodes in discriminating species of the large Asteraceae family. *BMC Evol. Biol.*, 10: 324.

Gregory, T.R. 2005, DNA barcoding does not compete with taxonomy. *Nature*, 434: 1067.

Hebert, P.D.N., Cywinska, A., Ball, S.L. and De Waard, J.R. 2003a, Biological identifications through DNA barcodes. *Proc. Roy. Soc. Biol. Sci. – Ser. B*, 270: 313–321.

Hebert, P.D.N., Penton, E.H., Burns, J.M., Janzen, D.H. and Hallwachs, W. 2004, Ten species in one: DNA barcoding reveals cryptic species in the neotropical skipper butterfly *Astraptes fulgerator*. *Proc. Natl Acad. Sci. USA.*, 101: 14812–14817.

Hebert, P.D.N., Ratnasingham, S. and DeWaard, J.R. 2003b, Barcoding animal life: Cytochrome c oxidase subunit 1 divergences among closely related species. *Proc. Roy. Soc. Biol. Sci. – Ser.* B, 270: S96-S99.

Hidayat, T., Pancoro, A. and Kusumawaty, D. 2012, Utility of *mat*K gene to assess evolutionary relationship of genus *Mangifera* (Anacardiaceae) in Indonesia and Thailand. *Biotropia*, 18: 74 – 80.

Hollingsworth, M.L., Clark, A., Forrest, L.L., Richardson, J., Pennington, T. R., Long, D.G., Cowan, R., Chase, M.W., Gaudeul M. and Hollingsworth, P.M. 2009, Selecting barcoding loci for plants: evaluation of seven candidate loci with species-level sampling in three divergent groups of land plants. *Molecular Ecology Resources*, 9: 439–457.

Hollingsworth, P.M., Graham, S.W. and Little, D.P. 2011, Choosing and using a plant DNA barcode. *PLoS One*, 6: e19254.

Ivanova, N.V., deWaard, J.R., Hajibabaei, M. and Hebert, P.D.N. 2006, *Protocols for high-volume DNA barcode analysis*, Draft submission to DNA working group - Consortium for the Barcode of Life, pp.24.

Jarret, R.L. 2008, DNA Barcoding in a crop genebank: The *Capsicum annuum* species complex. *The Open Biology Journal*, 1: 35-42.

Kojoma, M., Kurihara, K., Yamada, K., Sekita, S., Satake, M. and Iida, O. 2002, Genetic identification of cinnamon (*Cinnamomum* spp.) based on the *trnL–trnF* chloroplast DNA. *Planta Medica*, 68: 94–96.

Kress, W. J., Wurdack, K.J., Zimmer, E.A., Weigt, L.A. and Janzen, D.H. 2005, Use of DNA barcodes to identify flowering plants. *Proc. Natl. Acad. Sci. USA*, 102: 8369–8374.

Kress, W.J. and Erickson, D.L. 2007, A two-locus global DNA barcode for land plants: The coding *rbc*L gene complements the non-coding trnH-psbA spacer region. *PLoS One* 2: e508.

Kress, W.J. and Erickson, D.L. 2012, *DNA Barcodes: Methods and Protocols*, Methods in Molecular Biology, 858: Part 1, 3-8, Springer Publishing, DOI: 10.1007/978-1-61779-591-6_1.

Kress, W.J., Erickson, D.L., Jones, F.A., Swenson, N.G., Perez, R., *et al.*, 2009, Plant DNA barcodes and a community phylogeny of a tropical forest dynamics plot in Panama. *Proceedings of the National Academy of Sciences*, 106: 18621–18626.

Kress, W.J., Erickson, D.L., Swenson, N.G., Thompson, J., Uriarte, M., *et al.*, 2010, Advances in the use of DNA barcodes to build a community phylogeny for tropical trees in a Puerto Rican forest dynamics plot. *PLoS One*, 5: e15409.

Krishnamurthy, P.K. and Francis, R.A. 2012, A critical review on the utility of DNA barcoding in biodiversity conservation. *Biodiversity and Conservation*, 21: 1901-1919.

Lahaye, R., van der Bank, M., Bogarin, D., Warner, J., Pupulin, F., Gigot, G., Maurin, O., Duthoit, S., Barraclough, T.G., and Savolainen, V. 2008, DNA barcoding the floras of biodiversity hotspots. *Proc. Natl. Acad. Sci. USA*, 105: 2923–2928.

Ledford, H. 2008, Botanical identities: DNA barcoding for plants comes a step closer. *Nature*, 415: 616.

Li, F.W., Kuo, L.Y., Rothfels, C.J., Ebihara, A., Chiou, W.L., Windham, M.D. and Pryer, K.M. 2011b, *rbc*L and *mat*K earn two thumbs up as the core DNA barcode for ferns. *PLoS One*, 6: e26597.

Li, M., Au, K.Y., Lam, H., Cheng, L., Jiang, R.W., But, P.P.H. and Shaw, P.C. 2012, Identification of Baiying (Herba *Solani lyrati*) commodity and its toxic substitute Xungufeng (Herba *Aristolochiae mollissimae*) using DNA barcoding and chemical profiling techniques. *Food Chemistry*, 135: 1653–1658.

Li, M., Cao, H., But, P.P.H. and Shaw, P.C. 2011a, Identification of herbal medicinal materials using DNA barcodes. *Journal of Systematics and Evolution*, 49 (Special Issue: Plant DNA barcoding in China): 271–283.

Liu, J.I.E., Mo¨ ller, M., Gao, L-M., Zhang, D-Q. and Li, D-Z. 2010a, DNA barcoding for the discrimination of Eurasian yews (*Taxus* L., Taxaceae) and the discovery of cryptic species. *Molecular Ecology Resources*, 11: 89–100.

Liu, Y., Yan, H.F., Cao, T. and Ge, X.J., 2010b, Evaluation of 10 plant barcodes in Bryophyta (Mosses). *Journal of Systematics and Evolution*, 48: 36–46.

Miwa, H., Odrzykoski, I.J., Matsui, A., Hasegawa, M., Akiyama, H., *et al.*, 2009, Adaptive evolution of *rbc*L in *Conocephalum* (Hepaticae, bryophytes). *Gene*, 441: 169–175.

Mower, J., Touzet, P., Gummow, J., Delph, L. and Palmer, J. 2007, Extensive variation in synonymous substitution rates in mitochondrial genes of seed plants. *BMC Evolutionary Biology*, 7: 135.

Newmaster, S.G., Fazekas, A.J. and Ragupathy, S. 2006, DNA barcoding in land plants: Evaluation of *rbcL* in a multigene tiered approach. *Can. J. Bot.*, 84: 335–341.

Newmaster, S.G., Fazekas, A.J., Steeves, R.A.D. and Janovec, J. 2008, Testing candidate plant barcode regions in the Myristicaceae. *Molecular Ecology Resources*, 8: 480–490.

Njuguna, W. and Bassil, N.V. 2011, DNA barcoding: unsuccessful for species identification in *Fragaria* L. *Acta Hort.*, 918: 349-356.

Ogden, R., Mc Gough, H.N., Cowan, R.S., Chua, L., Groves, M., *et al.*, 2008, SNP based method for the genetic identification of ramin *Gonystylus* spp. timber and products: applied research meeting CITES enforcement needs. *Endangered Species Research*, 9: 255–261.

Pang, X.H., Song, J.Y., Zhu, Y.J., Xu, H.X., Huang, L.F. and Chen, S.L. 2010, Applying plant DNA barcodes for *Rosaceae* species identification. *Cladistics*, 26: 1-6.

Pennisi, E. 2007, Taxonomy. Wanted: a barcode for plants. *Science*, 318: 190-191.

Petit, R.J. and Excoffier, L. 2009, Gene flow and species delimitation. *Trends Ecol. Evol.*, 24: 386–393.

Pryer, K.M., Schuettpelz, E., Huiet, L., Grusz, A.L., Rothfels, C.J., Avent, T., Schwartz, D. and Windham, M.D. 2010, DNA barcoding exposes a case of mistaken identity in the fern horticultural trade. *Molecular Ecology Resources*, 10: 979–985.

Ratnasingham, S. and Hebert, P.D.N. 2007, BOLD: The Barcode of Life Data System (www.barcodinglife.org). *Mol. Ecol. Notes*, 7: 355-364.

Rubinoff, D., Cameron, S. and Will, K. 2006a, A genomic perspective on the shortcomings of mitochondrial DNA for "barcoding" identification. *J. Hered.*, 97: 581-594.

Rubinoff, D., Cameron, S. and Will, K. 2006b, Are plant DNA barcodes a search for the Holy Grail? *Trends Ecol. Evol.*, 21: 1–2.

Sass, C., Little, D.P., Stevenson, D.W. and Specht, C.D. 2007, DNA Barcoding in the Cycadales: Testing the potential of proposed barcoding markers for species identification of cycads. *PLoS One*, 2: e1154.

Schindel, D.E. and Miller, S.E. 2005, DNA barcoding - a useful tool for taxonomists. *Nature*, 435: 17.

Seberg, O. and Petersen, G. 2009, How many loci does it take to DNA barcode a *Crocus*? *PLoS One*, 4: e4598.

Smith, M.A., Fisher, B.L. and Hebert, P.D.N. 2005, DNA barcoding for effective biodiversity assessment of a hyperdiverse arthropod group: the ants of Madagascar. *Philos. Trans. R. Soc. London Ser. B*, 360: 1825–1834.

Song, J.Y., Yao, H., Li, Y., Li, X.W., Lin, Y.L., Liu, C., Han, J.P., Xie, C.X. and Chen, S.L. 2009, Authentication of the family Polygonaceae in Chinese pharmacopoeia by DNA barcoding technique. *J. Ethnopharmacol.*, 124: 434-439.

Srirama, R., Senthilkumar, U., Sreejayan, N., Ravikanth, G., Gurumurthy, B.R., *et al.*, 2010, Assessing species admixtures in raw drug trade of *Phyllanthus*, a hepato-protective plant using molecular tools.*J. Ethnopharmacol.*, 130: 208–215.

Steinke, D., Vences, M., Salzburger, W. and Meyer, A. 2005, TaxI: a software tool for DNA barcoding using distance methods. *Phil. Trans. R. Soc. B*, 360: 1975–1980.

Stoeckle, M.Y., Gamble, C.C., Kirpekar, R., Young, G., Ahmed, S. and Little, D.P. 2011, Commercial teas highlight plant DNA barcode identification successes and obstacles. *Sci. Rep.*, 1: 42.

Sun, Z. and Chen, S. 2012, Identification of cortex herbs using the DNA barcode nrITS2. *Journal of Natural Medicines*, DOI: 10.1007/s11418-012-0681-8 (In press).

Sun, Z., Song, J., Yao, H. and Han, J. 2012, Molecular identification of Cistanches Herba and its adulterants based on *nr*ITS2 sequence. *Journal of Medicinal Plants Research*, 6: 1041-1045.

Taberlet, P., Coissac, E., Pompanon, F., Gielly, L., Miquel, C., Valentini, A., Vermat, T., Corthier, G., Brochmann, C. and Willerslev, E. 2007, Power and limitations of the chloroplast *trnL* (UAA) intron for plant DNA barcoding. *Nucleic Acids Res.*, 35: e1–e8.

Valentini, A., Miquel, C. and Taberlet, P. 2010, DNA barcoding for honey biodiversity. *Diversity*, 2: 610–617.

Valentini, A., Miquel, C., Nawaz, M.A., Bellemain, E.V.A., Coissac, E., *et al.*, 2009, New perspectives in diet analysis based on DNA barcoding and parallel pyrosequencing: the *trn*L approach. *Molecular Ecology Resources*, 9: 51–60.

Van DeWiel, C.C.M., Van Der Schoot, J., Van Valkenburg, J.L.C.H., Duistermaat, H. and Smulders, M.J.M. 2009, DNA barcoding discriminates the noxious invasive plant species, floating pennywort (*Hydrocotyle ranunculoides* L.), from non-invasive relatives. *Molecular Ecology Resources*, 9: 1086–1091.

Vences, M., Thomas, M., Bonett, R.M. and Vieites, D.R. 2005, *Philos. Trans. Roy. Soc. London Ser-B*, 360: 1859–1868.

Vernooy, R., Haribabu, E., Muller, M.R., Vogel, J.H., Hebert, P.D.N., *et al.*, 2010, Barcoding life to conserve biological diversity: Beyond the taxonomic imperative. *PLoS Biol.*, 8: e1000417.

Ward, R.D., Zemlak, T.S., Innes, B.H., Last, P.R. and Hebert, P.D.N. 2005, DNA barcoding Australia's fish species. *Philos. Trans. R. Soc. London Ser. B*, 360: 1847–1857.

Will, K. and Rubinoff, D. 2004, Myth of the molecule: DNA barcodes for some species cannot replace morphology for identification and classification. *Cladistics*, 20: 47-55.

Wilson, O.E. 2004, The encyclopedia of life. *Trends Ecol. Evol.*, 18: 77-80.

Yao, H., Song, J., Liu, C., Luo, K., Han, J., Li, Y., Pang, X., Xu, H., Zhu, Y., Xiao, P. and Chen, S. 2010, Use of ITS2 region as the universal DNA barcode for plants and animals. *PLoS One*, 5: e13102.

Yao, H., Song, J.Y., Ma, X.Y., Liu, C., Li, Y., Xu, H.X., Han, J.P., Duan, L.S. and Chen, S.L. 2009, Identification of *Dendrobium* species by a candidate DNA barcode sequence: the chloroplast *psbA-trnH* intergenic region. *Planta Med.*, 75: 667-669.

Yu, J., Yan, H., Lu, Z. and Zhou, Z. 2011, Screening potential DNA barcode regions of chloroplast coding genome for *Citrus* and its related genera. *Scientia Agricultura Sinica*, 44: 341-348.

Zhang, J.M., Wang, J.X., Xia, T. and Zhou, S.L. 2009, DNA barcoding: Species delimitation in tree peonies. *Sci. China Ser-C: Life Sci.*, 52: 568-578.

Zhu, Y.J., Chen, S.L., Yao, H., Tan, R., Song, J.Y., Luo, K. and Lu, J. 2010, DNA barcoding the medicinal plants of the genus *Paris*. *Acta Pharm. Sin.*, 45: 376-382.

3

Genome Mapping and Gene Tagging for Marker Assisted Selection

R. Sujatha and K.V. Peter

In this era of molecular markers in genetic studies, genome analysis encompasses several stages of increasing resolution and sophistication, leading to the generation of high resolution chromosome maps. Since the entire plant kingdom cannot be covered under sequencing projects, molecular markers and their correlation to phenotypes provide us with requisite landmarks for elucidation of genetic variation.

A molecular marker is a site of heterozygosity for some type of DNA variation. Such a DNA marker in mapping analysis is analogous to a conventional heterozygous allelic pair. When they are used for mapping the genome, the identity of this DNA fragment is not important, but is simply used as a reference point to arrange and order the genetic information along the length of a chromosome. The molecular markers provide landmarks along the length of chromosome that facilitate the manipulation of chromosome segments in breeding and chromosome engineering programs.

Why to Map a Gene ?

- To study the genetic organization
- To know the gene/genes influencing a trait

- To locate a gene on a chromosome
- To aid in map based cloning of genes
- To aid in marker assisted selection (MAS)
- To study the evolutionary- relationship among the species

The first step is the assignment of trait specific genes and molecular markers to specific linkage groups. The next step involves calculating the map distances between gene and marker loci on a chromosome. Finally, physical mapping and sequencing provide the highest level of resolution in genome analysis. A fine structure map (saturated map) can be used to address fundamental and practical questions such as i) organization of functional vs. non-functional DNA regions, ii) elucidation of evolutionary processes and relationships, iii) systematic and taxonomic studies, iv) positional cloning, v) manipulation of genes and chromosome segments in breeding programs (marker assisted selection), etc.

1. The Basic Concepts

In genetics, gene is the unit of heredity transmitted from generation to generation during sexual or asexual reproduction. It exist in two or more alternative forms occupying the same site in the chromosome which is known as a locus. In molecular biology, gene is a segment of nucleic acid that encodes peptide or RNA. Since there are several thousand genes in an organism and the number of chromosomes is a few, several genes are located on the same chromosome. Two genes are said to be linked if they are located on the same chromosome. The group of genes which are said to be linked as they are physically located on the same chromosome is called a linkage group. Hence in a diploid organism, the number of linkage group is equal to the number of haploid set of chromosome which in turn is known as the genome.

As different chromosomes segregate independently during meiosis, the alleles of two genes located on different chromosomes also segregate independently. The chance that an allele at one locus co-inherits with an allele at another locus of the same parental origin is then 0.5 (i.e. 50 % - chance only) and such genes are unlinked. This chance increases if the genes are linked. But at the same time, all genes on the same chromosomes are not always co-inherited. During meiosis, the chromosome often breaks and then rejoins with the homologous chromosome and the phenomenon is known as crossing over and recombination respectively, due to which, new chromosomal combinations appear. The further the distance between two genes on the same chromosome, the more frequently there will be crossover, higher the number of recombination. Therefore, the recombination fraction is calculated from the proportion of recombinants in the gametes produced.

2. Genome Mapping vs. Gene Tagging – Aids to Marker Assisted Selection

Genome mapping is the determination of the relative positions and distances of the genes and other DNA sequences (molecular markers) of a linkage group on chromosomes. The entire set of linkage groups in an organism showing the specific order of linear arrangement of genes and the distance between them is known as the genome map of that organism. The techniques of genome mapping and linkage analysis were developed in 1911 by D.H Morgan and his graduate student Alfred H. Sturtevant and are used still today in much the same way but with far more advanced techniques.

The central idea of genome mapping is the phenomenon of linkage, breaking up of linkage by the crossing-over of chromosomes during meiosis and the concept that the frequency of recombination between two genes / markers can be used as a measure of the actual distance between them on a chromosome. For example, two genes that recombine often are far apart on a genetic map and two which rarely recombine are said to be linked and are very close together on a genetic map. Thus in genome mapping we are associating the gene of interest to a marker for which the linkage group is established and relative distance between the marker and the gene is known. In some cases the actual chromosomal location of the gene and the marker also is known.

Gene tagging can be done without an already prepared genome map. So in gene tagging, we are associating the gene of interest to a marker for which the linkage group or chromosome location is not established. When an easily detectable mutant is identified within a breeding population which is always co-inherited with a desirable trait, the mutant gene now known as the marker is considered as a tag for easy selection of the desired trait. But this will not give the extent of linkage of the marker with the gene of desired trait.

Alternatively the gene causing the mutation can be placed on a genetic map through a series of crosses that would establish its recombination frequency relative to other genes that had previously discovered and mapped. If the mutant gene is found to be in close proximity to the gene for the desired trait, the mutant gene or "marker" is said to be linked to it because the marker and the gene tend to co-segregate or co-inherit. In a breeding programme, this mutant gene could be used to detect the inheritance of the desired trait. If the mutant gene is expressed in the progeny, it is most likely that the progeny also has the desired trait due to its link to the mutant gene. This method of tracking the desired trait by selecting the marker gene is called indirect selection and was widely used by plant breeders whenever such genes are available.

As the availability of such tightly linked mutant genes are less, now molecular markers are used as tags for desired traits. These tags help in detecting the presence of gene and by selecting an individual containing this marker ensures that the gene of interest is also there in that progeny. This method is now popularly known as marker assisted selection (MAS). Here the DNA fragment acts as a pointer leading to the gene coding a specific trait. For example an F_2 population derived from crossing two inbreds may be genotyped with different marker systems as well as phenotyped. If a DNA fragment is segregating for genetic variants that affect a phenotypic trait, then the genotype of the marker in that region should be correlated with the phenotype. The closer the gene resides to the marker, the tighter that association will be, because fewer recombinations will have separated the marker and the gene loci. A variety of analytic techniques can be employed to scan the genome for significant associations and estimate QTL positions, effects, dominance relations and interactions. One must be applying these techniques with great care, however, as it is easy to produce statistical artefacts. However the success of mapping or tagging of a gene of interest depends on the accuracy of phenotyping.

3. Types of Genome Maps

3.1 Genetic/ linkage map

Historically, genetic / linkage maps were assembled by mapping genes with qualitative phenotypic differences to individual loci through controlled experimental crosses followed by calculation of map distance based on recombination frequencies. Determining the location of a gene or molecular marker on a genetic map relies on special cytogenetic stocks or linkage mapping, as measured by the recombination frequency in dihybrid and multihybrid crosses. In this case only grouping of the genes into different linkage groups will be there but to which chromosome a particular linkage group physically belongs will be difficult to say. For this, cytological techniques were used such as FISH (Fluorescent In Situ Hybridization) in which fluorescent labelling and hybridisation of any one of the member genes in the linkage group to a specific chromosome in a metaphase slide are tested. If there is hybridisation, it indicates the actual physical location of the whole set of genes to that chromosome. Then this was known as cytological maps.

Due to the limited number of morphological or physiological traits present in the initial maps, distances between loci were large and encompassed numerous unknown genes. These "gaps" were essentially inaccessible to linkage analysis because of the lack of available phenotypes assigned to those regions. The discovery and implementation of molecular markers provided large number of additional genetic markers able to fill in the gaps and provide a higher-resolution map in the process. But molecular markers may or may not correlate with phenotypic expression of a trait.

3.2 Physical maps

With the advent of sequence based molecular markers and whole genome sequencing, DNA molecules are directly examined to show the relative positions of the markers as well as to quantify the distances among markers in terms of the number of base pairs of DNA. Hence this is known as physical maps. Thus, physical mapping of the linked markers helps in association of the genetic distance to the physical distance between them.

There are different methods of physical mapping among which most common is the restriction mapping (good for small genomes). In this method, the genome is restricted by rare cutter restriction enzymes to give fragments of several hundred kilobases to several megabases. Thereafter the fragments are separated by pulse-field gel electrophoresis and analysed by southern hybridization using probes comprising of closely linked molecular markers. If probes representing two molecular markers hybridize to the same fragment, the fragment size is taken to be the maximum distance between the two markers. Thus, the physical distance between the two markers is mapped. Other methods include use of Sequence Tagged Site (STS) such as Expressed Sequence Tags (EST) and random genomic sequences.

4. Markers used in Genome Mapping

A marker is simply an assay, specific to a particular locus, that allows one to determine the genotype at that locus. The concept of genetic markers begins with Gregor Mendel, who used morphology (phenotype) based genetic markers in his experiment in the nineteenth century. Later, phenotype based genetic markers for *Drosophila* led to the establishment of the theory of genetic linkage. The limitations of phenotype based genetic markers led to the development of more general and useful direct DNA based markers that became known as molecular markers. Through the development of RFLP markers in the early 1980s, indirect selection in plant breeding using DNA markers became technically feasible. However, the laborious nature of the RFLP technique prevented a broad application of RFLP markers for marker assisted breeding. In the late eighties and the early nineties, molecular diagnostic methods based on PCR technology have been developed, such as RAPD, AFLP and microsatellites. These methods allowed an efficient detection of DNA fragments starting from small amounts of DNA. Many different marker technologies exist, and most maps are a combination of several different kinds of markers.

In principle, these technologies are interchangeable, although there are technical reasons for favoring a different marker system in one situation versus another. Several detailed review on the different marker systems are available (Gupta *et el.*, 2002, Agarwal *et al.*, 2008; Bhat *et al.*, 2010;) which discuss the merits and demerits of each type. A brief comparison of some of the widely used markers are provided in Table 1.

4.1 Cause of polymorphism in markers

Different marker systems detect the polymorphism in the DNA level which result due to various factors such as

- Mutation at restriction sites (RFLP, CAPS, AFLP)
- Mutation in PCR primer sites (RAPD, DAF, AP-PCR, SSR, VNTR, ISSR)
- Insertion or deletion between restriction sites (RFLP, CAPS, AFLP)
- Changes in the number of repeat unit between restriction sites or PCR primer sites: SSR, VNTR, ISSR
- Mutations at single nucleotides: SNP

5. Statistical Concepts Behind Mapping

5.1 Logarithm of the odds ratio

Grouping of markers into linkage groups is done based on the 'Log of the odds' (LOD) ratio. The term LOD was developed through the maximum likelihood approaches for linkage analysis by Mather as early as 1938. It is the logarithm of the ratio of probability of linkage verses probability of no linkage between two loci. A threshold has to be fixed for LOD value below which linkage is not considered significant. In many analyses, a significance level of LOD>3 is appropriate as an acceptance level of linkage between two loci. This value is equivalent to saying that the probability that two loci are linked has to be 1000 times ($\log 10^3$) more probable than the probability that they are not linked. Ratio will be calculated for each pair of loci and if it is >3, the loci will be grouped into the same linkage group. This will be repeated until every pair of markers are tested and assigned to the same or different linkage group. Markers belonging to different groups are inherited independently while markers in the same group are likely to be inherited together. Comparison of maps created from an array of LOD values often allows the researcher to determine the stability of putative linkage groupings. It is clear that any map only approximates reality and that map distances between markers will change as new markers become available.

5.2 Segregation distortion

Genetic markers show segregation distortion, which is used for testing the marker-trait association (Mackay and Powell, 2006). Segregation distortion between markers is the deviation of segregation ratios from the expected Mendelian ratios. This is tested based on chi square test. The genetic basis of the segregation distortion may be abortion of male or female gametes, selective fertilization of particular gametic genotypes, small population size etc. Hence before using the data for monogenic trait or QTL analysis, care needs to be exercised to ensure whether the data are suitable.

Examine whether the molecular marker whose association with the trait of interest being examined is exhibiting any segregation distortion, since this may lead to biased estimate of marker-trait association. Therefore, normally in a mapping population consisting of RILs, one needs to examine that all polymorphic markers segregate in a Mendelian 1 : 1 ratio and those which deviate significantly from this expected segregation are rejected and not used for any further analysis of marker-trait association. Similarly, the phenotypic data on the quantitative trait being examined should be tested for normality of distribution, before they are used for any further analysis and in case normality is not present, the data need to be transformed on a scale that will achieve normality of distribution.

5.3 Mapping functions

The map-unit is centiMorgan (cM) and the distance between two genes is determined by their recombination fraction. One cM is the distance over which, on average, one crossover occurs per meiosis. When considering the mapping of more than two points on the genetic map, it would be very handy if the distances on the map were additive. However, recombination fractions themselves are not additive. It depends on the phenomenon of interference.

Interference is the effect in which the occurrence of a crossover in a certain region reduces the probability of a crossover in the adjacent region. If there is complete interference, the event of a crossover in one region completely suppresses recombination in adjacent regions. In that case, the recombination fractions are additive. Also within small distances, recombination fractions are nearly additive. In general, with complete interference, and within small distances (below 15 cM), a mapping function is simply: distance (d) = r (recombination fraction)

If distances were not additive, it would be necessary to redo a genetic map each time when new loci are discovered. To avoid this problem, the distances on the genetic map are mapped using a mapping function. A mapping function translates recombination frequencies between two loci into a map distance in cM. A mapping function gives the relationship between the distance between two chromosomal locations on the genetic map and their recombination frequency. Two properties of a good mapping function are: (1)Distances are additive, i.e. the distance AC should be equal to AB + BC if the order is ABC and (2)A distance of more than 50 cM should translate into a recombination fraction of 50%. Two main mapping functions, Haldane and Kosambi, are available for this application (ref5) (Haldane 1919; Kosambi 1943). With no interference, the Haldane mapping function is appropriate: $d = -\frac{1}{2}\ln(1-2r)$ and given the map distance (d) the recombination fraction can be calculated as $r = \frac{1}{2}(1\text{-}e\text{-}2d)$. Kosambi's mapping function allows some interference: $d = \frac{1}{4}\ln[(1+2r)/(1-2r)]$.

6. How Mapping is Done?

The various steps involved are:

- Identification of divergent parents
- Generation of mapping population (F_2, BC, RILs, NILs)
- Identification of polymorphic markers (Phenotypic/ Protein/ DNA markers)

- Scoring of markers, Segregation analysis to establish linkage (software such as MAPMAKER, JOINMAP, iMAS)
- Identification of linkage group
- Detection and correction of errors
- Combine with other datasets and further refinement of the map
- Assigning linkage groups to chromosomes

6.1 Mapping population

The first step in producing a mapping population is selecting two genetically divergent parents (that will still produce viable progeny). Often one common cultivar and one wild parent are selected as they are likely to be the most divergent. The two selected parents are screened for polymorphism with the markers which are to be mapped to be sure that the progeny will produce recombinants. The mapping population is then produced by crossing the two parents to form an F_1 hybrid population which is selfed to produce an F_2 population or back crossed to produce BC_1 population which can be used for mapping. If parents are homozygous, the F_1 will be a uniform, heterozygous population with each plant containing one chromosome from each parent. During meiosis in F_1 plant, the homologous chromosomes may or may not cross-over and form recombinants. The expected ratio of phenotypes for the F_2 population is the classic Mendelian 1:2:1, it is the divergence from this ratio that determines the amount of linkage between genes. Inorder to classify the population according to this ratio it is always better to have co-dominant molecular markers such as RFLP , SSR etc.

Aside from F_2s, many other kinds of mapping populations may be used. Permanent mapping populations have been developed which obviate the need to make a new cross and generate a saturated marker map with each new QTL experiment. These populations also allow for replicated measurements on genetically identical individuals and can be shared among investigators. Recombinant inbred lines (RILs) and Near isogenic lines (NILs) are permanent population, in which independently derived F_2 or backcross individuals are repeatedly selfed (or otherwise inbred) until each line becomes fixed for one or the other parental alleles at nearly every locus. These F_2, BC_1 or populations derived from them are called structured populations.

6.2 Genotyping and Phenotyping of the mapping population

The segregating population is scored for the various molecular markers as well as the morphological trait values using an appropriate design. It should be ensured that the progenies are segregating for the character of interest to map the qualitative trait or the QTL. Accurate marker data and trait scoring data from a segregating population are vital for analysis. Genotyping errors (wrongly interpreted data from uncertain results) will change the map length considerably. It is shown that a 3% error rate in genotyping can double the genetic map length. For example, the AFLP technique generates highly informative gel images, in some cases containing up to 8000 informative data points per sequencing gel, manual scoring of which is very difficult and inaccurate. This huge information content has triggered the need for an image analysis software package that can convert these complex gel images into data sets which only contain the scores of the informative marker bands such as 'Keygene'. The software recognizes the different lane boundaries, identifies all the bands in the gel, corrects for mobility differences that may have occurred within the gel, and quantifies the intensity of all the bands. Subsequently, appropriate statistical analysis of the relative intensity of each band throughout the gel allows the automated scoring of the band. Moreover, extensive analysis of the relative intensities of the marker bands in segregating populations has demonstrated that the AFLP reactions are quantitative, and that co-dominant scoring (distinguishing homo- from heterozygosity) of the AFLP markers using the above described software can be achieved with a high level of confidence.

7. Framework map and Saturated map

Once a genetic linkage map of a chromosome is established for the first time using selected molecular markers and phenotypic traits, it is called a frame work map. For a chromosome map to be useful, it must be saturated with markers of several different types so that a variety of markers can be tried and assessed for their usefulness in detecting that specific trait. Hence into this map, gene locations which have been mapped in a similar manner can be integrated. Using large populations of different types (F_2, RILs BC_1etc) and greater number of molecular markers one can identify the marker which is tightly linked (<1cM) to a gene / QTL that governs the trait of interest. The genes can then be assessed for linkage with the closest markers on the map and indirect selection performed using them. Then it is known as a saturated map and the process is known as fine mapping. Marker thus identified should be further validated, ie., tested for the applicability and reliability in predicting the trait in related families and in different environments. It should also reveal sufficient polymorphism in different population derived from different parental genotype.

In saturated map, markers are available for each chromosome from one end to the other, and adjacent markers are spaced sufficiently close that recombination events only rarely occur between them. For practical purposes, this is generally considered to be less than 10 recombinations per 100 meioses, or a map distance of less than 10cM. In the model plant *Arabidopsis thaliana,* which has a particularly small genome, this requires as few as 50 markers. Several-fold more markers are needed for plant genomes such as wheat and maize

For example, earlier mapping was done using RFLP which is a codominant marker and very accurate. But since the technique is tedious, the availability of sufficient number of markers was less resulting in many gaps in the maps. Later the development of AFLP technique, combining the accuracy of RFLP with the easiness of PCR based technique, resulted in the generation of many genetic maps in different crops. These reports have demonstrated that majority of the amplified restriction fragments correspond to unique loci in the genome. In collaboration with laboratories throughout the world, reference AFLP linkage maps are being constructed using segregating populations which have been mapped before by using RFLP markers. This approach allows the integration of AFLP and RFLP linkage maps. Such reference RFLP/ AFLP maps have been or are being constructed in maize, wheat, barley, rice, oilseed rape, sunflower, potato, tomato, lettuce, pepper, cucumber, *Arabidopsis.*

8. Genetic and Physical Distances

There is no general relationship between genetic distance and physical distance (in base pairs). There is large variability between species for the average number of kilo base pairs (Kb) per centiMorgan. For humans, this average is about 1000 kb per cM. Even within chromosomes there is variation, with some regions having less crossovers, and therefore more Kb per cM, than other. The number of recombinations is not equal in the two sexes. It is usually lower in the heterogametic sex. In mammals, the female map is longer than the male map, as in females there are more recombinations for a certain stretch of DNA.

9. How to Attach a Tag on a Gene?

Linkage analysis in plants is typically conducted with experimental populations derived from a bi-parental cross. This can be used to find molecular markers closely associated with monogenic or polygenic traits.

9.1 Monogenic traits

For the identification of markers linked with monogenic traits, different approaches can be followed (Peleman,1999).

- Use of already completed gene map
- Create NILs / RILs for the trait of interest and several molecular markers
- Bulked segregant analysis

From an available saturated map of the crop select the molecular markers which are in tight linkage (<1cM) with the trait. This ensures that the selection for marker will also select for the trait of interest.

If such a map is not readily available, develop NILs / RILs for the trait of interest and screen a limited number of samples with a relatively large number of primer pairs. This way, many loci can be screened with a limited effort. The number of lanes per fingerprint can be limited by screening on set(s) of NILs/ RILs. Candidate markers that are identified in this way are then screened on a panel of phenotypically well characterised lines to confirm their linkage and to determine the predictive value of the markers.

Another efficient approach consists of the 'Bulked Segregant Analysis (BSA) method (Michelmore *et al.*, 1991). The method involves crossing two divergent parent for the trait of interest (eg: disease resistance) followed by grouping the segregants from the F_2 progeny into two, representing opposite edges of the variation range (e.g. bulk of resistant vs. bulk of susceptible plants after a disease resistance screen). DNA from all the members of each group (usually10-14 plants) is pooled together and the two bulks are then screened until a sufficient number of markers emerge. Differences in banding patterns found between the two bulks will reflect polymorphisms in regions tightly linked with the trait in question. Comparing the two bulks allows one to detect molecular markers linked with the analyzed trait.

This method can be used for both dominant and recessive monogenic traits. For dominant genes, 'cis' markers (linked with the trait of interest) will emerge from the screening, whereas 'trans' markers (linked with the opposite allele) will be identified for recessive traits.

9.2 Polygenic traits / quantitative trait loci (QTL)

Most traits of agricultural or evolutionary importance are controlled by multiple QTLs. Two of the most commonly used tools for dissecting complex traits are linkage analysis and association mapping.

The methods applied for characterization of polygenic traits differ due to difficulty in categorizing the individual plants of a mapping population into distinct phenotypic classes. This difficulty arises due to the fact that quantitative traits are governed by interaction of many genes as well as the significant role of environment in trait expression. To perform a whole-genome QTL scan, it is desirable to have a saturated map. QTLs are useful starting points for marker-assisted breeding, positional cloning, functional and evolutionary studies

9.2.1. Linkage analysis [Family-based linkage (FBL) mapping]

It is first necessary to obtain a progeny that are derived from a bi-parental cross segregating for the character of interest to map QTL. In 1961, Thoday suggested the use of the segregation of a qualitative marker to detect and estimate the effect of a linked polygene. This is the basis behind QTL mapping that has been used ever since.

The classical approach for the identification of loci involved in complex polygenic traits consists in the screening of a large number of individuals from a segregating population with a set of markers that are evenly distributed throughout the genome. Subsequently, statistical analysis is performed to identify regions in the genome that are involved in the trait. Initially determine whether a QTL is linked to a marker. For that, first partition the mapping population into distinct genotypic classes based on genotypes at the marker locus. Then using correlative statistics determine whether the individuals of one phenotype differ significantly, with individuals with other genotype with respect to the trait being measured. If the phenotype differ significantly it is interpreted that QTL affecting the trait is linked to the marker locus.

There are different methods of detecting the co-segregation of QTL alleles and linked marker alleles in pedigrees.

- Single marker analysis
- Simple Interval mapping
- Composite interval mapping

The earliest and simplest method for QTL analysis included single-marker regression or tests for independence in segregation of molecular marker and the trait of interest (Gupta, 2002). This method utilises t-test or ANOVA or linear regression to determine whether the co-segregation of QTL alleles and linked marker alleles are statistically significant or not. In other words, check if trait means for one marker genotype at any given locus differ significantly from the trait means for alternative marker genotypes. This will indicate if there is an association between the segregating markers and the character of interest. A serious limitation of this approach is that it may lead to underestimation of the effect of a linked QTL due to recombination between marker and QTL. That means, QTL with major effect but loosely linked cannot be distinguished from a QTL with minor effect and tight linkage with the marker. Also there is confounding of the effect of one QTL by many other QTLs that influence the trait.

Peleman (1999) reported that the BSA approach may also be useful for the identification of linked markers for quantitative traits.

A milestone in QTL analysis, therefore, was the QTL interval mapping. This approach needs a molecular genetic map and utilizes, in a single analysis, information from all the markers of a linkage group, thus permitting detection of QTL in each interval lying between any two flanking markers that individually may show no association with the trait.

Simple interval mapping (SIM) analyzes intervals between adjacent markers instead of single markers as in the previous method. This eliminates problems of recombination between marker and QTL and hence this method is statistically more powerful. Modifications of SIM have also been suggested in the form of composite interval mapping (CIM) and multiple interval mapping (MIM), which take care of some of the limitations of SIM. Composite interval mapping (marker regression approach) analyzes intervals between adjacent markers as well as additional markers unlinked to the interval markers to focus on the interval and eliminate confounding effects from other QTLs which are elsewhere in the genome ("background noise").

In all the above methods of QTL analysis, a mapping population is needed that is derived from a cross between two diverse parents, differing for the character of interest. This mapping population, which may consist of F_2, BC_1, doubled haploid (DH) lines or RILs, is used both for recording the data on the character of interest and also for genotyping individual plants/lines of the mapping population using the polymorphic molecular markers. These two sets of data are then used for conducting QTL analysis, whether it is the simple single-marker regression approach or the more sophisticated interval mapping.

In addition to locating chromosomal positions of QTL, estimating the sizes of identified QTL is also an important component in QTL mapping. The size of a QTL is usually measured by the proportion of the phenotypic variance contributed by the QTL. However, the genetic variance may be overestimated in a small line crossing experiment (Luo *et al.*, 2003).

The laborious nature of QTL analysis makes it difficult to precisely locate the QTL on the map and hence closely linked markers cannot be obtained, thereby preventing the broad scale application of indirect selection for quantitative traits. Hence the method of association mapping gained more importance recently for QTL mapping in plant genomes also. This was originally developed for human genome where a structured population will not be available.

9.2.2 Association mapping [Linkage disequilibrium (LD) mapping]

Although based on the same fundamental principles of genetic recombination as linkage analysis, association mapping examines this shared inheritance for a collection of individuals often with unobserved ancestry (Yu and Buckler, 2006). Association mapping exploits historical and evolutionary recombination at the population level. That is it relies on naturally-occurring recombinations. Such studies can look for marker-trait association in unstructured population such as complex pedigrees, samples of different varieties, or populations. This high-resolution method for mapping QTL based on the concept of linkage disequilibrium (LD) holds great promise for the dissection of complex genetic traits. It offers greater precision in QTL location than family-based linkage analysis and should therefore lead to more efficient marker-assisted selection, facilitate gene discovery and help to meet the challenge of connecting sequence diversity with heritable phenotypic differences.

However, for QTL detection, rather than location, Family-based linkage (FBL) mapping is usually more powerful. The two approaches are complimentary.

What is Linkage disequilibrium (LD)?

Linkage equilibrium and its opposite *viz.*, linkage disequilibrium, are terms used for the chance of co-inheritance of alleles at different loci (Bhat *et al.*, 2010). Alleles which are in random association are said to be in linkage equilibrium. The chance of finding one allele at one locus is independent of finding another allele at another locus. If there are two genes Aa and Bb, in the testcross progeny the chance of finding A allele with either the B-allele or the b-allele is 50%, then the two genes are in linkage equilibrium. Hence, if we look at the gamete-frequencies, then we speak of linkage equilibrium if the freq (AB) = freq (Ab) = freq (aB) = freq (ab). But when we consider a large random mating population this may not be true always. Here the genes may be in LD and the amount of disequilibrium is measured as

LD = freq (AB).freq (ab) – freq (Ab).freq (aB)

There are several reasons for this such as the actual physical linkage between the two genes, selection, migration (crossing) and mutation. If a new individual with AB gametes come into a population with ab gametes, then in the offspring there will be more AB and ab gametes than aB and Ab gametes if the genes are linked. LD is essential for mapping. We may expect full LD between linked genes within a family, as the number of recombinants is the result of meiosis event. Similarly, the same LD exists between a cross of inbred lines.

However, after a number of generations, the number of AB and ab gametes will approach that of the recombinant aB and Ab gametes, indicating linkage equilibrium, provided there is no selection. If the linkage is closer, this process will take longer. But ultimately, even if the distance between two genes is less than 1 cM, genes will become in linkage equilibrium. When there is equilibrium, the important consequence is that if we find a particular allele at one gene (e.g. a marker) we cannot say which allele at another gene (e.g. at a QTL) should be expected.

However, such statements are possible within families or across all families in a population if it was a recent cross from inbred lines, as in such cases there is LD. Population-wide LD exists in the case of selection or with linked loci short after crossing or migration, or when two genes are so close that hardly any recombinations occur.

Mapping based on LD

LD mapping in plants detects and locates QTL by the strength of the correlation between a trait and a marker (Mackay and Powell, 2006). Unlike FBL mapping, LD mapping does not require family or pedigree information and can be applied to a range of experimental and non-experimental populations. It relies on the decay of LD, initially present in a population, at a rate determined by the genetic distance between loci and the number of generations since it arose. Over a series of generations, in an unstructured population (a randomly mating population with no complicating factors such as population subdivision and immigration), only correlations between QTL and markers closely linked to the QTL will remain, facilitating fine mapping.

However, most populations have some degree of structure or subdivision and the simple relationship between strength of correlation and meiotic distance does not apply: correlations between unlinked loci often occur. Recently, methods of LD mapping which adjust marker-trait associations for these spurious associations have been introduced.

FBL mapping can be regarded as a special case of LD mapping in which LD is generated by establishing a population from a small number of founders in the recent past. An F2 population, for example, is derived from a single F_1 plant. The meiotic process and an appropriate experimental design ensure that the strength of the correlation between a marker and trait is proportional to the genetic distance of the marker from the QTL, with the correlation between unlinked loci being zero. If the F_2 was randomly mated for 100 generations, then the frequencies of non-recombinant chromosomes will vary resulting in a more precise location of QTL. In natural populations of crop plants, or among collections of cultivars, there have often been many rounds of historical recombinations. LD mapping exploits this historical recombination and provides opportunities for fine mapping that are difficult to achieve through family-based linkage analysis.

9.3 Software used

As the data generated using molecular markers are huge, mapping them manually into different linkage groups becomes impossible. Hence to calculate recombination values and map distances several software were developed based on the statistical principles mentioned earlier, which are then used to create genetic maps. Some of the commonly used soft ware used to map monogenic trait and QTL are given in Table 2.

10. Marker Assisted Selection (MAS)

This is the indirect selection method for early and quick selection of desirable genotypes using molecular markers from a breeding population which has great application in plant breeding programmes. Especially for traits for which the phenotypic tests are unreliable or expensive, markers can offer a solution. Before indirect selection can be applied, the genetic basis of the trait of interest needs to be elucidated (monogenic/polygenic) and markers linked to the gene(s) of interest have to be identified.

Once linked markers have been identified, the PCR based molecular markers can be converted into simple PCR assays, which allow screening of large number of plants for the trait of interest in a cost effective manner. A suitable linked DNA marker should allow the prediction of the phenotype in a broad range of the germplasm. But the occurrence of multiple alleles in the germplasm for a desired locus may sometimes complicate the identification of markers with a good predictive value in the germplasm.

MAS permit the breeder to make earlier decisions about the subsequent selections a while examining fewer plants. Early generation MAS is practiced in both qualitative and quantitative traits as it reduces the burden to carry the large number of genotypes for the next generation. Thus in later stages of breeding program resources can be used to focus on a fewer lines.

Marker Assisted Backcross Breeding (MAB) is now at the verge of becoming a standard application in modern plant breeding. Two different aspects can be distinguished in backcross breeding: (i) Selection for high recurrent parent genomic content: In this application, the DNA fingerprints are used to calculate the percentage of recurrent parent genome in each backcross individual, hereby taking the genome representation of the markers into account. (ii) Selection against linkage drag: When negative characteristics are linked with the trait that needs to be introgressed, molecular markers can be used to select for recombinants in the region. After phenotypic testing of these recombinants, individuals may be selected in which the region responsible for the linkage drag has been removed from the locus of interest.

Use of MAS for gene pyramiding is another major application especially in resistance breeding. This is the process of combining multiple disease resistance genes for specific races of a pathogen into single genotype to develop 'durable' disease resistance against different races. In conventional method, it is difficult to identify plants with one or two genes whereas with linked marker, one can determine the number of genes present in a plant. (Eg: blast and blight disease in rice). An added advantage in breeding for disease resistance behaviour is that this could be done in the absence of pathogen once marker information is available. Also the bioassay is of non-destructive type.

11. Achievements

11.1 MAS

Genetic linkage maps have been constructed for many crop and plant species, using a combination of polymorphic phenotypes and sequence polymorphisms. Such maps have served as framework within which genes controlling monogenic and polygenic traits can be located and subsequently isolated. Large number of reports are available about the occurrence of tightly linked markers for many of the genes/ QTLs controlling important agronomic traits, quality, biotic and abiotic stress tolerance etc. Marker - trait association has been successfully used for indirect marker assisted selection in cases like soybean cyst nematode (SCN) resistance, rice blast resistance, rice bacterial blight resistance etc.

In breeding for resistance to soybean cyst nematode (SCN; *Heterodera glycines*), a widely used phenotypic assay takes five weeks, extensive greenhouse space and 5–10 h of labour for every 100 plant samples processed. An SSR marker '*sat309*' was identified which was tightly *linked* (only 1–2cM) to the SCN resistance gene '*rhg1*'. Genotypic selections with *sat309* were 99% accurate in predicting lines, susceptible in subsequent greenhouse assays.

Pyramiding three bacterial blight resistance genes (Xa5, Xa13 and Xa21) into Indica rice cultivar PR106 (high-yielding but susceptible) was successfully carried out using markers RZ390/RG550; RG207; RG136, RG103,Pta248. The donor line used was IRBB62 with all three genes.

Another example is the quality improvement of maize kernel. The selection for selecting maize genotypes containing introgressed opaque 2 and endosperm modifier that gives hard kernel texture is very difficult in early days. Opaque 2 gene is recessive and endosperm modifier is dominant. Hence each conventional BC generation needs to be selfed to identify opaque 2 recessive gene. Along with homozygous opaque 2 gene, multiple endosperm modifier must be selected. With the use of three tightly linked SSR markers (*umc1066*, *phi057* and *phi112* which are internal repetitive sequences within the *opaque2* gene) individuals carrying the *opaque2* gene were selected in successive backcross generations, and hence, the time required was reduced to one-half.for these genes. A few other examples are provided as Table 3 which is an incomplete list.

11.2 Map based cloning

Map-based cloning, also called positional cloning, is the process of identifying the genetic basis of a mutant phenotype by looking for linkage to markers whose physical location in the genome is known. The first gene to be cloned on the basis of its physical position in crop plants was resistance to *Pseudomonas syringae pv. tomato* locus (Pto) of tomato.

Another example is the cloning of *Xa 21* gene from rice conferring resistance to bacterial blight disease. Initially 11 RFLP markers linked with blight resistance were identified among which one marker (RG103) co segregates with resistance locus in 11,100 F2 progenies. The BAC library was then screened using this to identify 16 overlapping clones. The susceptible variety (Taipei 309) was transformed with each of the 16 clones. Plants having a 9.6 kb clone showed disease resistance (Pathogen growth reduced to 10-100 fold) which was sequenced. Presence of a 3.1kb cDNA was confirmed by Northern blot and was designated the gene as Xa21.

But this method is generally restricted to organisms where the necessary tools, genetic map, physical map and or sequence of the genome are available. Also availability of saturated restriction and physical maps in crops containing large genome is very difficult. Each step is laborious, time consuming and requires significant investment.

12. Problems

At this juncture, a word of caution is also essential as the present success rate of MAS is limited. The main reasons include :

1. Resolution of the map and order of markers are critical milestones in genome mapping and they are largely dependent on population size. Only a large mapping population will include enough genetic information from many segregating gametes in order to produce a reliable map.
2. Only the QTLs of largest effect, and those closest to a marker locus, will show statistically reliable associations.
3. It may be difficult to estimate even the presence of QTL if they interact strongly in their effects.
4. The regions to which a QTL is localized can be quite large (several cM, where 1 cM can range from hundreds to thousands of kilobases). Such regions may contain many genes.

5. Mapping in out-breeding heterozygous perennial crops is still in infancy and is complicated by the absence of complete homozygosity in the parents. Hence generation of a classical F_2 is impractical. Because of the long vegetative phase, construction of an appropriate mapping population within a fixed time period is very difficult or in many cases impossible. In a progeny from this type of outbred parents, markers which segregate do not necessarily reveal polymorphism between the parents and up to four alleles per locus may segregate. Unlike the short-lived annual plant species, the development of specific genetic stocks and tester lines containing important traits is time consuming for perennial crops. Pedigrees available for many outbred tree species involve only two parents and their full-sib progeny or maternal half-sib families. However, each tree are highly heterozygous and could be considered equivalent to an F_1. There has therefore been a need to explore alternative approaches for mapping in such species. Association mapping may give better results in such crops if information on a large random mating population are available.

13. Future Prospects

Basic and exotic germplasms are potential sources of new and valuable genes contributing to productivity and adaptability. Identification of QTLs in the primary or secondary gene pools and their use in breeding will facilitate in improving the productivity of major crops. The advanced techniques like automation in genotyping, allele specific diagnostics and DNA microarrays should make MAS more powerful and effective.

Large insert libraries constructed in YACs (yeast artificial chromosome) and BACs (bacterial artificial chromosome) have been the tools of choice for generating high-resolution physical maps and identifying candidate genes during positional cloning.

References

Agarwal M, Shrivastava N and Padh H 2008. Advances in molecular marker techniques and their applications in plant sciences. *Plant Cell* Rep 27:617-631.

Bhat ZA, Dhillon WS, Rizwan Rasheed R, Javid Ahmad Bhat JA, Dar WA and Ganaie MY 2010. The role of molecular markers in improvement of fruit crops. *Not Sci Biol* 2 (2): 22-30 www.notulaebiologicae.ro.

Gupta PK 2002. Molecular markers and QTL analysis in crop plants. *Curr. Sci* 83(2):114-115.

Gupta PK, Varshney RK and Prasad M 2002. (In) Molecular Techniques in Crop Improvement (eds Jain SM, Ahloowalia BS and Brar DS), Kluwer, Dordrecht pp. 9–54.

Luo L, Mao YC and Xu SZ 2003. Correcting the bias in estimation of genetic variances contributed by individual QTL. *Genetica-*. 2003; 119(2): 107-113.

Michelmore RW, Paran I and Kesseli RV 1991. Identification of markers linked to disease-resistance genes by bulked segregant analysis: A rapid method to detect markers in specific genomic regions by using segregating populations. *Proc. Natl. Acad.* Sci. USA 88: 9828-9832

Peleman J 1999. Application of the AFLP® technique in marker assisted breeding. In: Gillet, E.M. (ed.) Which DNA Marker for Which Purpose? - Molecular Tools for Biodiversity. Institut für Forstgenetik und Forstpflanzenzüchtung, Universität Göttingen, Chapter 14

URLhttp://webdoc.sub.gwdg.de/ebook/y/1999/whichmarker/index.htm

Mackay I and Powell W 2006. Methods for linkage disequilibrium mapping in crops. *Trends in Plant Science* 12(2):57-63.

Yu J and Buckler ES 2006. Genetic association mapping and genome organization of maize. *Current Opinion in Biotechnology* 17:155–160. www.sciencedirect.com

4

Climate Change Research Amenities with Special Emphasis on CO_2 Fertilization Effect

Debashis Chakraborty and Sourav Saha

Climate Change: Understanding our Earth and Atmosphere

Our Mother Earth is the fifth largest planet in the solar system and the third from the sun. It is also the sole planet whose name comes from old English language only, and not from names of any other celestial bodies, whose names originated mostly from Greek/Roman mythology. Earth has a diameter of 6378 km, of which the earth crust (uppermost skin) is only 24 km. Earth is the only planet where water exists as liquid and 71% of its surface is covered by water. The earth's atmosphere is made up mostly of nitrogen (77%) and oxygen (21%) gases. CO_2 consists of a very small (trace) fraction (0.03%) of atmosphere, but is most vital for the life cycle. When earth was created, it has got much higher CO_2 than today, and all of them have been sequestered in rocks as carbonates and dissolved in oceans and a negligible amount is distributed in plant communities. This extremely small quantity of CO_2 is keeping life on earth, by maintaining earth's temperature at a comfortable 14°C (without CO_2, earth's temperature would have been -21°C). This effect is called the 'greenhouse effect', revolving which all issues of climate change exist. It is also to be noted that water vapour is also an important greenhouse gas (GHG). It is to also to be noted that in spite of substantial increase in solar constant, earth's average temperature has remained stable for billions of years. It is understood that this was possible through variation in CO_2, and it is the biological processes (Earth biosphere) that controlled the environment over years (Gaia hypothesis).

The atmosphere is a layer covering the Earth (50 km), of which the life sustains within a thin layer (<9 km). Atmosphere contains many greenhouse gases (any gas which exhibits greenhouse property-allow sunlight to enter the earth atmosphere, but trap the reflected light not to be escaped), of which CO_2, water vapour, methane and nitrous oxides (both natural and man-made) and aerosols (exclusively manmade). Since industrialization (1750-1850 AD), GHGs have increased on an average of 20-25%, but fossil fuels contributed significantly in increasing CO_2 in last 30 years. The IPCC report says that a natural exchange of 88-90 billion metric tons takes place between ocean and atmosphere, and 119-120 billion metric tons between atmosphere and vegetation+soils. The balance of CO_2 cycle is disturbed mostly by the fossil fuel (oil, natural gas, coal) consumption and industrial processes, which is irreversible (Fig. 1). Among the global anthropogenic GHGS, CO_2 from fossil fuel shares a whopping 56.6% of all GHGs on earth, 17.3% CO_2 comes from decay of biomass, and 2.8% from cement production and natural gas flaring; methane 14.3%, nitrous oxide 7.9%, 1.1% and hydrocarbons (HFCs, and PFCs) 1.1% and natural CO_2 as 2.8% (IPCC, 2007).

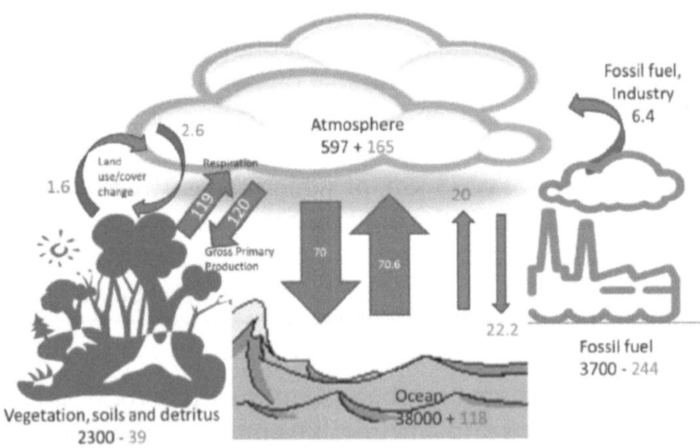

Fig. 1 : The carbon flux on Earth showing the reserve in GtC and fluxes in GtC yr^{-1}. The pre-industrial natural and the anthropogenic fluxes are represented by blue and red arrows, respectively (Adapted from Fig. 7.3, IPCC Fourth Assessment Report, 2007: Contribution of Working Group I)

Is Carbon-Di-Oxide Only to Blame?

Greenhouse gases other than CO_2:

1. Nitrous oxide: called 'laughing gas', it is colourless, non-inflammable, slightly sweet in odour and taste, produced by fertilizer use, fossil fuel consumption, animal waste management and through processes of nitrification and denitrification.
2. Ozone: It is the tropospheric ozone, which is important, created from atomic O with UV radiation
3. Methane: produced during production and transport of coal, natural oil; anaerobic decay of organic materials in landfills, wetlands and rice fields, by bovine animals in the process of their digestion. Methane has a very high potential in absorbing solar energy and has a life time of 21 years in the atmosphere before it breaks down into oxygen molecules and atoms. The methane concentration at Artic were measured as 1850 parts per billion (ppb), as against 745 ppb at the time of its discovery by Allessandro Volta in 1977.
4. Fluorinated gases: Hydrofluorocarbons, perfluorocarbons and sulfur hexafluorides emitted from a variety of industrial sources.

According to United Nations Framework Convention on Climate Change (UNFCCC), the direct non-CO_2 GHGS are methane (CH_4), nitrous oxide (N_2O), and the high global warming potential (high-GWP) gases including HFCs, PFCs, SF6 and NF3. Contribution by these non-CO_2 gases in global radiative forcing (change between balance between incoming and outgoing solar radiation) is commendable (Positive change refers warming, while negative means cooling of the atmosphere; IPCC, 2007) and stands as 28.4% as against CO_2 (71.6%). Among these non-CO_2 GHGS, CH_4 is responsible for 20.7%, N2O 6.9% and high GWPs 0.7%. Even though the radiative forcing of these non-CO_2 GHGS are significantly low, the global warming potential (GWP) of these gases are very high than CO_2; in respect to CO_2 (taken as 1), GWP of methane is 21, nitrous oxide 310, FCs as low as 140 but as high as 11,700, depending on their types, and sulphur hexafluoride as 23,900. According to EPA estimates, these non-CO_2 GHGs emissions on an equivalent CO_2 basis increased by 10% in 2005 from their earlier estimates in 1990 and are expected to increase by 45% between 2005 to 2030. Because of increase in GHGs, earth's natural radiative balance is disturbed; more heat is taken up by the atmosphere, increasing its temperature.

Rising CO_2 and Rising Temperature: Two Distinct Processes?

Whether the increase in CO_2 is natural or man-made, argument continues. The increase is also claimed to be natural (ROY Spencer's blog). 280+- 10 has been agreed as pre-industrial amount of CO_2 by scientific community. Nevertheless, there has been CO_2 concentration higher than this before (year to year fluctuations is very high), specially during 1942 (and all the observations were made through accurate chemical methods). However, anthropogenic factor is major cause in increase in atmospheric CO_2. Influence in temperature (warmer means increase) but effect is limited. And the source is fossil fuel (also called as human emissions). However, the human contribution to increasing temperature is still a debatable topic, as logically, only a few percent of atmospheric CO_2 is contributed by humans. If the exchange of CO_2 between atmosphere and oceans/vegetation is 150 GtC yr-1, which is 20% of the total CO_2 content in atmosphere, human emitted CO_2 has got only 1 in every 20 CO_2 molecules to be dissolved in oceans or captured by vegetation.

Concentration of CO_2 in March 2012 recorded as 394.45 ppm as against 392. 40 ppm in March 2011 (Tans and Keeling, 2012). The annual growth rate of CO_2 and their decadal average values (horizontal bars) clearly indicated increase in atmospheric CO_2 (Fig. 2). This growth is representative of the entire globe and the balance of CO_2 added to or removed from the atmosphere both by anthropogenic and natural processes.

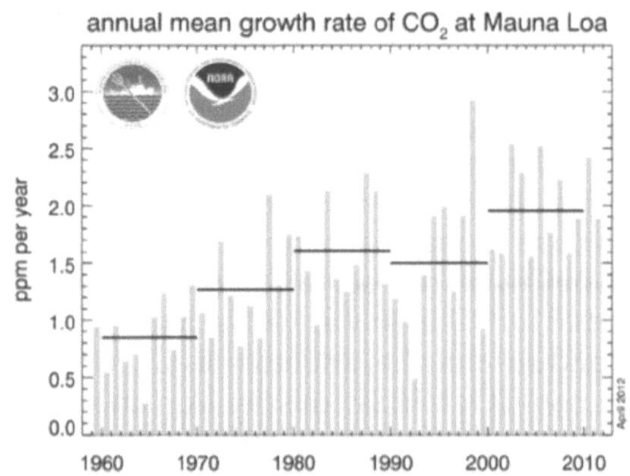

Fig. 2 : Annual mean growth rate of CO_2 at Mauna Loa, Hawaii. Horizontal bars indicate decadal average of growth rates (*Source:* Dr. Peter Trans and Dr. Ralph Keeling, Earth System Research Laboratory, NOAA, updated on April 2012)

Earth's lower atmosphere is getting warmed. Global mean surface temperature of Earth has been increased by 0.9 °F (or 0.5°C) between since 1880 and increased by 1°F (0.6°C) since 1970 (Fig. 3). It is also evidenced that eight warmest years since 1880 have passed since 2001 and the year 2011 was the 12th warmest year on record (Climate Research Unit, 2012).

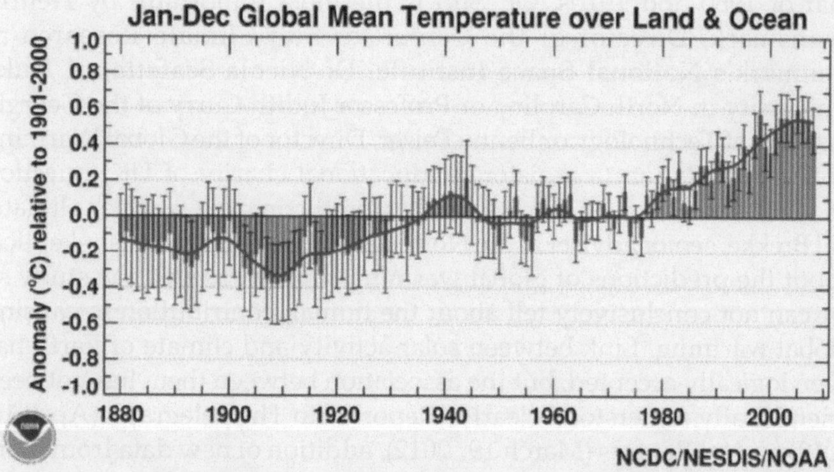

Fig. 3 : Annual average global surface temperature anomalies (*Source:* NOAA)

The Nongovernmental Panel on Climate Change (NIPCC), based at Arizona, USA initially created to provide an independent 'second' opinion about IPCC (Intergovernmental Panel on Climate Change, the official body endorsed by both UNEP and WMO with HQ at Geneva) 4th Assessment Draft Report (Finally released in 2007), has been looking on climate change from different perspective. The NIPCC has representative from 14 countries in the world (none from Govt. organizations), and opines that nature, rather than human activity has the key control over climate. According to their study, net effect of continued warming or increasing CO_2 has likely beneficial effects to plants and human beings (Idso and Fred Singer, 2009).

The rise in atmospheric CO_2 leading to rise in temperature or vice-versa has remained a puzzle to scientific community. The solar cycles (1 cycle approximates 11-year period), which is solar magnetic activities (appearance of sunspots), indicate that the cycle 25 will be weaker (grand minimum) compared to cycle 24, which starts from year 2014. This may have a significant effect on climate which might offset or mitigate the anthropogenic warming (Jones *et al.*, 2012). There has been a strong and increasing view of 'global cooling' due to sun's reduced activity and linked to the 'Dalton minimum' when the average temperature in Europe fell by 2°C (between 1790 and 1830). According to Don Easterbrook, Emeritus Professor of Geology, Western Washington University, global warming is only up to 2030, then cooling. Or take example of James Taylor, a Fellow of Heartland Institute, global cooling is already taking place; snow records at Rutgers University Global Snow Lab shows that last 10 years (2001-2010) snow exceeded that of 1960 and 1970s. Or 'Sun is the most important' by Henrik Svensmark, Director of the Center for Sun-Climate Research at Denmark's National Space Institute. Dr Nicola Scafetta, of Duke University in North Carolina or Professor Judith Curry of the Georgia Institute of Technology or Benny Peiser, Director of the Global Warming Policy Foundation (a registered educational charity of UK) question about the accuracy of climate models predicting the warmer climate. Pal Brekke, senior adviser at the Norwegian Space Centre is also skeptical about the predictions of global warming, and warns that the study so far can not conclusively tell about the human contribution in causing global warming. Link between solar activity and climate of Earth has been logically accepted, but the association between them has not been scientifically understood clearly. Reported in The Telegraph (April 10, 2012) and by Reuters (March 19, 2012), addition of new data from more than 400 weather stations in arctic regions, Russia and Canada to the global temperature datasets by Climate Research Unit and Met Office-Hadley Centre, UK showed that the global warming was more in past ten years than it was previously reported. The hottest among last 150 years were found 2005 and 2010, more than 1998, though not statistically significant.

The most recent study during drafting of this present article shows that increasing temperature or warming may trigger thawing of permafrost at the Arctic region, which is very sensitive to temperature (DeConto *et al.,* 2012). The thawing will release more of C from permafrost, giving s positive feedback by releasing more GHS in the atmosphere. The study implicates that the rise in temperature may also have led to an increase in amount of CO_2 in the atmosphere. This study reported about hyperthermals (sudden and extreme global climatic events) occurred nearly 55 million years ago, were linked to massive GHGs emission, leading to massive release of C from permafrost (melting) and 5°C rise in atmospheric temperature. This phenomenon is linked to today's situation, where global temperature rise is taking place associated with a rise in GHGs. According to the leading climate Scientist, James Hansen, a 'runaway greenhouse effect' (i.e. 'cannot be ruled out') is predicted in near future due to rapid increase rate of GHGs in the atmosphere, which will result in loss of oceans and all life on earth (Hansen, 2008). Thus, it is important to understand that there may be possibility of positive feedback of temperature rise to further release of C into atmosphere.

Population growth, industrial development, burning of fossil fuels and changing land use practices and all these man-made changes in nature, contribute to substantial increase in atmospheric CO_2. The increasing CO_2 concentration of earth's atmosphere and associated predictions of global warming have stimulated research programmes to determine the likely effects of future elevated CO_2 levels on agricultural productivity and on the functioning of natural ecosystems. The dramatic increase in global atmospheric carbon dioxide over the past century is hypothesized to have significant impacts on the earth system. The CO_2 concentration of the earth's atmosphere continues to increase in response to land-use change and the burning of fossil fuels. Being major determinants of crop growth and development increased atmospheric CO_2 and temperature can have significant impacts on the productivity of agro ecosystems, The increase in atmospheric CO_2 with concomitant increases in other GHGs is predicted to increase mean global temperatures by 3-4°C (Schlesinger and Mitchell 1985. Watson *et al.*, 1990). Since the industrial revolution, the combination of fossil fuel burning and deforestation has led to an increase of carbon dioxide (CO_2) concentration (26%) in the atmosphere (Keeling *et al.*, 1995). There is growing evidence of increase in atmospheric CO_2 concentration from a base value of 280 to a level of 379 ppm (IPCC, 2007) [See Fig. 4]. During the last 12 years, its increase has been 1.9 ppm year"1 and is predicted to be as high as 570 ppm by the middle of this century. Accordingly, as a consequence of this rise in CO_2, a warmer earth (rise in earth's average temperature from 2.0 to 4.5^{0}C; IPCC, 2007) is predicted. This might possibly exert in a significant impact on agricultural productivity (Aggarwal and Sinha, 1993). CO_2 concentration is predicted to double by the middle of next century if the CO_2 injection rate is left unabated. The concentration of atmospheric CO_2 is expected to reach 700 ppm by the end of the 21st century (Watson *et al.*, 1990).

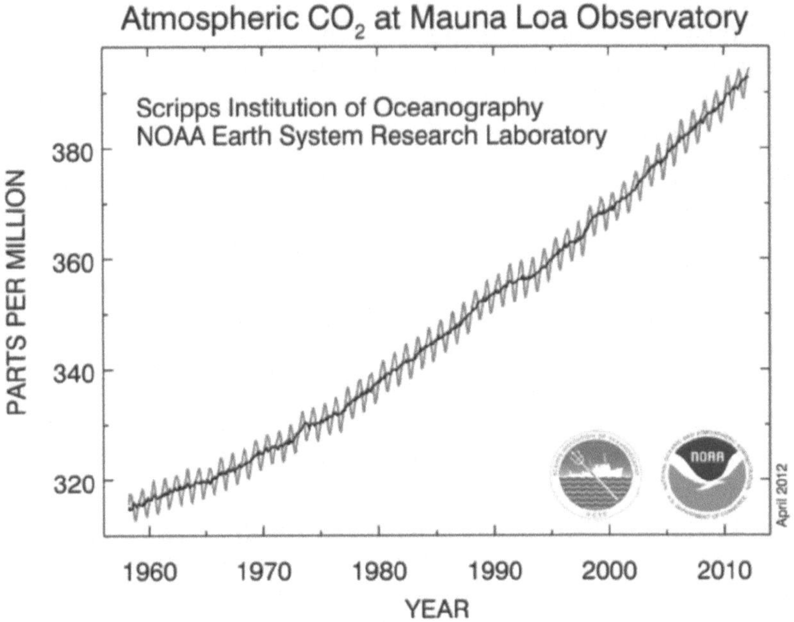

Fig. 4 : The Keeling Curve: Upward trend of CO_2 concentration in the atmosphere (*Source:* Earth System Research Laboratory, NOAA)

Fertilization Effect

Similar to many unresolved questions on increase in CO_2 and temperature and their global warming (or cooling) effect, possible impacts of increasing CO_2 level on plants are also ambiguous. A meta-analysis with 79 crops and wild plants showed more biomass and yields, but lower in N content (Jablonski *et al.,* 2001). This is more in cultivating crops than wild species, indicating a win-loss situation, a 'mixed blessing on agriculture' (More food output, but less nutritional quality). Most of the study reported positive effect on growth of plants under increased atmospheric CO_2.The increased plant growth due to elevated CO_2 level in atmosphere has been different for different photosynthetic types, C_3 and C_4 plants, the former showed to increase in more biomass (~40%) than the later (25%). As C_4 plants are already more efficient, the increased growth rate will be less. Keeping in view that only 5% of plants in the world are C_4 (rest are C_3 types), and only 1% of them have been economically explored (e.g. suragcane, corn, millets and sorghum), plenty of scope exist to look into the improved plant growth and related consequnces as a result of increase in atmospheric CO_2 level. Rising CO_2 level can have indirectly affect plant growth through change in air temperature, which is seen to rise with rise in CO_2, though the relation has not been convincingly established. Interestingly, CO_2 also causes plats to partially close the stomata (sinking effect), thus causing less water to be released (evapotranspiration) o the atmosphere, making the atmosphere 25% warmer than usual (Cao *et al.*, 2010). It is equally baffling of the effect light on plants under elevated CO_2, both quality and quantity. Other environmental factors like soil water or nutrient status will also play crucial growth as increase in growth needs more nutrients from soil as well. Interaction of O_3 with CO_2 will also be interesting, as O_3 has been reported to increase in lower atmosphere. Above all, growing importance of plant communities necessitates extensive study on effect of CO_2 on plant communities as a whole, both under natural and agro ecosystems.

History of Atmospheric CO_2 on Earth

Analysis of air entrapped in polar ice (~0.1 ml g-1 of ice) is possibly the best way of discovering atmospheric composition over past 100,000 yr. However, the enrichment of CO_2 in the trapped air has been partly understood, and so its reconstitution (Delmas *et al.*, 1980). It was suggested that CO_2 concentration was nearly 0.016% during the coldest part of last ice age (20,000-15,000 yr ago) (Delmas *et al.*, 1984). The NOAA paleoclimatology Programme, National Snow and Ice Data Centre, US and World Data Centres for Paleoclimatology and Glaciology jointly maintains archives of ice core data from poles and low latitude mountain glaciers and ice caps throughout the world. These data extending 415,000 years back shows that there has been ups and downs over the years and the last peak was attained 325,000 years back (~300 ppm), followed by a fall afterwards, and then again rise and fall. There were 4 ice ages before when the atmospheric CO_2 concentration went below 240 ppm. The Keeling curve (for last 50 years) can be successfully appended and linked to these dada without disturbing the pattern. but the unprecedented fact is the CO_2 concentration today, which has never achieved in the past, and that this unprecedented rise is caused by human activities. The circumstantial evidence is also strong. The CO_2 circulating between plants, animals and atmosphere becomes slightly radioactive (in the upper atmosphere by sun). However, it takes millions of years to get this radioactivity. CO_2 from fossil fuel cannot be traced radioactivity as their life is very short.

Charles David Keeling (1928-2005) of the Scripps Institution of Oceanography at California was the first to measure the atmospheric CO_2 in 1955 and thereafter till his death. It was too early in 1950s to think of any consequences of burning fossil fuel to the atmospheric CO_2 (CO_2 is one part of 2597 of Eath's atmosphere) and that plants would make use of it. Dr. Keeling started in taking air samples at Big Sur State Park in California and subsequently at Mauna Loa in Hawaii. Mauna Loa was selected because it was far from the major sources of CO_2 emission and thus, represents the average globe. The first measured value of CO_2 concentration was 315 ppm 1955, and steadily increasing (394 ppm on March 2012). The CO_2 level also rise and fall between seasons (black wigs in the picture), the highest and the lowest values are obtained during May and November. It is also clear that the rate of increase is also varying and higher today (2.14 ppm in 2005) than early years of measurement (0.71 ppm in 1960). There has also been slowing down of the rate immediately after the volcanic eruption at Mount Pinatubo of Philippines in 1991, producing a massive SO_2 cloud overcast, resulted in cooling of the atmosphere facilitating more CO_2 to dissolve to sea water. In recent times (after 1996), the seasonal fluctuations of CO_2 in the northern hemisphere became larger, indicating an early beginning of growing season due to global warming (northern hemisphere has the largest land mass and so the vegetation).

The Research: CO_2 Enrichment Experiments

The increasing CO_2 concentration of the atmosphere has led to major research efforts to determine what effects it will have on the vegetation of the Earth in the future (Dahlman *et al.*, 1985). This research generally has involved exposing plants to elevated (or subambient) levels of CO_2 for various lengths of time and observing the responses of the plants. History of CO_2 research on plants dates back to 1980s [a brief account may be found in Uprety *et al.*, 2007], but modern measurement at integrated soil-plant-atmosphere level at the plot scale was done by the Danish Ecophysiologist, Henrick Lundegårdh (1888-1969). His experimental CO_2 enrichment designs were similar to present-day FACE experiment, and published remarkable results between 1924 and 1930. The eddy covariance method has been used for CO_2 estimation at the field or ecosystems level (>1 ha) (Baldocchi, 2003). However, eddy technique was applied also at the plot level (Reicosky *et al.*, 1997). It is well accepted now that the enhanced CO_2 concentrations associated with global change will directly influence the process of photosynthesis in most species of plants. Early observations suggested that increased rates of photosynthesis would lead to increased growth rates of plants. There is a proliferation of experimental studies investigating the effects of CO_2 on plants. To understand the effects of elevated CO_2 on terrestrial ecosystems, most of the researchers use the facility of growth chambers, controlled environmental chambers, greenhouses, phytotrons, open top chambers (OTCs) and free air carbon dioxide enrichment (FACE). CO_2 released from the gas extraction pipes in the lakes will be blown through plastic ducts to a set of greenhouses with different CO_2 concentrations. From the greenhouses, the air will be blown by large fans to open-top chambers and/or through thinner perforated plastic sleeves (0.01 to 0.05 m^2 cross section) placed on the ground in the open. The effects of atmospheric CO_2 enrichment have been studied for more than a century in greenhouses, control environment chambers, OTCs and other elevated structures to confine the CO_2 gas around the experimental plants. Most of the agricultural scientists have used controlled environments and field based facilities, such as open top chambers (OTC), Free Air CO_2 Enrichment (FACE) facilities to study the crop responses to radiation, temperature, humidity and water availability.

Before 1980, most studies on the effects of elevated CO_2 were carried in controlled environmental facilities including leaf cuvettes, whole plant growth chambers and greenhouses. There are extensive, precisely controlled closed systems that continuously recondition and recirculate the air in controlled environment of growth cabinets, phytotrons, and other open field exposure systems, which is moderately more expensive and less precise than closed systems. During current decade, new approaches to CO_2 enrichment have potentially reduced the costs. Improved experimental design can help to characterize temporal dynamics of ecosystem carbon processes. Currently, most of the experiments using enrichment facilities follow a holistic approach to design experiments in ecosystems as natural as possible to observe their responses.

A number of CO_2 enrichment technologies have been developed so far as follow:

(1) Leaf Cuvettes (LCs)

LCs are designed for single leaf gas exchange measurements and can be used for measuring gas exchange as a part of system for studying the short time effect of elevated CO_2 levels on the CO_2 exchange processes in leaves. Cuvettes are specially designed to accommodate for variety of leaves like broad (e.g. soybean, sunflowers, alfalfa and most of dicotyledons), narrow (e.g. cereals crops like wheat, rice), conifers (e.g. pines) or needle-like (e.g. junipers) and even fruits and pods of plants (Plate 1). The modern cuvettes are integrated with temperature through PAR (photosynthetically active radiation) and IR (Infrared) sensors and light controlling systems (light emitting diode or LED-a semiconductor light source). Nowadays, leaf cuvettes are available for CO_2 monitoring of cacti, mosses, ferns, lichens etc. However, there are concerns about the possible diffusion leaks of especially water vapour through chamber gaskets of the standard clamp-on leaf chambers (cuvettes) (Rodeghiero *et al.,* 2007). Authors suggested improvement in cuvette design necessary for reliable estimates as plants in natural environment are often characterized by low photosynthetic capacities.

Leaf cuvettes may be two types (Uprety *et al.,* 2006). The open exhaust system is used to determine leaf photosynthesis and transpiration rates. It analyses the change of CO_2 and water vapour fluxes, once a packet of air of know composition is passed over the leaf. An infrared gas analyzer (IRGA) may be used to measure in CO_2 flux. In the second type, the change in CO_2 and water vapour is monitored through the rate of injection of the same in a closed chamber to maintain a balance (a set point). A detailed account of these two systems can be found in Sinclair and Horie (1989) and Allen (1991). An improved version of leaf cuvettes was used for individual evaluations of gas exchange parameters and independent control of air current of upper (abaxial) and lower (adaxial) leaf surfaces (Yasutake *et al.,* 2000). Recent studies on CO_2 exchange by using leaf cuvettes can be found in wheat (Tazoe *et al.,* 2009; Wall *et al.,* 2011), sorghum (Prasad *et al.,* 2009), rice (Pederson *et al.,* 2009), Barley (Wall *et al.,* 2011), hibiscus and pepper (Stimler et al, 2010), eucalyptus (Kristine *et al.,* 2011 Barton *et al.,* 2012), eucalyptus and other trees (Wilkinson *et al.,* 2008).

(2) Gas Exchange Canopy Chambers

Leaf cuvettes are comfortable at the leaf level, and can be successfully used to learn the CO_2 and vapour exchange to small plants and even trees at the field level. However, scaling the observations to whole-plant level remains a challenge. Large (gas exchange) canopy chambers can be used for the whole plant or even trees, and these can be open (steady-state) or closed (transient-state) types (Denmead, 1984; Reicosky *et al.,* 1990; Livingston and Hutchinson, 1995). In the open system, the gas concentrations in the entry and exit airstreams are monitored, while maintaining a constant circulation inside the chamber. This system has the advantage of continuous monitoring the plant response, but has many limitations. Apart from the change in radiation quality inside the chamber (mostly diffuse component of radiation coming through the wall of the chamber), there is always an overpressure inside the chamber to keep the structure inflated, and thus, the air flow will be quite different than the natural wind systems in the field. The solar radiation coming through the chambers walls is modified with portion of diffuse radiation being the maximum. As the system is open, the canopy has to be closed for a large time during the course of measurement, thereby significantly modifying the microclimate inside the chambers. As the chamber is covered by plastic/mylar sheets, the chamber acts as a greenhouse with increased intensity of long-wave radiation and simultaneous increase in air temperature inside the chamber. The humidity inside the chambers often remains higher while CO_2 concentration may go down due to net photosynthesis and plant respiration. Due to stirring of air inside through fans, aerodynamic condition is also impaired. In order to overcome these problems, a non-steady-state through-flow system was designed (Muller *et al.*, 2009). The movable upper part of the chamber (6) is made of transparent (95% PAR transmittance) polypropylene film secured to adjustable aluminum frame while the base frame (7) is inserted 5 cm deep into the soil (Plate 2). A tube (1) supplies the ambient air to chamber through a blower (2), while four radial blowers at the base mix the incoming air and produce a wind of desirable speed inside the chamber. An impeller anemometer is fitted in a tube (4) which measures the rate of bulk air-flow. The tube is connected to the air-conditioning unit (5), which is responsible for cooling and drying of incoming air. A slight overpressure is maintained inside the chamber to avoid any leakage as well as keeping the structure stable. The open top chamber, a special type of steady state is separately discussed later.

These transient types are only meant for ambient level of CO_2 and water vapour and no flow measurement or climate conditioning (elevated CO_2 etc.) required. These types are portable, easily moving and can be automated, but small (<9 m3 in volume) and mostly limited to small plants (mostly herbaceous). The canopy is covered for a small period of time (few minutes only) while the measurements are taken, thereby not modifying the microclimate inside the chamber. The gas fluxes are monitored through simple regression of gas concentration over time or by other methods and has been reviewed (e.g. Jarvis *et al.*, 1971; Ricosky *et al.*, 1990; Wagner and Ricosky, 1992; Wagner *et al.*, 1997). However, as these are not automated, night observations become difficult. An automated closed-system chamber has been proposed to overcome this limitation (Steduto *et al.*, 2002).

A transient type chamber to cover an adult tree (~20 m3 in volume) yet easily movable has been recently successfully designed (Perez-Priego *et al.*, 2010; Fig. 5). The top (two trapezoidal-shaped windows) and the four walls are hinged, and can be opened and closed at will. The air inside the chamber is circulated through a vacuum pump, while another pump is used to sample the air flow inside the chamber and pass to the IRGA for analysis of CO_2 and water vapour.

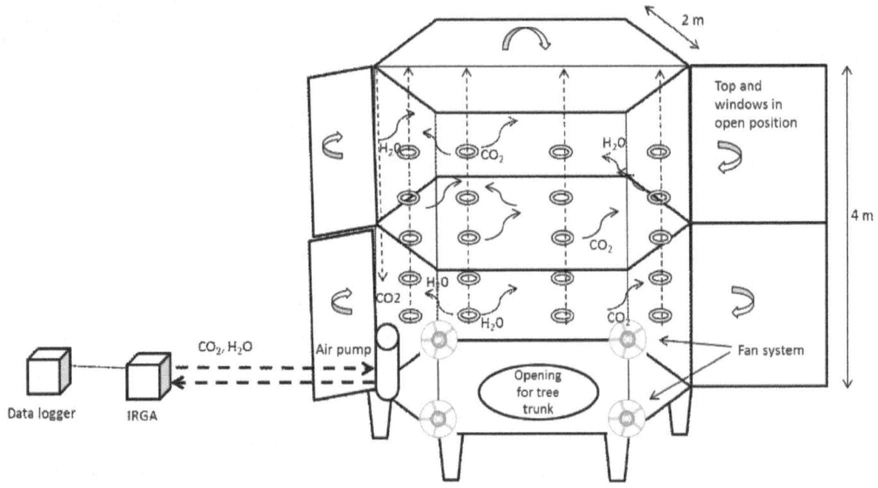

Fig. 5 : The transient-type closed chamber used for studying gas exchange on an olive trees; aerrows indicate air flow (*Source:* Perez-Priego *et al.*, 2010) CR-23X is the data logger and LI-840 is IRGA

A special type of chamber is also to monitor effect of CO_2 and temperature on spruce trees. These whole-tree chambers can be found are found at Flakaliden in Northern Sweden, with CO_2 and temperature control, are used for studies of the impacts of elevated air temperature and atmospheric CO_2 concentrations on 40-year-old Norway spruce trees (Plate 3).

The National Phytotron Facility at the Indian Agricultural Research Institute, New Delhi, established in 1997 is the first facility of its kind in Asia and 4th in the World after Australia, Canada and the US (Plate 4). This was a result of joint effort by the FAO, the United Nations Development Programme, the Indian Council of Agricultural Research and the Department of Science and Technology, Government of India. Built at a cost of Rs 15 crore, this foundation has a core facility of 22 °C, humidity, light and carbon dioxide controlled chambers, backed by 10 greenhouses to study the live responses of plants under controlled conditions and the possible impact of climate change and greenhouse gases. It has a self-contained area of 2700 m^2, housing 22 growth chambers and green houses. The facility is made available for plant scientists belonging to ICAR, CSIR, SAUs, traditional universities and agro-industries to understand the complicated interaction of physiochemical environments and living systems, especially the plants and their pathogens. This controlled environment research facility overcomes environmental constraints like seasons and locations and facilitates the faster development of better varieties and efficient crop management programmes, particularly in the rain-fed and other low yield areas.

(3) Soil Plant Atmosphere Research (SPAR) Units

Artificially lit growth chambers have low light intensities and the quality of light is also undesirable, compared to the open field light available from the sun. In order to overcome these limitations, soil-plant-atmosphere research (SPAR) system was first developed at the USDA-SEA, Coastal Plains Soil and Water Conservation Research Centre, Florence, SC (Phene *et al.*, 1978). This system fundamentally consists of a base soil bin covered with an acrylic aerial chamber, secured and sealed to the top of the bin (Fig. 6). Each unit has a pressure-regulated water outlet, a flowmeter and a remote controlled irrigation system. The aerial chamber for plants is separated from the soil bin by plastic sheets. The plastic sheets are also sealed around each plant and well as the edge of the bin to minimized gas exchange between soil and the aerial chambers with plants. Each aerial chamber is made of materials with high optical clarity like plexiglass, teflon film, mylar sheets, polycarbonate films or simple acrylic sheets, and can accommodate several rows of plants, where the temperature and the air flow is controlled. The mixing of air inside the chamber is auto-adjusted through mass-flow controllers. A lateral side panel gives access to the plants. Soil temperature is also thermostatically controlled by brine flowing through copper tubing placed around each soil bin or electrical heaters. Dew point temperature is controlled by chilled water heat exchangers. Soil and roots of the bin can be totally exposed at one side, by removing an exterior lateral side panel. Provisions are made to place sensors without exposing the soil and roots, and ceramic candles are embedded for draining excess water or maintain desirable soil water potential through irrigation. The whole system is computerized to collect data automatically. All the micrometeorological parameters including solar radiation, net radiation, leaf and soil temperatures, CO_2 assimilation, transpiration and relative humidity are monitored simultaneously, making SPAR units as vital tools in studying the interactive effect of CO_2 and other environmental variables which are likely to change concomitantly in future like ozone (Olszyk *et al.*, 2002), air temperature (Allen *et al.*, 2003) and ultraviolet radiation (Zhao *et al.,* 2003). SPAR units are designed to gather process-level quantitative data to understand the interactions of environmental variables including CO_2 with plant growth, which can further be used in simulation modeling. State-of-art SPAR facility can now be available in many research centers across globe, one of them is at the Mississippi State University, USA, (Plate 5) and another at USDA-ARS (Plate 6) capable of producing data from small plant scale simplicity to the field scale complexity. Although solar radiation variation inside other greenhouses has been studied in details, the same inside the SPAR unit has not been focused (Kim *et al.,* 2004). An ideal SPAR system should allow >90% of the photosysthetically active radiation part of solar radiation, the soil deep enough to promote unrestricted root growth, controlling system of diurnal and seasonal variation of environmental variables including air-temperature and gas exchange. These should also support in growing plants through its entire life cycles (annuals) or multiple years (perennials). The among-chamber and within chamber variability can be minimized by taking appropriate experimental protocols and considering positional effects inside a chamber during sampling (Fleisher *et al.,* 2009). The SPAR system has an advantage over the FACE technology (discussed later) is the temperature forcing, in which temperature can set from the optimum value to the point where reproductive process fails. SPAR climate research has indicated that the manifestations of global climate change on plants will occur mostly through the rise in atmospheric temperature, rather than increase in CO_2 levels (Allen and Boote, 2000; Prasad *et al.,* 2002; 2003).

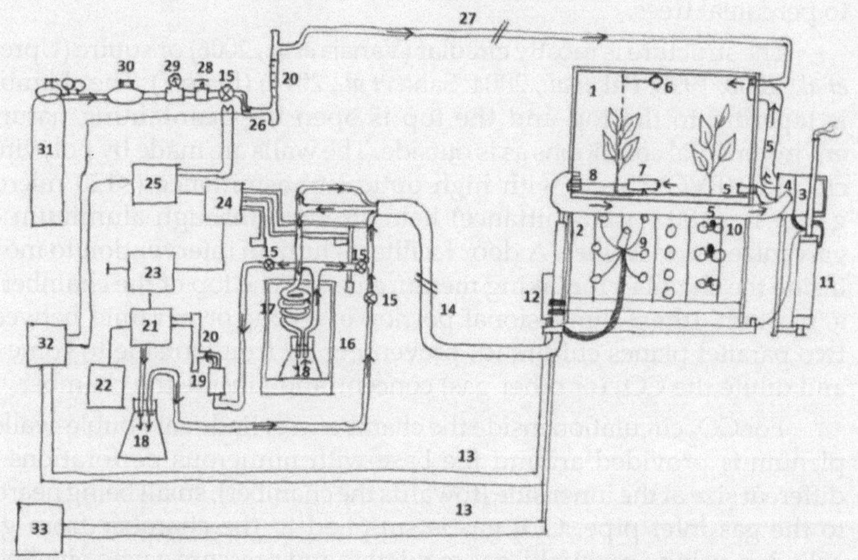

Fig. 6 : SPAR system (Redrawn after Phene *et al.,* 1978) [1. Upper plexiglass plant growth chamber; 2. Steel soil bin; 3. Air-conditioner; 4. Heater; 5. Inlet and outlet air-tubes; 6. Net radiometer; 7. Psychrometer; 8. Vacuum pump; 9. Soil moisture sensors; 10. Tensiometers; 11. Transpiration measurement; 12. Electrical junction box; 13. Wires for electrical connection; 14. Underground plastic tubes for air samples; 15. Valves; 16. Refrigerator for removing water from air sample; 17. Water condensing copper coil; 18. Water trap; 19. Drying of air samples; 20. Rotameter; 21. CO_2 infrared analyzer; 22. CO_2 level strip chart recorder; 23. CO_2 control system; 24. CO_2 sample scanner; 25. Timer for injecting CO_2; 26. CO_2 temperature measuring thermocouple; 27. Plastic tube to inject CO2 to the growth chamber; 28. Manual gas flow valve; 29. Pressure gauge; 30. Pressure regulator; 31. CO_2 tank; 32. Integrating DDAS for measurement; 33. Calculator based DDAS measurement and control

(4) Open Top Chambers

The first open top chamber is believed to be designed by Heagle at the USDA-ARS Air Quality-Plant Growth and Development Research Unit, to assess the impact of air pollutions on plants (Heagle *et al.,* 1973). The design was further modified by Rogers and his colleagues (Rogers *et al.,* 1983) to study effect of CO_2 on plants. Broadly categorized as steady-state gas exchange chambers, Open Top Chambers (OTCs) are small green houses, extensively used to study the plant response to elevated CO_2 in the field. Nowadays OTCs are also used to study the effect of air pollution, especially the ozone on plants. Over the years, OTCs have been modified to accommodate plants of different architectures starting from bushy plants to tall orchards, annual crops to perennial trees.

The structure is mostly circular (Vanaja *et al.*, 2006) or squire (Uprety *et al.*, 2000; 1995: Pal *et al.*, 2004: Saha *et al.*, 2011) (Plate 7). The chamber is tapering to the top and the top is open for maintaining natural environmental conditions as is outside. The walls are made by polyvinyl chloride (PVC) sheets with high optical transmittance (<120 micron gauge for >90% transmittance) held strongly through aluminum or galvanized iron frames. A door facilitates human intervention to move inside the chamber for taking measurements. The top of the chamber is a 'frustum' (the 3-dimensional portion of a cone or pyramid between two parallel planes cutting it), preventing air from outside to come in and dilute the CO_2 (or other gas) concentration inside the chamber.

For CO_2 circulation inside the chamber, a cylindrical double-walled plenum is provided around the base with numerous perforations of different size at the inner side (towards the chamber), small being nearest to the gas inlet pipe. CO_2 gas is supplied to the chamber from gas cylinders using a manifold, gas regulators and pressure gauze pipelines. Air mixed with CO_2 are drawn inside the chambers by blowers and thoroughly mixed inside the chamber and slowly move upward creating a CO_2 enriched environment over the canopy. Blowers are also helpful in keeping the inside temperature close to the outside of the chamber. In a comprehensive first account about the functioning of OTCs, Leadley and Drake reported usefulness of OTCs in terms of its simple design and construction, but substantial modification of micro-climate inside the chamber (Leadley and Drake, 1993). The authors observed substantial spatio-temporal variation in CO_2 concentration inside the chamber, along with 3 °C increase in temperature and 20% reduction in light intensity. The relative humity has been reported higher inside the chamber other than the top frustum due to plant transpiration and restricted mixing of air. The open to chambers have not been used successfully for large plants and are limited to mostly herbaceous plants.

CO_2 gas is supplied to the chamber from gas cylinders using a manifold, gas regulators and pressure gauze pipelines. Gas cylinders are fitted in a row with thick copper outlet pipe. CO_2 gas is released from cylinders to chamber through a manifold and underground pipe and injected into the chamber along with the ambient air using a blower. Air blowers (diameter ~30 cm) are used for maintaining desired concentration of CO_2 into chamber tunnel. Blower also helped in maintaining inside air temperature closure to that of outside ambient atmosphere.

Uprety *et al.*, (1995) designed and developed OTCs for South Asian climatic conditions. These chambers with frustum are being used by South Asian CO_2 research network for multi country multi disciplinary CO_2 crop response studies. It has 5 sub systems:

(i) Supply of pure and high concentration of CO_2

(ii) System of valves regulators and flow meters

(iii) CO_2 controlled chambers

(iv) Appropriate gas analyzer with feedback control

(v) Computed data acquisition and programming.

Open-top chambers are relatively inexpensive to build because they consist simply of an aluminum frame covered by panels of polyvinyl chloride plastic film. By using open-top chambers, we could study the reaction of the ecosystem in situ and accommodate some plots with larger plants, like trees: the height of an open-top chamber may be increased with time as the tree grows. According to the experience of open-top chambers users in the temperate zone, the micro-environment can be expected to be modified as follows: CO_2 can be kept at the target concentration of ± 50 ppmv, temperature increases up to 3°C in the summer during midday hours, solar radiation is reduced by ± 20%, wind velocity is also reduced while relative humidity is increased (Leadley & Drake, 1993). Heagle *et al.* (1979) found less than 1°C differential between inside and outside air temperature in winter wheat. Relative humidity is always found to be higher than the external air (Olszyk *et al.*, 1982: Saha *et al.*, 2011). Recently the results from cross validation study between open top chamber and Eddy covariance method shows net ecosystem CO_2 exchange was 8% higher in open top chambers regardless of the large difference of microclimate between these two, under identical conditions (Sabinadore *et al.*, 2003). It is proposed to attach a cooling system in OTC that can remove the bias of overestimating crop growth response to elevated CO_2 (Van oijen *et al.*, 1999).

CO_2 Supply and Monitoring System for OTC

The plenum at the base of OTCs is provided for carbon dioxide circulation in the OTCs. The 100% CO_2 gas of commercial grade is used to elevate CO_2 levels within the chambers. Four to six CO_2 gas cylinders of 45 -60 kg capacity each are generally used and the gas is released to the chambers through a manifold fitted with copper tubing. Within each chamber the copper tubing is fitted with solenoid valve and rotameters to regulate the gas supply. The normal air from the air compressor is mixed with CO_2, which is then pumped into CO_2 circulating pipe. The CO_2 is released into the OTCs through perforated GI pipe fitted at the base of OTC chamber. The uniformity of the CO_2 is maintained by pumping CO_2 gas diluted with air by air compressor. The air is sampled from the centre point of the chamber through a coiled copper tube which can be adjusted at different heights as the crop grows. The equipment for monitoring and controlling the CO_2 in OTCs is fully automatic and the desired CO_2 level can be maintained throughout the experimental period in the OTCs with the help of this system. The system basically consists of CO_2 analyzer, pump to draw the sample from OTCs, the valves and meters to control and regulate the CO_2 and air flow, air samplers from each chamber, CO_2 gas cylinders for supply of CO_2 gas, air compressor to maintain the uniformity of CO_2 gas at set ppm in the chamber, the Program Logic Control (PLC) and Supervisory Control And Data Acquisition (SCADA) platform and computers. The non-dispersive infrared (NDIR) gas analyzer, which is microprocessor based system with single beam optical system is used used for measuring the concentration of CO_2 of the air sample, drawn from OTCs. The PLC and SCADA continuously monitor and control the desired CO_2 level. The software facilitates to set different concentrations of CO_2 in different OTCs at a time and also modify the time to analyze the air samples drawn from each OTC. Thus this facility continuously records and displays the actual CO_2 concentration, temperature and relative humidity of OTCs.

CO_2 Supply and Monitoring System of South Asian type OTC in IARI

Pure CO_2 gas (99.7%, v/v CO_2 and <10 ppm CO) procured from M/S Gas Associates, New Delhi, are released from commercial grade cylinder fitted with a regulator (DURA, make ESAB, India) through solenoid valve and PV tubing connected to an air-exhaust blower at the base. Continuous injection of pure CO_2 into plenum of OTCs is done where it is mixed with air from air compressor before entering into the chamber. The air sample is drawn periodically into a CO_2 sensor (NDIR, make Topak, USA) and the set level of CO_2 is maintained with the PLC and SCADA (Winlog software (Make SELCO Italy). The CO_2 data logging, control and operation are performed by a computer through DOIP (digital input and output module) on real time basis. During experimental period, temperature and humidity inside all chambers are recorded with the help of sensors (Model TRH 511, Ambetronics, Switzerland) fitted in the middle of each OTC and data logger (Model

TC 800D, Ambetronics, Switzerland). In elevated CO_2 OTC chambers, CO_2 levels are needed to be maintained from 6 AM to 6 PM after the emergence of the crop. For the chamber-controls of OTC without any external supply of CO_2, similar circulation of air through the blower is needed. Chambers are needed to wash frequently with a gentle stream of water to remove the dust, if any for maintaining its transparency.

Advantages of OTC

a) Inexpensive and easy to maintain.

b) Principal advantage of OTC is the ability to provide control over the atmospheric variations to which plants are exposed while maintaining more or less the same soil conditions of the field setting

c) Climate in OTC tracks the dynamic change in temperature, light, rainfall as experienced in unclosed area without costly and complex environmental controls characteristics of closed chambers.

d) OTC subjects the whole plant to the treatment without restricting the root growth.

Limitations in OTC Research

a) Not fully natural

b) Limitation of size and height for tall crops or forest species, unless the system is expandable to accommodate the anticipated future size of the plants during the lifetime of the experiment.

The inside OTC environment can be significantly differing compared to outside as listed in Table 1.

Table 1 : Summary of ways that the open-top chamber inside environment can differ from outside field

Parameter	Differential response from outside
Solar radiation	0.7- 0.9 of outside
Air movement	Much less in daytime and higher at night; turbulent intensity (variation of wind speed) much reduced both day and night
Solar radiation	30% lower to 3% higher; more variable within a plot, e.g., plants in the shade of a south wall (in northern hemisphere) receive less than those within the reflected rays off a north wall; more diffuse
Long-wave radiation and Air vapor pressure	Increased (150 W m^{-2})
Air temperature	Typically increased 0.5 to 2.5°C
Foliage temperature	Often increased similar to air temperature, but can decrease
Transpiration	10-30% lower
Rainfall	Lower
Dew formation	Less on leaves inside
Pollutants	O_3 concentrations lower
Insect populations	Lower

(5) Free Air CO_2 Enrichment Technology (FACE)

The FACE system provides a unique opportunity to examine the effects on plant growth at elevated CO_2 and other environmental conditions without any direct perturbation of microclimate. Development of FACE program had following objectives:

a) To maintain sufficiently steady CO_2 concentration so that fluctuations in CO_2 cause no perceptible effect on plants grown within the environment

b) To provide reliable field data on the response of crops and other plant communities to develop and evaluate mechanistic plant growth models, which can contribute to predict the effects of future CO_2 concentration and climate changes on plant growth and water use.

c) To contribute for an evaluation of global CO_2 fluxes and carbon balances between vegetation and rising atmospheric CO_2,

The FACE, by contrast, is characterized by a series of vertical vent pipes, placed circularly around the plot, which release CO_2 towards the centre of the ring. CO_2 concentration, wind velocity and wind direction are continuously measured and the information collected are used by a computer-controlled systems to maintain elevated concentration of CO_2 throughout the plot (Allen *et al.*, 1991). Maricopa FACE, the first such system developed in Arizona, had arrays of 32 vertical vent pipes, with each vent pipe containing multiple gas emitter ports. Each vent pipe in an array is connected to a common 22 m diameter torridly distribution plenum through an individually controlled valve. The proportional integrative differential (PID) adjust the amount of gas in to the plenum and another algorithm considers both wind direction and speed and controls the vent pipes that emit CO_2 enriched air.

South Asian FACE is also based on the principle of injecting additional CO_2 in open field suitably so as to attain a predetermined elevated level of gas concentration with uniform distribution inside FACE ring under varying metrological conditions of wind, temperature and humidity. Main components of this facility are:

a) CO_2 storage and distribution system

b) FACE ring (Plenum), through which CO_2 enriched air is injected into open field

c) Different sensors and actuators used for monitoring the environment and controlling the operations

d) An electronic system for controlling these operations. CO_2 concentration is measured at the center of the array with the help of an IRGA.

The PID valve controls quantity of CO_2 to be released in to plenum arms, which depends on the voltage (0- I0 V) applied to it. To evaluate proper operation of PID valve, a feedback signal is also monitored to know the actual status of the valve. This signal (0-5 V) corresponds to the fully closed and fully open conditions of PID valve. Fumigation of CO_2 in to FACE ring from the plenum is made at the crop canopy height. The height of the plenum could be adjusted from time to time to match with the height of the crop canopy by resting the plenum on height adjustable stands. Plenum could be raised up to 150 cm from ground level in steps of 5 cm. Two banks of 25 cylinders in parallel are used for CO_2 storage and dissemination under constant pressure. A buffer tank is installed after the cylinder bank, which took care of the pressure variations in supply line due to decreasing level of CO_2 in the cylinders during Mid - FACE operation. Control system measured in every second the data from meteorological sensors (wind speed and direction) along with CO_2 concentration. Based on these inputs, control system, through a specially developed PID algorithm, controlled flow rate of CO_2 in the plenum with the help of PID controller valve.

Direction of wind decides which on/off valve is to be opened to release CO_2. In case the wind velocity is less than 0.5 m/s (which is taken as the case of calm air), valves are opened in a cyclic order as no dispersion of CO_2 occurs in such conditions. Thus a predetermined constant level of enriched CO_2 is maintained at the center of plenum. South Asian FACE is expected to include temperature enrichment device in future (Fig. 7). Here CO_2 requirement is 10- 15 times higher than OTC

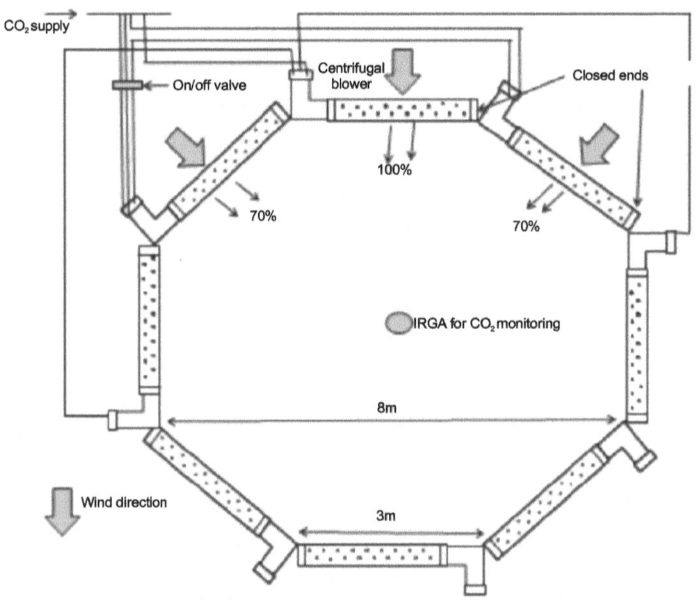

Fig. 7 : Octagonal ring of South Asian Free air CO_2 enrichment (FACE) technology

Advantages of FACE

1. The free air CO_2 enrichment (FACE) approach is that data can be obtained under conditions with unquestioned realism-for both absolute and relative responses.
2. The second advantage of the FACE approach is that the large plot size also allows many cooperators to make many complementary measurements on the same plant material, which is a big aid to data interpretation.

3. The third advantage of the FACE approach is that it is less expensive per unit of high-CO_2-grown plant material. FACE has been criticized because the prodigious quantities of CO_2 required make it expensive.

However, FACE also has numerous important limitations, likewise.

1. Levels of some environmental variables, such as temperature, light, and humidity, cannot easily be controlled on a field scale. Moreover, it would be technically very difficult to even impose differential levels of these variables in the field, e.g. to raise the temperature of a plot a few ^0C above ambient.
2. Technical difficulties to decrease the CO_2 concentration in a field plot below ambient in order to study the effects of low CO_2 levels (or subambient levels of other gases, such as air pollutants)
3. The minimum FACE ring diameter would be about 6 m (for 1-m-tall vegetation), so that even the smallest possible FACE plots would still be large compared with the smallest possible OTCs. Considering also that costs go up roughly in direct proportion to enriched plot area (actually diameter for FACE) and to the degree of enrichment above ambient, for studies that require very high concentrations of CO_2, other approaches than FACE with a smaller scale may be fiscally necessary
4. CO_2-enriched through vent-pipes has the potential to cause microclimate perturbations ("blower effect") under very stable and calm atmospheric conditions as during still nights.
5. FACE experiments typically impose a steep increase in CO_2 concentrations at the beginning of the experiment. This abrupt change in environmental conditions may induce different responses of plants and ecosystem processes that have grown under normal CO_2 for decades. In particular, enhanced photosynthesis induces an elevation in nitrogen- demand, which often leads to nutrient stress and consequent down-regulation of photosynthesis.
6. Even though FACE experiments are sufficiently large to capture most critical ecosystem processes, they are still like an island within the surrounding ecosystem (Hendrey & Miglietta 2006).

Generally the CO_2 concentration- using OTC is maintained at 700 ppm while in FACE concentrations between 550 and 600 ppm are more typical. The main aspects of the OTC and FACE techniques are summarized in Table 2.

Table 2 : Main aspects of the OTC and FACE techniques used for CO_2 enrichment

FACE (Free Air CO_2 Enrichment)	OTC (Open Top Chamber)
Diameter (8-30 m)	Diameter (~1 m)
Side walls absent	Presence of side walls
CO_2 distributed by a ring of vertical pipes	CO_2 distributed by a circular tube
Computer-controlled system that adjusts CO_2 flow rate	UV-B are not transmitted through the walls
Forest canopy fully developed	No wind from outside
Species can compete	Increased temperature
High costs	No dispersion of pathogen and pests
Blower effect under still wind conditions	Plant-atmosphere coupling is altered
Steep increase in CO_2	Forest-ecosystem processes cannot be studied
Undisturbed rooting	Rooting barrier

Advantages and Limitations of FACE over OTC

1. In the OTC the presence of side walls limits CO_2 consumption but induces a significant impact on the microclimate, altering air flow, intercepting rainfall, restricting access to insect pollinators and pests, increasing air temperature and water vapour humidity and lowering transmittance on sunny days (Leadley and Drake 1993, Long *et al.*,2004). The transparent OTC walls are constructed in order to create a wind barrier and thereby make it easier to control the concentration of CO_2 (or other gas of interest) inside. Of course, an assumption is being made that the effects of air movement on plant growth are less important than those of CO_2 (or other gas) or at least that meaningful relative comparisons can be made between plants grown in treatment chambers and those from suitable control chambers; however, air movement itself can affect plant growth. The environment inside an open-top chamber can differ from that of a field outside as listed in Table 2 (Kimball *et al.*, 1994). In OTC, the wind is removed, preventing wind effects and dispersal of pathogens and pests. The OTC walls

also form a partial barrier between the crop and the cold sky. Because the walls are warmer than the sky, more downward long-wave radiation is emitted by the walls than the sky and then subsequently absorbed by the crop .the degree of the effect depending upon the optical properties of the chamber walls for long-wave radiation.In FACE experiments, microclimate is minimally affected, but large quantities of CO_2 are required to compensate the CO_2 that diffuses away from the plot, especially under windy conditions. An additional limitation of OTC is the presence of a rooting barrier that prevents roots from exploiting soil outside the chamber and vice-versa, eventually inducing feedback inhibition on photosynthesis and production (Long *et al.*, 2004).

2. While OTC is made of inexpensive materials and requires low amounts of CO_2, FACE requires a high investment in instrumentation, building material, CO_2 and transport.
3. The major limitations of OTC in a forest environment are: the influence of trees and stand development patterns, the lack of an ecosystem prospective, scaling issues and absence of boundary layers. Trees and forests are very well coupled to the atmosphere, but this coupling is often greatly perturbed when enclosed in chambers (Lee and Jarvis 1996).
4. In the OTC method, ventilation disables the natural coupling between vegetation and atmosphere. The applied artificial turbulence alters the exchanging process between canopy and atmosphere, by means of periodic irruptions of air in the canopy instead of a continuous mixing as it occurs in a natural environment. In FACE experiments, the components of the plant-soil nutrient cycle can be integrated, species can compete for resources, and a forest canopy may fully develop (Norby *et al.*, 1999).
5. Finally experiments in both OTCs and FACE are used showed that the relative growth responses to elevated CO_2 are not significantly different between OTCs and FACE, but the absolute growth of the cotton was 30% greater inside OTCs, while that of the wheat was about the same for both OTCs and FACE. Thus, for many studies the FACE approach is preferred because both absolute and relative responses to elevated CO_2 can be obtained reliably. Furthermore, the large plot size supports & multidisciplinary teams with much

destructive sampling, and there is an economy of scale. Yet, OTCs remain a workable alternative in some experiments that appear technically difficult or too expensive with FACE (Kimball *et al.*, 1997).

A list of various FACE experimental setups is listed in Table 3. The physiological and agronomic traits under the FACE and Chambers are listed (Table 4).

(6) Free Air Temperature Enrichment Technology (FATE)

FATE components include two 1500-watt IR lamps (positioned at about 1.2 m above ground), which can homogeneously irradiate 40-50 cm path of vegetation at an angle of 40° above ground. These lamps are regulated by proportional action controller, which modulates IR flux density (frequency, 10Hz) to obtain a preset target differential (T) between heated and a corresponding unheated plot. IR lamps, used in simulating temperature, are made of tungsten filament and can irradiate at 2000°C. Filter is used to selectively cut off visible light so that light can be continued in night. It also removes wavelength to which phytochrome is sensitive so that photomorphogenesis can be avoided. FATE irradiation is uniform in space and time.

(7) Temperature Gradient Chambers (TGCs)

To study the interaction of temperature increase with elevated CO_2 or the effect of elevated temperature only, Temperature Gradient Chambers (TGCs) are constructed for maintaining different CO_2 concentrations. Earlier TGCs (50 cm high, 10m long) are plastic green house with an air inlet at one end and an exhaust fan at other end. TGCs in IARI (3.6 m height x 18 m length) are constructed as commercial green houses with steel pipes (diameter- 19 mm) are covered with 0.1 mm thick UV transparent PVC film. TGC is in north south orientation to obtain fairly uniform spatial distribution of direct solar beam in summer and to minimize the influence of shading of neighboring chambers in winter. Air inside the chambers is naturally heated through sunrays and exhaust fans are installed at other ends to create the temperature gradient. In advanced systems, ventilation rate is controlled automatically according to fluctuation in air-temperature inside the chambers due to variation in solar radiation, based on the temperature measured over the canopy periodically (Plate 8). CO_2 is supplied through the longitudinal tubes along the inside wall of the chamber. TGCs are

Table 3 : List of existing FACE experiments in the world

Country	Places	Name of experiment	Crop/ Ecosystem (ppm)	Level of CO_2	Plot diameter (m)	Contact person	Since when started	Comments
Australia	Horsum	AGFACE (Australian Grains Free Air CO_2 Enrichment Facility	Wheat-Pulse	550	16	Glenn Fitzgerald: glenn.fitzgerald@dpi.vic.gov.auau Maichel Tausz: michael.tausz@unimelb.edu.au	2007	Dryland cropping systems
	Pontville, Tasmania	TasFACE	Native grassland	550	1.5	Marc Hovenden: Mark.Hovenden@utas.edu.au	2002	Also increase in temperature (use of IR lamps)
	Walpeup	AGFACE	Wheat	550	4	Glenn Fitzgerald: glenn.fitzgerald@dpi.vic.gov.auau	2008	Dryland
	Yabulu, Queens land	OzFACE	Natural tropical Savanna	370, 460, 550	15	Chris Stokes	2001	Soil C is also monitored
China	Wuxi Experi ment Station, Jiangsu Province	China FACE	Rice-wheat	Ambient + 200	12	Jianguo Zhu: jgzhu@issas.ac.cn	2001	-

Contd.

Germany	Braunschweig, DEU	FAL Change FACE	Winter Barley, Winter wheat, Corn, Sugar beet, Grass, Cover crops	550	20	Hans Joachim Weigel: hans. weigel@vti.bund.de	1999	N dose and water application rates
	Giessen, DEU	GiFACE	Grassland	Ambient +20%	8	Ludger Gruenhage: Ludger. Gruenhage@bot2.bio.uni-giessen.de	1998	Oldest running FACE Experiment
India	New Delhi	FACE						
	Hyderabad	FACE						
Italy	Genomics Research centre	Durum FACE	Durum wheat	570	14	Luigi Cattivelli: Luigi.cattivelli@entecra.it	2011	Rainfed wheat
Japan	Tsukuba, Ibaraki	Tsukuba FACE	Rice	Ambient +200	17	Toshihiro Hasegawa: Thase@affrc.go.jp	2010	Whole devoted to rice paddy
New Zealand	Bulls	New Zealand FACE	Grazed pasture grassland	475	12	Paul Newton: paul.newton@agresearch.co.nz	2008	Temperature effect also added
Switzerland	Hofstetten, CHE	Web FACE	Mature temperate forest	600	80	Christian Korner: ch.koerner@unibas.ch	2000	Only experiments on mature forest ecosystem

Contd.

United Kingdom	Bangor	Bangor FACE	Deciduous Forest	Ambient +200	8	Douglas Godbold: d.l.godbold@ bangor.ac.uk	?	
United States	Minne sota	BioCON	Grassland	550	20	Peter Reich: preich@umn.edu	1998	Both C3 and C4 grasses
	Cheye nne, WY	PHACE	Native rangeland (Prairie)	600	3.4	Jack Morgan: Jack. Morgan@ars.usda. gov	2006	Both temperature and CO2 effect
	North Carolina	Duke FACE	Pine Forest	Am bient + 200	30	Ram Oren: ramoren@duke.edu	1996	Data available on request
	California	Jasper Ridge Global Change Experiment	Grassland	Double the ambient	2	Chris Field: chris@jasper. stanford.edu	1997	Warming, increase precipitation and N deposition also studied along with CO2 increase in natural ecosystem
	Tennessee	ORNL FACE	Deciduous forest	~40% higher than ambient (also diurnal variation)	25	Rich Norby: norbyrj@ornl.gov	1998	An user facility: Open to all
	Wisconsin	ASPEN FACE	Aspen forest	Ambient +200	30	Mark Kubiske: mkubiske@fs.fed.us		
	Illinois	SoyFACE	Soybean	550	20	Don Ort: d-ort@life. uiuc.edu	2001	Large experimental plots
		Sky Oaks	Chaparral ecosystem	550	16	Walter Oechel: oechel@sunstroke. sdsu.edu	?	Part of a Global Change Research Group

designed to grow wheat to study the responses of specific genotypes to variable temperature and CO~ concentrations. TGCs are generally covered with 6 mm UV stable polythene sheet. Outlet plenum is a box covered with PVC sheet. Fan is mounted on the roof pointing outwards. This film is elastic and reduces the damage of water and hail. Temperature difference between inlet and outlet doesn't go more than 5°C. Temperature measured by differential thermocouples is used to set the required fan speed electronically. CO_2 concentration is continuously monitored at the outlet fan by an IRGA. Airflow needed to maintain temperature gradient is sufficient to prevent any measurable difference in CO_2 concentration between modules (~10 ppm). There are provisions for monitoring both CO_2 and temperature effect inside the TGTs.

Table 4 : Summary of meta-analysis results from FACE and chamber techniques and literature reviews. (a): Ainsworth & Long 2004; (b): Curtis & Wang 1998; (c): Long *et al.*, 2004; (d): Drake *et al.*,1997; (e): Leakey *et al.*,2009.

Parameters	FACE	Chambers
Stomatal conductance	-20% a	-20% b
Above-ground dry-matter	28% [a]	28.8% [b]
Light saturated CO_2 uptake	47% [a]	31% [b]
Yield	17% [a]	28-35% [b]
Starch content	83% [c]	67.6% [b]
Leaf Nitrogen	-4% [a]	-15 % [b]
Photosynthesis	30% [c]	53% b
Rubisco Activity	-20% [c]	-20% [b, d]
Dark respiration	0% [e]	15-20% [b, d]

A brief summary of various CO_2 enrichment methods are listed in Table 5.

Table 5 : Comparison between CO_2 enrichment methods

CO_2 Enrichment methods	Advantages	Disadvantages
Leaf Cuvettes	Single leaf response study Specialized study on single factor approach	No whole plant response, Low light Higher constant day night temperature Higher relative humidity
SPAR system	High light similar to natural condition Control of temperature and humidity	Chamber effects Higher temperature and relative humidity Wind gradients Limited replications Expensive per unit
CO_2-TGTs	Suitable established CO_2 and temperature gradients for interaction studies	Possible chamber effects Gradients may be variable
Open top chambers	*in situ* study on crops and natural vegetation Near natural environment Regulates elevated CO_2	Gradients of wind and humidity produce chamber effects Many chambers need to deal with natural variability of agro eco systems
FACE	Closed to natural environmental conditions Large area of plant material	Complex controls Large are needed Cost may be higher

Major Global Facilities

SoyFACE

The SoyFACE (Soybean Free Air Concentration Enrichment) is facility at the University of Illinois for growing crops under elevated CO_2 facility in situ (i.e. actual field growing conditions) (Plate 9). This facility utilizes the elevated O_3 effect on soybean plants also, and how to cope up with this elevated CO_2 or O_3 conditions in future (strategic research) It also seeks to exploit the genotypes or genes to increase or sustain yields under the changes atmospheric conditions with elevated CO_2, O_3, temperature and drought stress singly or in combination. Research also involves the C sequestration potential of soil and how the soil quality will be changed, and finally the effect on pathogens. This facility allows exposure of plants to elevated CO_2 and O_3 (level which is expected at 2050) under complete natural conditions (i.e., not confined or under any boundary). The wind speed and direction is monitored at the centre of FACE rings, and the release of gas is automated to keep the concentration at the desired level. The gas is released through small perforation of the tubes around the circumference of the experimental field. Thus the microenvironment is not changed (as happens in the closed or semi-closed chambers).

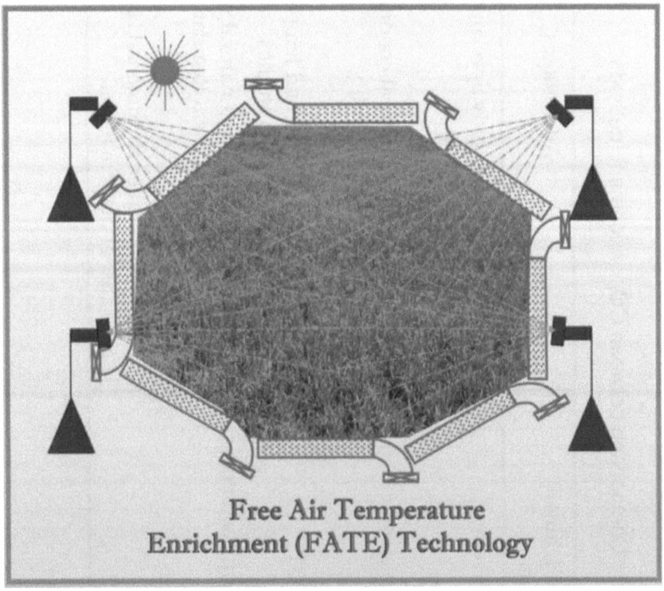

Fig. 8 : The Free-Air-Temperature-Enrichment technology (*Source :* Sheel Biotech limited)

The Swiss Crane Project

The Swiss Canopy Crane Project: In an attempt to understand the effect of enrichment of atmospheric CO_2 on mature forest, a FACE system was developed for long canopy trees (>30 m height), called web-FACE (Plate 9). In this system, a liquefied CO_2 is vaporized and supplied at a pressure of 450 kPa above the ambient, and released from thin black plastic tubes of a 8.5 km network (similar to the microtubes used for drip irrigation systems) into the crown of trees. The CO_2 concentration at tree canopy is measured automatically and the access to the canopy is facilitated by freestanding tower crane for measurement. The spatial and temporal monitoring of CO_2

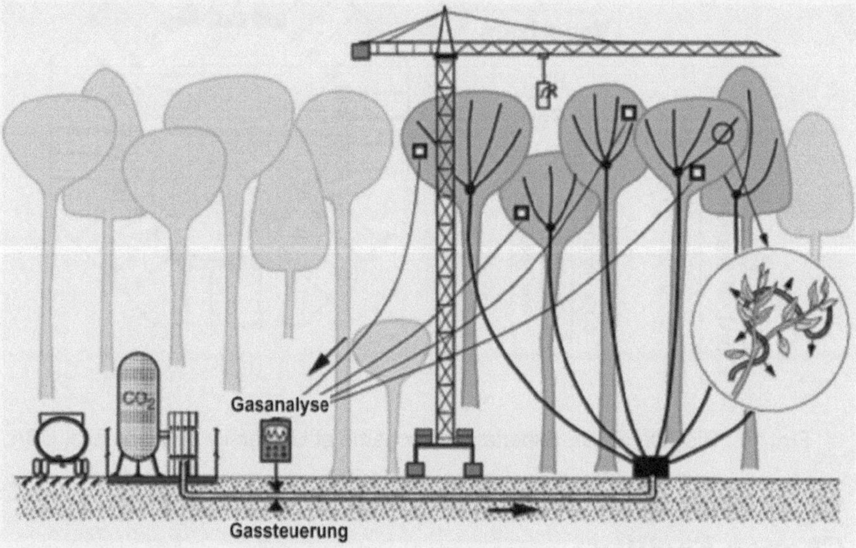

Fig. 9 : CO_2 release and monitoring at a temperate forest at Hofstetten, Switzerland.

BioCON

BioCON Experiment at University of Minnesota: The FACE system used in the BioCON experiment was developed by Brrokhaven National Laboratory, U.S. Department of Energy. It has six fumigation circles, three with elevated CO_2 (550 ppm) and three with ambient (360 ppm) levels (Fig. 10). For monitoring wind speed and direction, sensors are placed at the centre of each circle. These sensors also transmit information about the CO" concentration above the crop canopy to a computer and the necessary release of CO_2 when and from which side of circle is controlled. The whole operation of CO_2 release in done at the fan houses. Here the CO_2 is mixed with ambient air and released by a large fan to the underground pipe, called 'pleum', which runs at an octagonal shape around the plot circle. This specific experiment runs in a completely randomized design, CO_2 treatment is the whole-plot factor, while sub-plot factor may be N treatment, crop species or anything with randomly assigned replicates (Plate 10).

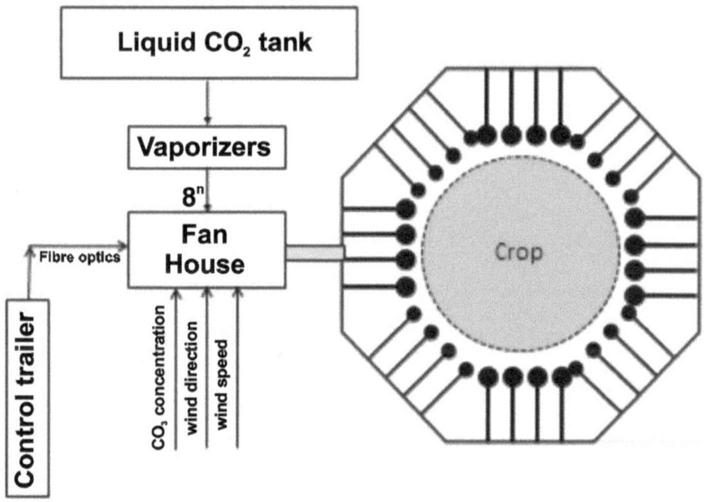

Fig. 10 : BioCON FACE Experimental scheme at University of Minnesota, USA

The Atmospheric Radiation Measurement Climate Change Facility

Popularly known as ARM facility funded by US department of energy. It is a user facility for study of climate change research by the scientific community all-round the globe. It has two mobile and one aerial facility, and a data archive. This facility has three field sites for collecting climatic data - Cluster of sites at Southern Great Plains centering Oklahama (*in situ* and remote sensing facilities), North Slope of Alaska (data on cloud and radiative processes over high latitudes), and Tropical Western Pacific at Papua New Guinea, Nauru Island and Darwin Australia (monitoring climate change variability like El Nino, ENSO, southern oscillation and sea surface temperature over pacific ocean).

The Primary Industries Climate Challenges Centre (PICCC)

This is a collaborative venture between the University of Melbourne and the Department of Primary Industries, Victoria. It combines the resources and activities of its partners to lead research, development and extension related to both mitigation and adaptation of climate change at both the enterprise and landscape level. PICCC do not fund for direct research but facilitates in developing collaborative projects with PICCC as one of the partners. The proposals must be linked to primary industries research related to climate change.

References

Allen, L.H. Jr and Boote, K.J. 2000. Crop ecosystem response to climate change: soybean. In: Climate change and global crop productivity (K.R. Reddy and H.F. Hodges Eds.), pp. 133-160, CAB International, Wallingford, Oxon, U.K. and New York.

Allen L.H. Jr., Pan, D., Boote, K.J., Pickering, N.B., and Jones, J.W. 2003. Carbon dioxide and temperature effects on evapotranspiration and water use efficiency of soybean *Agron. J.*, 95 : 1071-1081

Allen, L.H. Jr., Bisbal, E.C., Boote, K.J., and Jones, P.H. 1991. Soybean dry matter allocation under subambient and superambient levels of carbon dioxide. *Agron J.* 83 : 875-883.

Baldocchi, D.D. 2003. Assessing the eddy covariance technique for evaluating carbon dioxide exchange rates of ecosystems: past, present and future. *Global Change Biology* 9 : 479-492.

Barton, C.V.M., Duursma, R.A., Medlyn, B.E., Ellsworth, D.S., Eamus, D., Tissue, D.T., Adams, M.A., Conroy, J., Crous, K.Y., Liberloo, M., Löw, M., Linder, S., and McMurtrie, R.E. 2012. Effects of elevated atmospheric [CO_2] on instantaneous transpiration efficiency at leaf and canopy scales in Eucalyptus saligna. *Global Change Biology*, 18 (2) : 585-595.

Cao, L., Bala, G., Caldeira, K., Nemanid, R., and Ban-Weiss, G. 2010. Importance of carbon dioxide physiological forcing to future climate change. *Proceedings of the National Academy of Sciences*, May 5, 2010. doi: 10.1073/pnas.0913000107.

CRU, 2012. Climate Monitor Online. Climate Research Unit, School of Environmental Sciences

Crous, K.Y., Zaragoza-Castells, J., Ellsworth, D.S., Duursma, R.A., Löw, M. , Tissue, D.T.,and Atkin, O.K. 2011. Light inhibition of leaf respiration in field? Grown Eucalyptus saligna in whole? Tree chambers under elevated atmospheric CO_2 and summer drought. *Plant, Cell and Environment,* 35(5) : 966-981.

Denmead, O.T. 1984. Plant physiological methods for studying evapotranspiration: problems of telling the forest from the trees. *Agricultural Water Management*, 8 (1-3) (1984), pp. 167-189

Delmas, R.J., Ascencio, J.M. and Legrand, M. 1984. Polar ice evidence that atmospheric CO_2 20,000 yr BP was 50% of present. *Nature*, 284 : 155 - 157

DeConto,R.M., Galeotti, S., Pagani, M., Tracy, D., Schaefer, K., Zhang, T., Pollard, D., and Beerling, D. 2012. Past extreme warming events linked to massive carbon release from thawing permafrost. *Nature*, 484 : 87-91.

D.C. Reicosky, S.W. Wagner, Devine and O.J., 1990. Methods of calculating carbon dioxide exchange rates for maize and soybean using a portable field chamber. *Photosynthetica*, 24 (1990), pp. 22-38

Fleisher, D.H., Timlin, D.J., Yang, Y., Reddy, V.R., and Reddy, K.R. 2009. Uniformity of soil-plant-atmosphere-research chambers. *Transactions of the ASABE*, 52(5) : 1721-1731.

Hansen, James. 2008 Climate threat for planet: Implications for energy policy, and intergenerational justice. Bjerknes Lecture, American Geophysical Union, San Francisco, California, December 17, 2008.

Heagle, A.S., D.E. Body, and Heck. W.W. 1973. An open-top field chamber to assess the impact of air pollution on plants. *J. Environ. Qual.* 2 : 365-368.

Idso C. and Fred Singer, S. 2009. Climate Change Reconsidered: 2009 Report of the Nongovernmental Panel on Climate Change (NIPCC), Chicago, IL: The Heartland Institute, 2009. 739 p.

IPCC 2007.Climate Change 2007: Synthesis Report, Pachauri, R.K. and Reisinger, A. (Eds.), IPCC, Geneva, Switzerland. pp 104.

Jablonskil, L.M., Wang, X.,and Curtis, P.S. 2002. Plant reproduction under elevated CO_2 conditions: a meta-analysis of reports on 79 crop and wild species. *New Phytologist*, 156(1) : 9-26.

Jones, G. S., M. Lockwood, and Stott P. A. (2012), What influence will future solar activity changes over the 21st century have on projected global near surface temperature changes?, *J. Geophys. Res.*, doi:10.1029/2011JD017013, in press.Faculty of Science, University of East Anglia, Norwich NR4 7TJ, UK

Keren Stimler, Stephen A. Montzka, Joseph A. Berry, Yinon Rudich,and Dan Yakir, 2010. Relationships between carbonyl sulfide (COS) and CO2 during leaf gas exchange. *New Phytologist*, 186 (4) : 869-878.

Kim, S.Y., Reddy, V.R., Baker, J.T. Gitz, D.C.,and Timlina, D.J. 2004. Quantification of photosynthetically active radiation inside sunlit growth chambers. *Agriculture and Forest Meteorology*, 126(1-2) : 117-127.

K.R. Reddy, J.J. Read, J.T. Baker, J.M. McKinion, L. Tarpley, H.F. Hodges, and Reddy, V.R. 2001. Soil-plant-atmosphere-research (SPAR) facility-a tool for plant research and modeling. *Biotronics*, 30 : 27-50.

Livingston, G.P.,and Hutchinson, G.L., 1995. Enclosure-based measurement of trace gas-exchange: applications and sources of error. In: Matson, P.A., Harris, R.C. (Eds.), Biogenic Trace Gases: Measuring Emissions from Soil and Water. Blackwell Scientific Publication, London, pp. 14-51.

Leadley P.W. and Drake, B.G. 1993. Open top chambers for exposing plant canopies to elevated CO2 concentration and for measuring net gas exchange. *Plant Ecology*, 104-105(1) : 3-15.

Müller, J., Eschenröder, A.,and Muller, W.D. 2009. Through-flow chamber CO2/H2O canopy gas exchange system-Construction, microclimate, errors, and measurements in a barley (*Hordeum vulgare* L.) field. *Agriculture and Forest Meteorology*, 149(2) : 214-229.

NEIC, 2002. National Energy Information Centre, Energy Information Administration, Washington DC.

Olszyk, D., Tingey, D., Wise, C.,and Davis, E. 2002. CO_2 and O_3 alter photosynthesis and water vapor exchange for Pinus ponderosa needles *Phyton-Annales Rei Botanicae*, 42 (3) : 121-134.

Pedersen,O., Rich, S.M.,and Colmer, T.D. 2009. Surviving floods: leaf gas films improve O_2 and CO_2 exchange, root aeration, and growth of completely submerged rice. *The Plant Journal*, 58(1) : 147-156.

P.G. Jarvis, J. Catsky, T.E. Eckart, D. Koch and Koller D. General principles of gasometric methods and the main aspects of installation design. Z. Sestak, J. Catsky, P.G. Jarvis (Eds.), Plant Photosynthesis Production: Manual of Methods, Dr. W. Junk N. V. Publishers, The Hague, pp. 49-110.

Phene, C.J., Baker, D.N., Lambert, J.R., Parsons, J.E.,and McKinion, J.N. 1978. SPAR - A soil-plant-atmosphere research system. *Transactions of the ASAE*, 21(5) : 924-938.

Prasad, K.V.V, Boote, K.J., Allen, L.H. Jr and Tomas, J.M.G. 2002. Effect of elevated temperature and carbon dioxide on seed-set and yield of kidney bean (*Phaseolus vulgaris* L.), *Global Change Biology*, 8 : 710-721.

Prasad, K.V.V, Boote, K.J., Allen, L.H. Jr and Tomas, J.M.G. 2003. Supra-optimal temperatures are detrimental to peanut (*Arachis hypogaea* L.) reproductive processes and yield at both ambient and elevated carbon dioxide. *Global Change Biology*, 9: 1775-1787.

Prasad, P.V. V., Vu, J.C.V., Boote, K. J. and Allen, L. H. 2009. Enhancement in leaf photosynthesis and upregulation of Rubisco in the C? sorghum plant at elevated growth carbon dioxide and temperature occur at early stages of leaf ontogeny. *Functional Plant Biology*, 36 (9) : 761-769.

Reicosky, D.C., Dugas,W.A., and Torbert H.A. 1997. Tillage- induced soil carbon dioxide loss from different cropping systems. *Soil and Tillage Research*, 41(1-2) : 105-118.

Rogers, H.H., G.E. Bingham, J.D. Cure, J.M. Smith, and K.A. Surano. 1983. Responses of Selected Plant Species to Elevated Carbon Dioxide in the Field. *J. Environ. Qual.* 12:569-574.

Rodeghiero M, Niinemets U. and Cescatti A. 2007. Major diffusion leaks of clamp-on leaf cuvettes still unaccounted: how erroneous are the estimates of Farquhar *et al.* model parameters? *Plant Cell Environ.* 30(8):1006-22.

Sinclair, T.R. and Horie, H. 1989. Leaf nitrogen, photosysthesis and crop radiation use efficiency: A review. *Crop Science*, 29 : 90-98.

Stedutoa, P., Çetinkökùa, Ö., Albrizioa, R.,and Kanberb, R. 2002. Automated closed-system canopy-chamber for continuous field-crop monitoring of CO2 and H2O fluxes. *Agricultural and Forest Meteorology*, 111(3) : 171-186.

S.W. Wagner and Reicosky D.C. 1992. Closed-chamber effects on leaf temperature, canopy photosynthesis, and evapotranspiration *Agronomy Journal*, 84 (4) : 731-738.

S.W. Wagner, D.C. Reicosky, and Saamuel R. 1997. Alessi. Regression models for calculating gas fluxes measured with a closed chamber. *Agronomy Journal*, 89: 279-284.

Tans and Keeling, 2012. Dr. Pieter Tans, NOAA/ESRL (www.esrl.noaa.gov/gmd/ccgg/trends/) and Dr. Ralph Keeling, Scripps Institution of Oceanography (scrippsco2.ucsd.edu/).

Tazoe, Y., von Caemmerer, S, Badger, M.R., and Evans, J.R.2009. Light and CO_2 do not affect the mesophyll conductance to CO_2 diffusion in wheat leaves. *Journal of Experimental Botany*. 60(8) : 2291-2301.

Uprety DC, Garg SC, Bisht BS, Maini HK, Dwivedi N, Paswan G, Raj A,and Saxena DC 2006. Carbon dioxide enrichment technologies for crop response studies. *Journal of Scientific and Industrial Research* 65 (11): 859-866.

Uprety, D.C., Dwivedi, N., Mohan, R.,and Paswan, G. 2001. Effect of Elevated CO_2. Concentration on Leaf Structure of Brassica Juncea under Water Stress. *Biologia Plantarum*, 44(1) : 149-152.

Uprety, D.C., Abrol, Y.P. and Reddy V.R. 2007. Crop response to elevated CO2. In: Biodiversity and its significance (Tandon, P., Abrol, Y.P. and Kumaria, Suman Eds), I.K. International Publishing House, New Delhi. pp 291-331.

Vanaja, M., Maheswari, M., Ratnakumar, P.,and Ramkrishna, Y.S. 2006. Monitoring and controlling of CO_2 concentrations in open top chambers for better understanding of plants response to elevated CO_2 levels. *Indian Journal of Radio and Space Physics*. 35: 193-197.

Wall, G.W., Kimball, B.A., White, J.W., Ottman, M.J. 2011. Gas exchange and water relations of spring wheat under full-season infrared warming. *Global Change Biology*, 17 (6) : 2113-2133.

Wall, G.W., Garciab, R.L., Wechsung, F., and Kimball, B.A., 2011. Elevated atmospheric CO_2 and drought effects on leaf gas exchange properties of barley. *Agriculture Ecosystems and Environment*. 144 (1) : 390-404.

Wilkinson, M.J., Monson. R.K., Trahan, N., Lee, S., Brown, E., Jackson, R.B., Polley, H.W., Fay, P.A.,and FALL, R., 2008. Leaf isoprene emission rate as a function of atmospheric CO2 concentration. *Global Change Biology*, 15(5) : 1189-1200.

Yasutake, D., Kitano, M., Hamakoga, M., Araki, T., nagasuga, K., and Suzuki, Y. 2000. An improved cuvette for individual evaluations of gas exchange parameters on adaxial and abaxial leaf surfaces under different air currents. *Biotronics*, 29 : 33-42.

Zhao, D., Reddy, K.R., Kakani, V.G., Read, J.J.,and Sullivan, J.H. 2003. Growth and physiological responses of cotton (*gossypium hirsutum* l.) to elevated carbon dioxide and ultraviolet-b radiation under controlled environmental conditions. *Plant Cell Environ.*, 26 (5) : 771-782

5

Advances in Molecular Breeding of Perennial Spice Crops

K. Nirmal Babu, V.A. Parthasarathy and K.V. Peter

Understanding biology and genetics at molecular level has become very important for better understanding and manipulation of genome architecture and this creates new opportunities to streamline plant breeding. Breeding is simply defined as the selective mating of individuals of a population to isolate or combine desired morphological, physiological or genetic traits such as appearance, yield, and disease resistance. This is performed with the assistance of identifiable traits.

Molecular markers have contributed significantly in this respect and have been widely used in plant science in a number of ways, including genetic fingerprinting, diagnostics, identification of duplicates and selecting core collections, determination of genetic distances, genome analysis, identification of markers associated with desirable breeding traits which are useful in marker assisted breeding, genomics and development of transgenics. Use of these markers also significantly reduces breeding time and cycles required for crop improvement. This is reflected in the fact that complete molecular and genetic maps have now been constructed for at least 20 major crop species and many more have been characterized at less defined level and the genomes of a few model plant species were completely sequenced.

Molecular breeding is essentially based on the use of molecular markers in association with linkage maps and genomics to select plants with desirable traits based on a genetic assay(s), tagging and isolation of useful genes. This can make plant breeding more precise, rapid and cost effective in comparison to phenotypic selection. A large number of environmentally insensitive genetic markers which adequately follow the inheritance of important agronomic traits coupled with genomics applications have dramatically enhanced the efficiency of plant breeding. A number of breeders are now using markers to increase the effectiveness of selection in breeding and to significantly shorten the development time of varieties. New advances in automated technology enabled new approaches in marker assisted breeding or Breeding by Design. The advances in applied genomics and the availability of enormous marker data provide us with the tools to determine the genetic basis for traits of agronomical importance in many field and horticultural crops. Also, methods for assessing the allelic variation at these agronomically important loci are now available. This combined knowledge will eventually allow the breeder to combine favorable alleles at all these loci in a controlled manner, leading to superior varieties. Extending on these capabilities, the understanding of the genetics basis of agronomically important characters and the allelic variation, would enable the breeder to design superior breeding lines '*in silico*'.

Applications of DNA Markers in Breeding Programes

Different stages of impact that can be distinguished in Marker Assisted Breeding are viz, Marker Assisted Selection (MAS), creation and identification of novel varieties, removal of linkage drag, pyramiding resistance genes, breeding of polygenic traits, introduction of novel characteristics and effective exploitation of (exotic) germplasm.

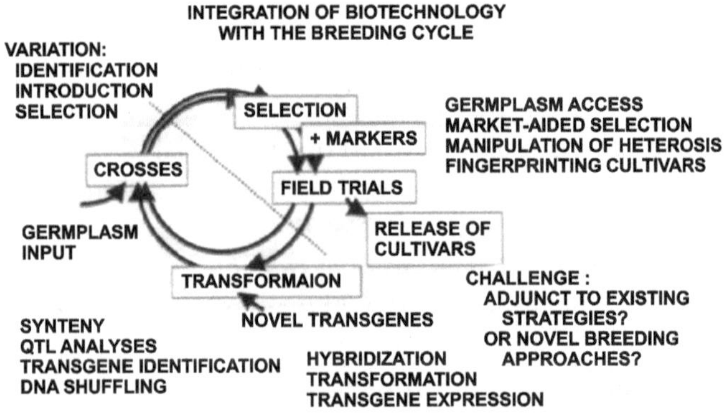

Molecular characterization, estimation of genetic diversity, finger printing and mapping strategies for perennials

Genetic diversity

The classification and identification of living organisms have provided a wealth of information about the organisms, phylogentics and their diversity. This was further strengthened with the advent of efficient DNA marker and DNA, RNA or Protein sequence information combined with bioinformatics.

Genetic maps

Plant breeding and genetics rely on genetic maps. These are made by analysis of segregating plant populations derived by crossing parents with contrasting characteristics.

Association mapping

A complementary approach is to analyse Linkage Disequilibrium in natural populations where large structured mapping populations do not exist.

Pseudo – Test Cross

Most linkage maps in plants have been obtained from segregating populations derived from crosses between inbred lines. Such populations are generally not available in trees and are difficult to obtain due to genetic load and time constraints. Grattapaglia and Sederoff (1994) used a Pseudo – Test Cross approach as a Mapping Strategy to construct Genetic Linkage Maps in out breeding perennial forest trees like Eucalyptus as well as in any highly heterozygous sexually reproducing living organisms. In this method, crosses are made between two heterozygous individuals and many single-doses RAPD markers will be heterozygous in one parent, null in the other and therefore segregate 1: 1 in their F1 progeny following a 'pseudo' testcross configuration and can be used for making linkage groups and preparation of frame work maps with ' MAPMAKER' soft ware. This strategy is ideally suited for marker assisted strategy in perennial sexually reproducing spice crops with long pre bearing period.

Candidate Genetic loci

Candidate gene based association mapping using single nucleotide polymorphisms (SNPs) is another approach used in mapping of genes controlling complex traits. The approach is to detect polymorphic DNA markers that are in linkage disequilibrium with the polymorphism with 'character' causing site, as opposed to the 'character' variant itself. The initial association studies will focus on candidate gene regions for which a number of single nucleotide polymorphisms (SNPs) have been previously identified, as opposed to genome wide scans.

Molecular Breeding of Perennial Spice Crops

Spices are, low volume high value, cash crops which influence the economy of the countries in which they are grown. The important spice crops of interest are black pepper, cardamom, vanilla, ginger, turmeric. Most of them are perennial polyploids with long pre bearing period. They are essentially open pollinated heterozygous populations. The seed progenies segregate and are not uniform hence are propagated vegetatively for commercial purposes. Ginger has no seed propagation and turmeric has rare seed progeny. The polyploid nature and long pre bearing period makes breeding of these crops cumbersome, time consuming and understanding of their genetics difficult. The perennial and heterozygous nature with long pre bearing period makes conventional molecular approaches not suitable for Spices crops. Under such circumstances population approaches like association mapping, use of pseudo crosses and SNPs in candidate gene approaches will help. Any reduction in breeding time coupled with precision breeding will help in developing high yielding high quality varieties resistant to major diseases and pests will not only help in increasing the production productivity of spices but also help in production of clean organic spices which is so important to compete in international markets. Use of molecular markers to augment breeding is of recent origin in spices.

Black pepper (*Piper nigrum* L)

Black pepper, native to India, is the most important spice of the world. Conserving and fingerprinting the genetic diversity in pepper and development of resistant varieties to '*Phytophthora* Foot rot', pollubeetle (*Longitarsus nigripennis*), nematodes (*Radopholus similes, Meloidogyne incognita*) will increase the production productivity of black pepper. Use of molecular markers to augment breeding is of recent origin in black pepper. Most of the studies are confined to molecular profiling of germplasm and genetic fidelity studies.

Molecular Characterization and Diversity

Jaramillo and Manos (2001) used Internal Transcribed Spacers (ITS) of nuclear ribosomal DNA and elucidated phylogeny of 51 species of *Piper.*

Biochemical (isozymes) and Molecular marker (RAPD, AFLP and ISSR) polymorphism was used for assessment of genetic diversity in black pepper, characterize important cultivars, varieties and related species to develop finger prints and to study the inter relationships

Gaia *et al.* (2003) studied the genetic diversity in 78 clones of *Piper nigrum from South America* using 8 isoenzyme systems and concluded that most of the clones of *P. nigrum* L. were quite similar and homogeneous. Lebot *et al.* (1999) used AFLP profiling of various morpho and chemotypes of Hawaiian cultivars of *Piper methysticum* and concluded that Kava in Hawaii has extremely narrow genetic base and morphological and phytochemical variation is controlled by very few genes and different morphotypes are most likely somatic mutants from a common clonal source introduced by Polynesians during early settlement.

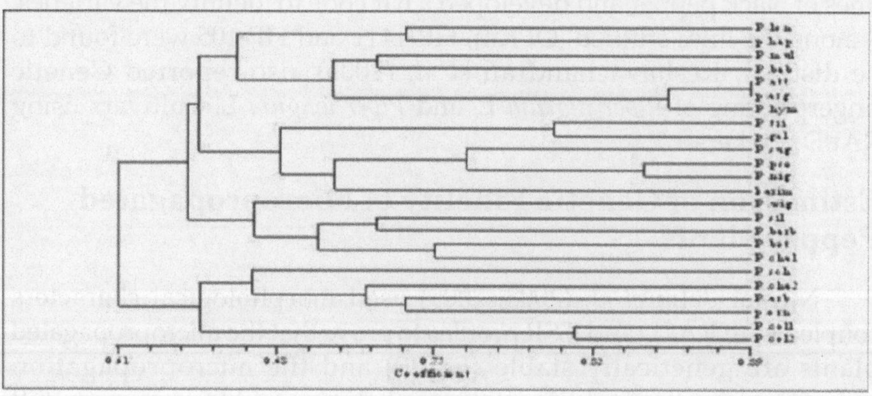

Fig 1 : Phylogenic tree constructed based on RAPD and ISSR polymorphism showing inter relationships among the *Piper* species (Nirmal Babu *et al.*, 2003)

Nirmal Babu *et al.* (2003) studied molecular inter relationships between 24 Pipe*r species* using RAPD and ISSR profiles. The study indicated that *P. pseudonigrum, and P. sugandhi* are closest to *P. nigrum,* the cultivated black pepper (Fig.1). Babu *et al.* (2003) studied RAPD and AFLP polymorphism among nine species of *Piper*. The exotic species *P. colubrinum* and *P. arboreum* were found to be highly distinct.

Nirmal Babu *et al.* (2003) also conducted similar studies on fourteen major cultivars and ten released varieties Cluster analysis indicated that there are distinct differences between most of the cultivars and varieties. Sreekara and Subhakara are nearest to each other while Panniyur 1 is farthest. Pradeepkumar *et al.* (2003) studied 24 black pepper (*P.nigrum*) accessions, including 9 advanced cultivars and 13 landraces using DNA markers. Good variability was observed among *P. nigrum* cultivars.

Nazeem *et al.* (2005) used both RAPD and AFLP analysis to assess the variability and inter- relationships among 49 cultivars of black pepper. Two selections from the variety Karimunda, namely Sreekara and Subhakara, together formed a single cluster with almost 92 percent similarity. But Panniyur 1 and Panniyur 3 which are the progenies of the same parentage were distinctly dissimilar. Nisha *et al.* (2007) after AFLP analysis of thirty popular and agronomically important cultivars of black pepper grouped the accessions into three major clusters and four diverse cultivars with only 30% similarity. Karimunda, a widely grown and popular cultivar was unique in the fingerprint profiles obtained.

Sreedevi *et al.* (2005) used RAPD to characterize 7 new high yielding lines of black pepper and developed a bar code to identify the varieties. Among the lines studied OPKm, HP-1411 and HP-105 were found to be distinct. Keshavachandran *et al.* (2005) also reported Genetic fingerprinting of *Piper nigrum* L. and *Piper longum* L. Cultivars using RAPD markers.

Estimation of Genetic Fidelity of Micropropagated Pepper plants

Nirmal Babu *et al.* (2003, 2005) used morphological characters coupled with RAPD and ISSR profiles to prove that the micropropagated plants are genetically stable (Fig. 2) and the micropropagation technology can be used for commercial cloning of black pepper. IISR has an accredited test laboratory By the DBT for virus diagnostics and genetic fidelity testing of tissue cultures plants of spices.

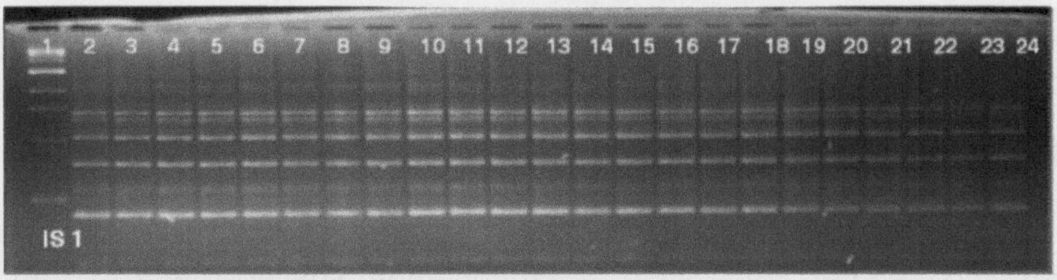

Fig. 2 : ISSR profiles of Micropropagated plants of pepper cv. Subhakara indicating genetic uniformity (Nirmal Babu *et al.*, 2005)

Development of Mapping Population and Preliminary Molecular Maps

Two mapping population of 200 segregating progenies were developed between Panniyur 1 X Subhakara for preparation of preliminary genetic map and P 24 X Subhakara to tag *Phytophthora* resistance in black pepper (Nirmal Babu *et al.*, 2003, 2011). RAPD and ISSR profiling of 96 progenies of first mapping population was done along with their parents and over 200 polymorphic markers segregating in the population were scored. Preparation of frame work molecular map is in progress using pseudo test cross approach. This progeny segregates for *Phytophthora* resistance. With an objective to tagging this character, the mapping population was phenotyped (screened) for *Phytophthora* resistance using leaf screening method and a few resistant lines were observed from both crossed and selfed progenies. The three different methods of screening i.e. Leaf lesion index, Stem lesion length, Depth penetration index in both selfed and crossed progenies indicated that the resistance segregated in the nearest ratio of ratio of 1 resistant: 15 susceptible which is a ratio controlled by genes with duplicate functions. Preparation of frame work molecular map is in progress using pseudo test cross approach. This progeny segregates for *Phytophthora* resistance.

Identification of Hybrids and Markers Linked to Important Characters

Identification of hybrids

Isoenzymes were utilized to examine the hybrid nature of plants derived from inter specific crosses. Sasikumar *et al.* (1999) used isozyme analysis to characterize inter specific hybrids between *Piper nigrum.* x *P. attenuatum* and *P. nigrum* x *P. Barberi* to identify the true hybrids by the presence of Hybrid specific as well as male parent specific bands.

Johnson *et al.* (2005) used male parent-specific RAPD markers for identification of hybrids in black pepper (*Piper nigrum* L.)

Markers Linked to Phytophthora Resistance

In an effort to identifying RAPD markers linked to *Phytophthora* resistance, eleven lines each of *Phytophthora* susceptible and tolerant cultivars were used for RAPD profiling. The study indicated that in general the tolerant lines formed a cluster of their own and a few suscepts were grouped with tolerant lines. One marker at 700 bp region was found to be present in tolerant lines and absent in suscepts (Nirmal Babu *et al.*, 2003) Fig. 3.

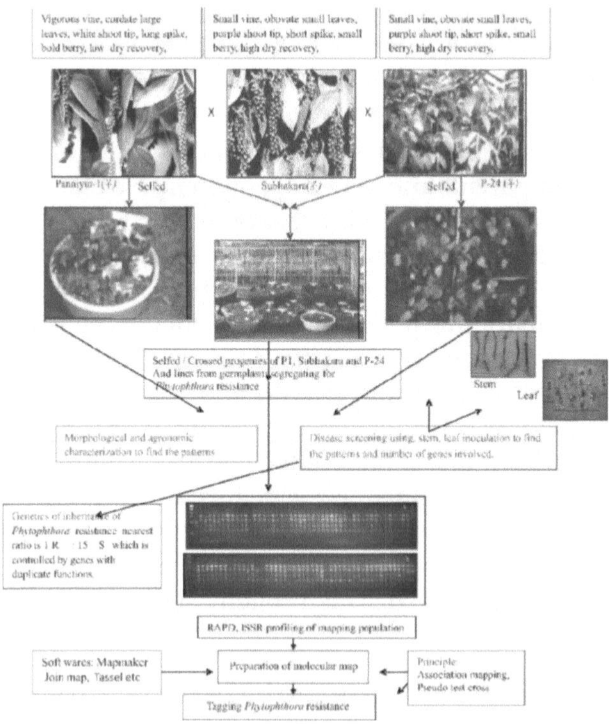

Fig 3 : Strategy for preparation of molecular map and tagging of *Phytophthora* resistance in Black pepper

Identification of adulterants

Dhanya *et al.* (2007) reported simple protocol for extraction of high quality genomic DNA from dried black pepper. This protocol can be used for characterizing and checking adulteration in traded black pepper.

Development of SCAR Marker for Phytophthora Resistance in Black pepper (*Piper nigrum* L.)

Sheji Chandran *et al.* (2006) identified a RAPD marker associated with *Phytophthora* resistance and converted it in to SCAR marker (Fig.4). A unique band 360 base pair appeared in all the *Phytophthora* resistant lines but absent in the susceptible lines. This marker can be used for screening black pepper germplasm at seed ling stage itself.

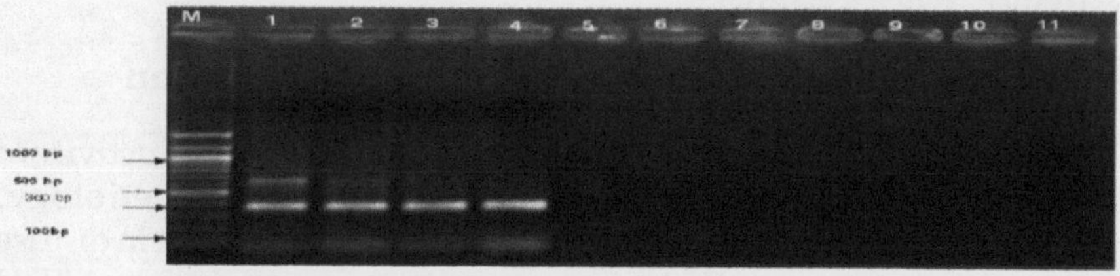

Fig 4 : Differential amplification of SCAR among susceptible and resistant lines of black pepper (Sheji Chandran *et al.*, 2006) Lane M. 100 bp ladder, 1- 4 *Phytophthora* resistant lines, 5-11 *Phytophthora* susceptible lines.

Comparative genomics, Tagging and Isolation of Candidate genes

In perennial polyploids like black pepper conventional mapping strategies are laborious. Under such circumstances, genomics approach and using information from other sources with comparisons at heterologus genomes or genes give us the necessary leads for tagging. Candidate genes responsible for pathogenesis can also be identified from sequence information available on r genes and information available from *Arabidopsis* genome. Preliminary work on isolation of genes responsible for agronomically important characters, especially for biotic and a biotic stresses was done. A few putative genomic and c DNA fragments associated with resistance like *β-1,3-glucanase, Osmotin, CMV coat protein, methyl glutaryl CoA reductase, pea lectin* and *tomoto protease inhibitor* genes are isolated from black pepper and *Piper colubrinum* using genomic approaches, substantive hybridization, differential display, heterologous probes etc. Programmes on isolating, cloning and validating of full length genes is in progress (Girija *et al.*, 2005, Sujatha *et al.*, 2005, Bhat *et al.*, 2005, Dicto and Manjusha 2005, Johnson *et al.*, 2011)

Reports on genome wide analysis of NBS-LRR-encoding genes in *Arabidopsis* are available. A PCR based candidate gene approach with degenerate primers for isolating Resistance gene candidates. This was used to isolate R gene candidates in Black pepper using degenerate primers designed from the conserved Ploop and GLPL motifs of NBS region of known R genes. A 500 Bp fragment was amplified. Comparative genomics has already made much headway for solanaceous crops to which capsicum belongs where much information on *Phytophthora* resistance is available. Information available on mapping of heterologous loci for example, *Ph-2* locus controlling partial resistance to *Phytophthora infestance* in tomato, *Rps 1*-k in soybean, *Phyto.5.2* from capsicum can also be used and are being tested to tag phytophthora resistance in black pepper.

Small Cardamom, *Elettaria cardamomum* Maton

Small Cardamom is also native to India. India. The productivity of cardamom is hampered by various diseases of viral etiology. Development of Virus resistant varieties is an important input in disease management strategy and being a cross pollinated crop, clones are ideal for generating true to type planting material from high yielding high quality genotypes.

Molecular Characterization and Diversity

Elettaria cardamoum Maton is a monotypic genus in India but a few related genera like *Amomum, Aframomum, Alpinia* and *Hedychium,* most of which co exist with small cardamom. Nirmal Babu *et al.* (2011) studied about 11 genera viz., *Amomum, Aframomum, Alpinia,, Hedychium* and *Elettaria* and 96 collections of cultivated cardamom were studied for RAPD, PCR - RFLP and ISSR polymorphism to study their relationships and diversity. The study indicated that *Amomum subulatum* is closest to cultivated cardamom among the genera studied. There were no duplicates among cultivated cardamom genotypes studied (Fig 5). The study indicated a clear divergence in Kerala and Karnataka collections with comparatively less divergence within a population and varieties developed by hybridization have higher divergence. The clustering of NKE (Natural Katte Escapes) lines indicate their possible origin as segregating progeny from populations of narrow genetic diversity. Controlled breeding rather than selection from open pollinated progeny is to be preferred in cardamom to bring more variability in germplasm.

Tagging of Agronomically Important Characters

Development of structured mapping populations for tagging of Katte and Rhizome rot resistance are also in progress in cardamom involving katte resistant lines and rhizome rot resistant lines and the F2 populations are in the process of field planting.

Tagging of Katte Resistance

One putative RAPD marker was also identified associated with Katte Resistance.

Identification of adulterants

Syamkumar *et al.* (2005) reported standardization of protocol for the isolation of amplifiable genomic DNA from dried capsules of traded cardamom. This protocol will help in the PCR-based characterization of different grades of commercially traded cardamom and to identify adulterants if any. RAPD profiling coupled with biochemical characterization was used to characterize different international brands of cardamom to determine their quality.

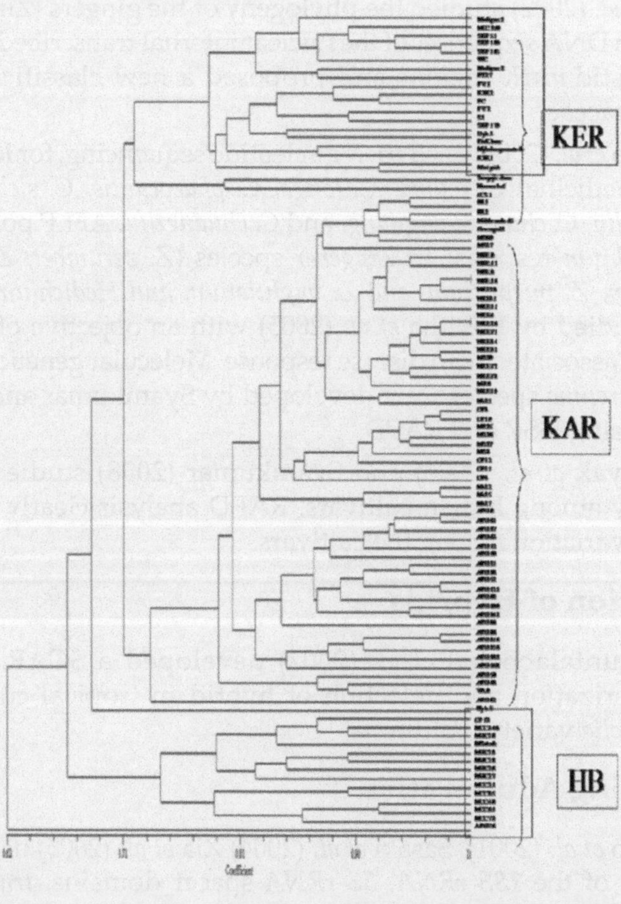

Fig 5 : Dendrogram of diversity among the cardamom germplasm collections from different regions

Turmeric (*Curcuma longa L*)

Turmeric is cultivated extensively in tropical regions of the world. India is the largest producer, exporter, and consumer of turmeric. Rhizome rot is the most important production constraint.

Molecular Characterization and Diversity

Chase (2004) and Ngamriabsakul *et al.* (2003) studied the phylogeny and relationships in Monocots based on analysis of DNA sequence data of seven genes representing all three genomes in Zingiberales and nuclear ribosomal DNA (ITS1, 5.8S and ITS2) and chloroplast DNA (*trn*L (UAA) 5[prime prime or minute] exon to *trn*F (GAA) and suggested that the tribe *Zingibereae* as well as the genus *Curcuma* are monophyletic. Kress *et al.* (2002) studied the phylogeny of the gingers (Zingiberaceae) based on DNA sequences of the nuclear internal transcribed spacer (ITS) and plastid *matK* regions and proposed a new classification of the Zingiberaceae.

Cao *et al.* (2003) used *trn K* nucleotide sequencing, for identification of six medicinal *Curcuma- C. longa, C. phaeocaulis, C. sichuanensis, C. chuanyujin, C. chuanhuangjiang,* and *C. chuanezhu*. AFLP polymorphism and *Pythium* response in *Zingeber* species (*Z. zerumbet, Z. cernum, Z. neesanum, Z. purpurium, and Z. wightianum and Hedichium flavescence)* were studied by Kavitha *et al.* (2005) with an objective of identifying markers associated with disease response. Molecular genetic fingerprints of 15 *Curcuma* species were developed by Syamkumar and Sasikumar (2007) using ISSR and RAPD

Nayak *et al.* (2006) and Syamkumar (2008) studied molecular diversity among Indian cultivars. RAPD analysis clearly showed the genetic variation among the cultivars.

Detection of Hybrids

Anuntalabochai *et al.* (2007) developed a SCAR marker for characterization and detection of hybrid in tropical cut flower, *C. alismatifolia* variety 'Patumma'

Checking Adulteration

Cao *et al.* (2001), Sasaki *et al.* (2004) Xia *et al.* (2005) used sequence analysis of the *18S rRNA, 5S rRNA* spacer domains, *trnK* gene and chemical fingerprints for quality control and authentication of Chinese and Japanese *Curcuma* drugs. Detection of extraneous *Curcuma* sp.

contamination of powdered samples of turmeric using random amplified polymorphic DNA was reported by Sasikumar *et al.* (2004). The analysis of 3 market samples of turmeric powder from Kerala, India, revealed the prevalence of wild species *C. zedoaria* powder over the *C. longa* (the common culinary turmeric) powder, although the curcumin levels of the samples were in accordance with the quality standards prescribed for the commodity.

Genetic Fidelity and Identification of Somaclonal Variants

Nirmal Babu *et al.* (2003) studied morphological and molecular variations among micropropagated and callus regenerated plants and found variations in both but with higher percentage of variation in callus regenerated somaclones. *In vitro* plants developed through microrhizome exhibited least amount of variations. They inferred that this is due to the accumulated vegetative mutations (mosaic) in turmeric. The genetic fidelity studies of turmeric germplasm conserved in *in vitro* genebank using RAPD profiling showed their genetic integrity (Ravindran *et al.* 2004). Salvi *et al.* (2003) and Praveen (2005) Sumathi (2007) analyzed turmeric somaclones using RAPD and concluded that plants regenerated using shoot tips showed genetic stability whereas, callus derived and inflorescence derived plants showed somaclonal variation.

Genetic stability of 12 month old *in vitro* conserved *C. longa* cv. Prathibha plants were assessed using RAPD by Tyagi *et al.* (2007) and no significant variation was observed.

Ginger, *Zingiber officinale* Rosc

Ginger is the third most important spice originated in South Asia. There is no seed set in ginger leading to limited variability and this hampers Crop improvement programs. Rhizome rot caused by *Pythium aphanidermatum,* bacterial wilt caused by *Ralstonia solanacearum* are the major diseases affecting ginger. Due to lack of seed set no direct molecular breeding could be done in ginger but Biotechnological tools can play an important role in ginger improvement.

Molecular Characterization and Diversity

Very little information is available on molecular characterization in ginger. Ninety six accessions of ginger were analysed using RAPD and the detected polymorphism is moderate to low (Sasikumar and Zachariah, 2003).

Molecular Characterization of Traded Ginger for Checking Adulteration

A simple and rapid method for isolating good quality DNA with fairly good yields from mature rhizome tissues of turmeric and ginger has been perfected by Shyam kumar *et al.* (2003). Isolated DNA was amenable to restriction digestion and PCR amplification. Genetic profiling of traded ginger from India and China using twenty RAPD primers and fifteen ISSR primers gave consistent amplification pattern. Significant variation was observed between the produces from the two countries (Sasikukumar unpublished).

Jiang *et al.* (2006) used metabolic profiling and phylogenetic analysis for authentication of ginger.

Genetic Fidelity Testing and Identification of Somaclonal Variants

Studies on RAPD profiling within the replicates of *in vitro* conserved and cryo preserved lines of ginger using RAPD primers, did not detect any polymorphism indicating the genetic stability (Ravindran *et al.,* 2004).

Nirmal Babu *et al.* (2003) used RAPD for detecting genetic variability of selected 'variants' among micropropagated and callus regenerated plants. They observed differences in RAPD profiles (Fig 6). This may be because of the accumulated natural mutations in ginger which has resulted in many varieties and cultivars even without sexual reproduction. Sumathi (2007) also reported higher number of variations in RAPD profiles in callus regenerated plants.

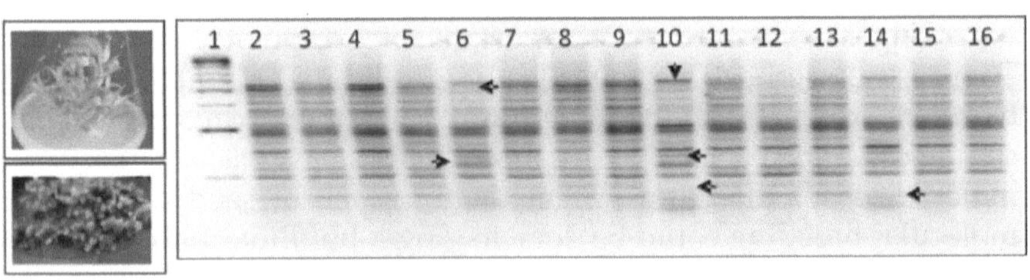

Fig. 6 : RAPD polymorphism in micropropagated plants of ginger. (Nirmal Babu *et al.* 2003)

Detection of Pathogens

Kumar and Anandaraj (2006) developed an efficient DNA isolation protocol and PCR based detection of bacterial pathogen in soil. This

PCR based method using universal *Ralstonia solanacearum* specific primer offer a rapid method for unambiguous detection of this pathogen at a concentration of 10^3-10^4 cells per gram of soil.

Tagging and Isolating Candidate Genes for Resistance

In the absence of sexual reproduction in Ginger no mapping population could be developed. Using genomic information from related and heterologus genomes or genes will give us the necessary leads for tagging.

Candidate genes approach using degenerate primers and functional genomics is more suitable for Zinger improvement. Aswati and Thomas (2006, 2007) reported isolation, characterization and expression of resistance gene candidates (RGCs) using degenerate primers based on conserved motifs from the NBS domains of plant resistance (R) genes . Swetha and Subramanian (2007) reported isolation and molecular analysis of R-gene in resistant *Zingiber officinale* (ginger) varieties against *Fusarium oxysporum.* They observed that the R-gene is present only resistant varieties. These cloned R-genes provide a new resource of molecular markers for marker assisted selection (MAS) and rapid identification of *Fusarium* yellows resistant ginger varieties.

Kavitha and Thomas (2008) reported *Zingiber zerumbet* (L) Smith, a wild species related to the cultivated ginger is a potential resistance donor for soft rot disease in ginger caused by *Pythium aphanidermatum* Fitzp. AFLP analysis of *Z. zerumbet* accessions revealed a high genetic diversity in *Z. zerumbet* unexpected for a clonal species. They studied the genetic diversity and *P. aphanidermatum* resistance of 74 *Z. zerumbet* accessions belonging to 15 populations. Eight accessions were found to be immune to the infection.

Chen *et al.* (2005) - using primers designed from the conserved regions of monocot mannose-binding lectins - isolated. cloned and characterized a mannose-binding lectin form cDNA derived from ginger rhizomes. Yua *et al.* (2008) reported isolation and functional characterization of β-eudesmol synthase from *Zingiber zerumbet* Smith. They identified a new sesquiterpene synthase gene (ZSS2) from *Zingiber zerumbet* Smith. Huang *et al.* (2007) reported molecular cloning and characterization of Violaxanthin de- epoxidase (VDE) key enzyme of xanthophyll cycle plays an important role in protecting photosynthesis apparatus from the damage of excessive light in ginger. Cloning and expression analysis of PR5 in *Curcuma amada* and *Zingiber officinale* was reported by Prasath *et al.,* 2012. Isolation of resistance genes from such related genera will help in ginger improvement via transgenic approaches.

Vanilla (*Vanilla planifolia* Andr)

Vanilla - a native of central America is the natural source of vanillin. It is introduced to India and there is very limited variability in the cultivated form of this spice. Vanilla, a cross pollinated crop of hybrid origin is known to have many meiotic and post-meiotic chromosomal abnormalities. As a result the segregating seedling progenies can give various cytotypes and genetically variant types.

Molecular Characterization and Genetic Diversity

The selfed progenies, somaclones and interspecific hybrids of *V. planifolia* obtained through seed culture have shown variations in morphological characters such as plant height, internodal length, leaf length and leaf breadth. Variations were evident in *in vitro* cultures, as well as after planting out (Minoo, 2002). In the absence of many classical phenotypic markers, in perennial crops like vanilla, molecular markers such as RAPD and AFLP, were used to estimate genetic similarities and interrelation ships in cultivars seed progenies, somaclones and interspecific hybrids.

Minoo *et al.* (2007) studied isoenzyme, RAPD, AFLP polymorphism supplemented by morphological characters to study the existing variability in cultivated vanilla, species interrelation ships, identification of interspecific hybrids and fingerprinting of important genotypes (Minoo *et al.* 2006). The study indicated that there is limited variability among the cultivated collections of *V. planifolia* grown in India with Mauritius collection is most divergent. Among the species studied, *V. tahitensis* is most closely related to a *V. planifolia* collection from while *V. andamanica* is the most distant. In spite of superficial morphological similarity, *V. andamanica* is not closely related to *V. planifolia* or *V. tahitensis,* and its collections are the most divergent from all other species studied forming a separate and unique cluster.

There is considerable variability among the eight different collections of *V. andamanica,* supporting the probability that this species did originate in the Andaman Islands where sexual reproduction is likely. AFLP profiling (Fig. 7) indicted that there are significant variations among the selfed seed progenies of *V. planifolia*. This variation further increased when the plant regeneration is through callus or when grown in colchicines containing medium. The progenies obtained from inter specific crosses between *V. planifolia* and *V. aphylla* are truly hybrids and *in vitro* technology can be used for generation of variability and its utility in crop improvement. (Minoo *et al.* 2006).

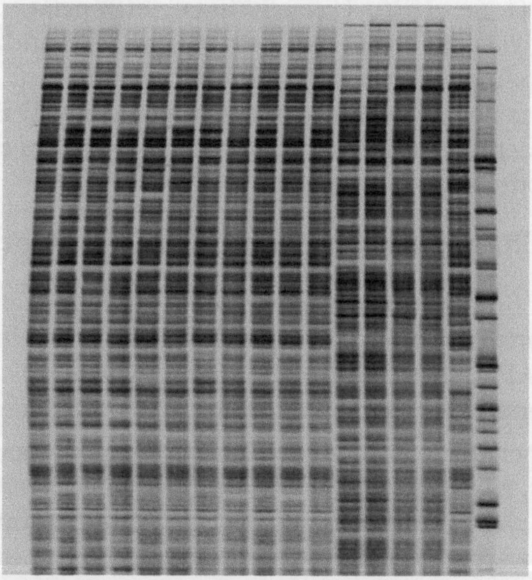

Fig. 7 : AFLP profiles of vanilla seedling progenies indicating genetic variability and interspecific hybrids indicating their hybrid nature (lanes 1-10 : Seedling progenies of *V.planifolia*, 14-16: Interspecific hybrids of *V.planifolia* and *V.aphylla*) 1 : V1, 2 : V2, 3 : V4, 4 : V6, 5 : V7, 6 : V8, 7 : V10, 8 : V11, 9 : V12, 10 : V24, 11 : *V.planifolia,* 12 : *V.aphylla1,* 13 : *V.aphylla2* , 14 : VH1, 15 : VH4, 16 : VH5, 17 : Water control. . Arrows indicate species specific bands

Tree Spices

Cinnamon, Clove, Nutmeg, etc are some of the important tree spices grown in India. In these perennial tree crops, identification and clonal multiplication of high-yielding 'elite' genotypes becomes a priority due to long pre bearing period. Much variation is observed in nutmeg being a dioecius and seed propagated plant especially for fruit characters. In cinnamon also variations have been reported for morphological, chemical and bark characters. However, in clove limited introduction and due to the naturally self pollinating nature has limited the variability and the genetic base available is very narrow.

Molecular Diversity in Cinnamon

Murashige *et al.* (2002) studied genetic variation among four cinnamon species viz. *Cinnamomum cassia*, *C. zeylanicum*, *C. burmannii* and *C. sieboldii* using nucleotide sequences of chloroplast DNA for correct identification of *Cinnamomum* species. Judy (2005 studied eleven species of Cinnamomum viz. *C. verum, C. cassia, C .malabatrum, C. riparium, C. macrocarpum, C. perottitii, C. weightii, C. citronella, C. tamala, C. sulphuratum, C. glaucens and C. camphora* using RAPD and reported that cultivated cinnamon *C. verum* is closely related to *C. malabatrum*. Assessment of genetic variability in cassia using RAPD indicated that there is a considerable variation among the progenies of cassia in India where it's introduced.

Molecular Characterization of Myristica Germplasm

Molecular fingerprinting of selected high yielding elite accessions of nutmeg (*Myristica fragrans)* with high sabinene and low myristicin contents, by RAPD, ISSR and rDNA-RFLP markers was done Unique bands were identified for some important elite lines (Sheeja *et al.*, 2006).

Molecular characterization of wild and related genera of *Myristica viz: M. fragrans, M. malabarica, M. beddomi, M. amygdalina, M. fatua, M. andamanica M. prainii* and *Knema andamanica.* and a number of species specific bands were identified. This will help in purity estimation and identification of adulterants.

Molecular Characterization for Identification of Sex Linked Markers

Shibu *et al.* (2000) and Ganeshaiah *z* (2000) identified sex specific DNA markers for identifying the productive female trees in nutmeg. Najeebudeen (2005) reported a male specific band of 450 bp from *Myristica fragrans* .

Future Perspectives

In general the crop improvement programmes in spices is aimed at selection and breeding of high yielding quality genotypes. Appropriate usage of molecular marker technology in conjunction with conventional methods can aid to a great extent in meeting this objective. Molecular markers are being used in short listing of core collections of genetic resources of spices which can be effectively employed in directional breeding programmes. Identification of markers associated with sex segregation will help in reducing breeding time especially in perennial tree spices. Development of finger prints not only help in maintaining genetic purity but also in certification of genuine planting materials of elite genotypes and checking adulteration in traded sample especially in pharma industry. Generating molecular maps will help in tagging of genes of agronomic interest and maker assisted selection. Isolation of useful genes using genomic approaches will ultimately lead to gene pyramiding through transgenic approaches. Comparative genomics has already made much headway for solanaceous crops. This information can be effectively utilized in developing *Phytophthora* resistance in Black pepper. Similarly the information generated at International Network on Improvement of Banana and Plantain (INIBAP) will help in better understanding of other related sub families like Zigiberaceae to which cardamom belongs.

References

Aarts MG, Hekkert B, Holub EB, Beynon JL, Stiekema WJ, and Pereira A 1998 Identification of R-gene homo logoues DNA fragments genetically linked to disease resistance loci in *Arabidopsis thaliana*. *Mol Plant Microbe Interact*, 11:251-258.

Anuntalabhochai, S., Sitthiphroma, S., Thongtaksinb, W., Sanguansermsric, M. and Cutlerd, R.W. 2007. Hybrid detection and characterization of *Curcuma spp*. using sequence characterized DNA markers. *Scientia Horticulturae*, 111 : 389 - 393

Aswati Nair R and Thomas G 2006 Isolation, characterization and expression studies of resistance gene candidates (RGCs) from *Zingiber* spp. *Theor Appl Genet* 116:123-134.

Aswati Nair R and Thomas G 2007 Evaluation of resistance gene (R-gene) specific primer sets and characterization of resistance gene candidates in ginger (*Zingiber officinale* Rosc.) *Current Science* 93 (1): 61-66.

Bhat AI, Haresh PS and Madhubala R 2005 Sequencing of coat protein gene of an isolate of Cucumber Mosaic Virus infecting black pepper in India *J. Plant Biochem. and Biotech*. 14: 37- 40.

Cao H, Sasaki Y, Fushimi H, Komatsu K and Cao H 2001 Molecular analysis of medicinally used Chinese and Japanese *Curcuma* based on 18S rRNA and trnK gene sequences. *Biol. Pharm. Bull*. 24(12) : 1389-1394.

Dicto J. and Manjusha S. 2005. Identification of elicitor-induced PR5 gene homologue in *Piper colubrinum* Link by suppression subtractive hybridization. *Journal of Current Science*, Vol. 88(4) : 25

Girija D, Beena PS and Nazeem PA 2005a Molecular cloning of a cDNA fragment encoding the defense related protein β-1,3-glucanase in black pepper (*P. nigrum* L.) p. 81-82. Proceedings of the Kerala Science Congress, January 2005, KFRI, Peechi, Kerala.

Grattapaglia, D and Sederoff, R. 1994 Genetic Linkage Maps of *Eucalyptus grandis* and *Eucalyptus urophylla* using a Pseudo - Test Cross: Mapping Strategy and RAPD markers. *Genetics* 137: 1121- 1137.

Jaramillo MA and Manos PS 2001 Phylogeny and patterns of floral diversity in the genus *Piper* (Piperaceae). *American Journal of Botany*, 88 : 706-716.

Kavitha PG and Thomas G 2008 Defense transcriptome profiling of *Zingiber zerumbet* (L.) Smith by mRNA differential display, *J. Biosci*. 33(1).

Kress, W.J., Prince, L.M. and Williams, K.J. 2002 The phylogeny and a new classification of the gingers (Zingiberaceae): evidence from molecular data. *American J. of Botany*, 89:1682-1696.

Minoo D 2002 Seedling and somaclonal variation and their characterization in Vanilla. Ph.D Thesis. Calicut University, Kerala, India.

Minoo Divakaran, Nirmal Babu K, Ravindran PN and Peter KV 2006 Inter specific hybridization in vanilla and molecular characterization of hybrids and selfed progenies using RAPD and AFLP markers. *Scientia Horticulturae*. 108:414-422.

Nirmal Babu K *et al.*, 2003. Molecular characterization and preparation of molecular maps in black pepper. Final report, National Agricultural Technology Project, ICAR, New Delhi. pp. 50.

Praveen K 2005 Variability in Somaclones of Turmeric (*Curcuma Longa* L.)Ph.D Thesis. Calicut University, Kerala, India.

Quirin EA, Ogundiwin EA, Prince JP, Mazourek M, Briggs MO, Chlanda TS, Kim KT, Falise M, Kang BC and Jahn MM 2005 Development of sequence characterized amplified region (SCAR) primers for the detection of *Phyto.5.2*, a major QTL for resistance to *Phytophthora capsici* Leon. in pepper. *Theor Appl Genet* 110:605–612.

Sasaki, Y., Fushimi, H., Cao, H., Cai, S.Q. and Komatsu, K. 2002 Sequence analysis of Chinese and Japanese Curcuma drugs on the 18S rRNA gene and trnK gene and the application of amplification-refractory mutation system analysis for their authentication. *Biol. Pharm. Bull.*, 25(12):1593-9.

Sasikimar B and Zachariah TJ 2003 Organization of ginger and turmeric germplasm based on molecular characterization, Final report, ICAR Adhoc Project, IISR, Calicut.

Sasikumar, B., Zachariah, T. J., Syamkumar, S., Remya, R., 2004. PCR based detection of adulteration in the market samples of turmeric powder. *Food Biotechnology* 18(3): 299-306.

Sheji Chandran, Smitha KS, Renu Susan George, Ishwara Bhat A and Anandaraj M. 2006. Development of SCAR marker for locating *Phytophthora* resistance in black pepper (*Piper nigrum* L). Indian Phytopathological Society, Southern Zone Meeting, Kasargod.(Abstract. P.24)

Singh HP, Parthasarathy VA and Nirmal Babu K (Eds) 2011 Advances in Horticulture Biotechnology - *Vol 3 - Molecular Markers and Marker Assisted Selection – Fruit Crops, Plantation Crops and Spices*, Westville Publishing House, New Delhi. pp.330.

Singh HP, Parthasarathy VA and Nirmal Babu K (Eds) 2011 Advances in Horticulture Biotechnology - *Vol 5 – Gene Cloning and Transgenics*, Westville Publishing House, New Delhi. pp. 566.

Sreedevi M, Syamkumar S and Sasikumar B 2005 Molecular and morphological characterization of new promising black pepper (*Piper nigrum* L.) lines. *J. Spices Aromatic Crops* 14:1-9.

Sumathi V 2007 Studies on Somaclonal Variation in Zingeberacious Crops, Ph. D Thesis, University of Calicut, Kerala, India.

Syamkumar S, Mrudula Jose and Sasikumar B 2005. Isolation and PCR amplification of DNA from traded cardamom *(Elettaria cardamomum* M.*). Plant Molecular Biology Reporter*, 23: 1a – 1e.

Syamkumar, S. and Sasikumar, B. 2007. Molecular marker based genetic diversity analysis of *Curcuma* species from India. *Scientia Horticulturae*, 112: 235-241.

6

Molecular Stress Biology of Spice Plants- A New Frontier of Plant Improvement

E V Soniya and Sweda Sreekumar

The harmony between human beings and plants is one of the evergreen reasons that maintains and supports life on earth. This fact clearly explains why any hazard to the life of plants prominently affects man. One important feature distinguishing plants from other complex multicellular organisms is that plants are sessile and thus have to endure environmental challenges (Zhu, 2001).

Crop productivity is limited by stress factors. The stresses which pose major threats to survival of plants belong to two categories- abiotic and biotic stress. Drought, salinity, high or low temperatures, light, deficient or excess nutrients, heavy metals, pollutants etc., either individually or in combination causes abiotic stresses. Biotic stress refers to any stress provided by living organisms, be it unicellular or multicellular. Plants facing a variety of biotic and abiotic stresses are able to protect themselves by means of tolerance mechanisms which result from changes at the molecular level. They are often confronted by a multitude of stresses in their natural environment. To overcome this, plants have evolved various defence strategies, which involve comprehensive reprogramming of the cellular metabolism. This includes a multilayered inducible immune system (Jones and Dangl, 2006).

Several approaches like synseed technology, protoplast fusion and development of novel transgenics have great significance in conservation and increasing the production and productivity of spices (Nirmal Babu *et al.*, 2005). At molecular level, plants react to stress conditions by up-regulation or down-regulation of gene expression that produces several alterations in protein synthesis and thus their biological functions (Deeba *et al.*, 2012). Understanding the molecular basis aids in developing selection strategies for enhancing tolerance. Plant genome mapping and molecular biology techniques offer a new opportunity for understanding the genetics of stress-resistance genes and their contribution to plant performance under stress (Sairam and Tyagi, 2004).Secondary metabolites are involved in protective functions in response to both biotic and abiotic stress conditions (Ramakrishna and Ravishankar, 2011).

An intensive study on molecular stress biology is inevitable for developing improved plant varieties which resist stress, thus leading to a better yield. Though such studies on various plants resulted in a large body of information, most of the works were focussed on food crops like cereals and vegetables as they are one of the major requirements for sustaining human life. Not to mention, food crops always remain an area of priority. But spice crops which are inevitable in providing flavour and aroma to food making it more appetizing have not been much focussed on. To date, information on this aspect is limited.

Importance of Spices

A spice is an aromatic substance of plant origin, primarily from various parts of the plant like dried seed, fruit, root, flowers or bark used in very small quantities as a food additive for flavor, color, or as a preservative. Spice products are used in traditional and local medicine. Many spices are also used for religious rituals in Asia and in cosmetics, perfumery and liquorices in other parts of the world. Besides adding zing to the food, they help in digestion. About 109 spices are cultivated, most of which are concentrated in the Mediterranean region and Asia, from where they most probably originated.

The different spice crops used in several parts of the world are listed in Table 1.

Perennial spice crops like allspice, cardamom, black pepper, cinnamon, clove and nutmeg provide livelihood security to a large number of people in Asia and the Pacific region. Black pepper is one of the oldest spices used by man. It is a perennial climber grown for its fruits used as spice and medicine (Ravindran, 2001). Some studies suggest that spices like cinnamon may be effective in the treatment of cancer (Hyeon *et al.*, 2003; Nishida *et al.*, 2003).

Table 1 : List of Spices as per International Standard Organization(ISO)-Document 676

Ajowan/ ajwain	*Trachyspermum am mi*
Allspice/ Jamaica pepper/ Pimento	*Pimenta dioica*
Angelica /Garden angelica	*Angelica archangelica*
Aniseed myrtle/ anise myrtle	*Syzygium anisatum*
Aniseed/anise	*Pimpinella anisum*
Aridan	*Tetrapleura tetraptera*
Asafoetida	*Ferula asafoetida*
Ashanti Pepper/ West African pepper	*Piper guineense*
Bay laurel/ Bayleaves/ laurel	*Laurus nobilis*
Bergamot	*Mentha citrata*
Bitter fennel	*Foeniculum vulgare*
Black caraway/ Carum bulbocastanum	*Bunium persicum*
Black cardamom/ Large cardamom	*Amomum subulatum/A.costatum*
Black cumin/ blackseed	*Nigella sativa*
Black pepper	*Piper nigrum*
Brazilian pepper/ Pink pepper	*Schinus terebinthifolius*
Camboge	*Garcinia cambogia*
Capers	*Capparis spinosa*
Caraway	*Carum carvi*
Cardamom	*Elettaria cardamomum*
Cassia/ Chinese cassia	*Cinnamomum aromaticum*
Celery/ Garden celery	*Apium graveolens*
Chervil	*Anthriscus cerefolium*
Chilli pepper	*Capsicum annuum*
Chinese cinnamon/ Chinese cassia	*Cinnamomum cassia*
Chinese ginger	*Boesenbergia rotunda*
Chive	*Allium schoenoprasum*
Cinnamon	*Cinnamomum zeylanicum*
Cinnamon myrtle	*Backhousia myrtifolia*
Clove	*Eugenia caryophyllata*
Common sage/garden sage	*Salvia officinalis*
Coriander	*Eugenia caryophyllata*

Contd.

Cubeb/ tailed pepper	*Piper cubeba*
Cumin	*Cuminum cyminum*
Curry leaf/ sweet neem leaf/Kari	*Murraya koenigii*
Dill	*Anethum graveolens*
Epazote/ Wormseed	*Dysphania ambrosioides / chenopodium ambrosioides*
Eucalyptus	*Eucalyptus citriodora*
Fenugreek	*Trigonella foenumgraecum*
Galangal	*Kaempferia galanga*
Garlic	*Allium sativum*
Garlic mustard	*Alliaria petiolata*
Ginger	*Zingiber officinale*
Greater galangal	*Alpinia galangal*
Guinea pepper/African pepper	*Xylopia aethiopica*
Horse-radish	*Armoracia rusticana*
Hyssop	*Hyssopus officinalis*
Indian leek/ Chinese chive	*Allium tuberosum*
Indian Long pepper	*Piper longum L*
Japanese mint/ field mint/ Corn mint	*Mentha arvensis*
Javanese long pepper	*Piper retrofractum/P. officinarum*
Juniper berry	*Juniperus communis*
Kokum/ kokam	*Garcinia indica*
Laurel/ True laurel/Bay leaf/ Sweet flag	*Laurus nobilis*
Lavender	*Lavandula angustifolia*
Leek, winter leek	*Allium porrum*
Lemon balm/balm/Melissa	*Melissa officinalis*
Lemon grass	*Cymbopogon citratus*
Lemon verbena/ lemon beebrush	*Aloysia citrodora*
Lesser galangal	*Alpinia officinarum*
Licorice/ Liquorice	*Glycyrrhiza glabra*
Lovage/ Garden lovage	*Levisticum officinale*
Mace	*Myristica fragrans Houtt.*
Malabar tamarind/ gambooge	*Garcinia gummi-gutta*
Mango ginger	*Curcuma amada*
Marjoram/Sweet marjoram	*Origanum majorana/ Marjoram hortensis*
Melegueta pepper /Grain of paradise	*Aframomum melegueta*
Mexican Oregano	*Lippia graveolens*

Mustard (black)	*Brassica nigra*
Mustard (white)	*Sinapis alba*
Nutmeg	*Myristica fragrans*
Onion	*Allium cepa*
Oregano/ Origan	*Origanum vulgare*
Pandan wangi	*Pandanus amaryllifolius Roxb.*
Parsley	*Petroselinum crispum*
Peppermint	*Mentha piperita*
Perilla/Beefsteakplant/ Chinese basil	*Perilla frutescens*
Poppy seed	*Papaver somniferum*
Red pepper/ Sweet pepper	*Capsicum annum*
Rosemary	*Rosemary officinalis*
Saffron	*Crocus sativus*
Senegal pepper/ Grains of Selim/Kani pepper	*Xylopia aethiopica*
Sesame	*Sesamum indicum*
Shallots (small onion)	*Allium ascalonicum*
Spearmint/ garden mint	*Mentha spicata*
Star anise	*Illicium verum*
Stony leek/ Japanese bunching onion	*Allium fistulosum*
Sumach/ Sumac	*Rhus coriaria/ R. glabra*
Summer savory	*Satureja hortensis*
Sweet basil	*Ocimum basilicum*
Sweet fennel	*Foeniculum vulgare*
Sweet flag	*Acrocus calamus Linn.*
Szechuan pepper/ Japanese pepper	*Zanthoxylum piperitum*
Tamarind	*Tamarindus indica*
Tarragon/ Estragon	*Artemisia dracunculus*
Tejpat/ Indian cassia	*Cinnamomum tamala*
Thyme/ common thyme	*Thymus vulgaris*
Turmeric	*Curcuma longa*
Vanilla	*Vanilla planifolia*
West Indian Bay	*Pimenta racemosa*
White pepper	*Piper nigrum*
Wild thyme/ creeping thyme	*Thymus serpyllum*
Winter savory	*Satureja montana*
Zedoary/ White turmeric	*Curcuma zedoaria*

Source : Peter K.V. (2012) Susheela, R (2006)

Along with coriander, seed spices like cumin, fennel and fenugreek are extensively used for culinary and medicinal purposes and as condiments (Peter, 2012). Fennel and cumin are also aromatic crops, as they provide essential oils used in the perfumery and confectionary industries. Cumin has a long history, referring almost 5000 years back to the ancient Egyptian civilization, where it was used both as a spice and a preservative in mummification. The seeds of cumin are antispasmodic, carminative and stimulant used in the treatment of flatulence and bloating (Chevallier, 1996).

Water management is crucial for flowering and fruit set of these spices. Excess moisture leads to diseases and pest problems. Scarcity of water also affects growth and yield. Similar is the case with high or low temperature, salinity and other abiotic and biotic stresses. Negligence in the research pertaining to molecular biology associated with stress may eventually result in the limitation of culturing many precious spice crops. One of the most important hurdles on the road to save them from diminishing is the different stress components. The solution to tackle this problem may arrive at our hands once the researchers are ready to focus on the responses of these plants subjected to different abiotic and biotic stresses. More research pertaining to the effect of different stresses on the various spices would be appreciable, although some studies based on this cannot be neglected.

Salinity

Plant growth and development are adversely affected by high salt concentration in soil. This affects seed germination, growth and vigour, flowering and fruit set ultimately leading to a low economic yield and quality of produce. Almost half of the world's irrigated lands are affected by salinity (Zhu, 2001). Mangal *et al.* (1986) reported reduced fennel seed production with increased salinity. An excess of salt in soil decreases the osmotic potential creating a water stress and severe ion toxicity in plants. Salts interact with mineral nutrition thus causing nutrient imbalances and deficiencies. All these lead to molecular damage, arresting plant growth and resulting in death (McCue and Hanson, 1990). Results of the study conducted by Ghamarnia *et al.* (2012) showed that seed yield, oil yield and water use efficiency of black cumin decreased rapidly as a result of increased of water salinity during irrigation.

Both salt-tolerant and salt-sensitive varieties exist within a particular crop species. Identification of regulatory genes and molecular markers linked to tolerance traits is an effective tool for selecting cultivars with improved tolerance. Genes involved in conferring salt-tolerance are often isolated from the salt-tolerant species and utilized for understanding the mechanisms which distinguish them from their salt-sensitive counterparts.

Most of the experimental approaches to isolate genes whose expression is influenced by salinity in plants involve screening cDNA libraries constructed from mRNA populations isolated from plants or cells exposed to high salinity. Differential screening is done with probes derived from mRNA isolated from salt-stressed and non-stressed plant tissues. The mRNA that corresponds to genes, preferentially expressed in the salt stressed plants is isolated and characterized. Also, cDNAs which encode proteins which enhance salt-tolerance are selected using salt-sensitive mutants of yeast which are rescued following transformation with plant cDNA libraries.

Cloned plant genes and transgenic plants have become a standard tool in plant-stress biology. These technologies are mainly applied to model systems and have greatly enlarged the knowledge of mechanisms of tolerance. The various abiotic stresses cause changes in plant processes at all levels of organization (morphological, physiological, biochemical and molecular). Genetic dissection of salt tolerance in *Arabidopsis* established the involvement of the SOS pathway in response to ionic aspect of salt stress. The molecular analysis of stress responses has arrived at a stage where research can build upon a large collection of characterized genes. Identification of quantitative trait loci for abiotic stress resistance may be an effective analytical tool (Sairam and Tyagi, 2004).

Water Stress

Moisture stress is one of the major abiotic factors that limit agricultural productivity. Drought stress occurs when the available water in the soil is reduced to such critical levels and atmospheric conditions add to continuous loss of water (Xu *et al.,* 2010). Shrinking of cells leads to loss of turgor, osmotic stress and a potential change of membrane potentials. Upon severe water loss from the cells, membrane disintegration and abolition of metabolic processes occur (Mahajan and Tuteja, 2006).

Plants experiencing moisture stress have reduced physiological processes like photosynthesis, transpiration and reduced biomass and reduced yield. If there is extreme moisture stress, particularly at early growth stages, plants may not recover from moisture stress and will die. However, if the level of stress is not severe, plants deploy different strategies to maintain fitness. Some of the adaptations to moisture stress include increases in certain hormone concentrations to regulate stomatal conductance limiting transpirational water loss. In addition, moisture-stressed plants tend to increase root growth to absorb more water from the soil. Moisture-stressed plants have smaller canopies, are more susceptible to insect injury, and suffer heavier yield loss compared to unstressed plants (Peterson and Higley, 2001). Although optimal moisture supply may enhance the level of tolerance to insect injury, excess moisture can be harmful for plant growth (Guidi and Soldatinit, 1997; Linkemer *et al.*, 1998). Spices like cumin need low water for growth cycle and are resistant to drought (Dehaghi and Filabi, 2011). Sepehr *et al.*, (2012) reported that flooding stress in cumin seeds decreased nitrogen and phosphorus concentrations in roots and shoots thereby increasing the activity of nitrate reductase enzyme.

Vivek *et al.* (communicated) isolated a stress-responsive CDPK gene, ZoCDPK1 from a stress cDNA generated from ginger, using rapid amplification of cDNA ends (RLM-RACE) - PCR technique and characterized its role in stress tolerance. Their results suggested that ZoCDPK1 functions in the positive regulation of the signaling pathways that are involved in the response to salinity and drought stress in ginger.

High Temperature

High temperature alters membrane properties (Sangwan *et al.*, 2002) and also reduces or inactivates enzyme activity through protein denaturation (Kampinga *et al.*, 1995). It induces a series of characteristic cellular and metabolic responses in plants (Weis and Berry, 1988). Heat stress causes rapid synthesis of molecular chaperones called heat shock proteins (HSP) accompanied by a reduction in the synthesis of other proteins (Czarnecka *et al.*, 1984). HSPs prevent the denaturation and aggregation of target proteins, and facilitating protein refolding thereby providing thermotolerance to organisms (Perez *et al.*, 2009).

In the study conducted by Lu *et al.* (2013), suppression subtractive hybridization (SSH) technology, one of the most powerful and commonly used methods for identifying differentially expressed transcripts (Wang *et al.*, 2011, 2012; Dipnarayan *et al.*, 2011; Prabu *et al.*, 2011; Zhang *et al.*, 2011), was employed to compare gene expression between thermally stressed *P. ternata* plants and those not subjected to heat shock to allow modeling of *P. ternata* gene expression in response to high temperature. They found that the expression of PtSAD gene was downregulated at beginning of heat stress, leading to speculation that the level of saturated fatty acids may have increased in cell membranes of *P. ternata* in response to heat stress.

According to Dang *et al.* (2012), *CaWRKY40*, a transcription factor identified from pepper (*Capsicum annuum*) may be used as a tool in genetic engineering of simultaneous disease and heat stress resistance of crops adapted to high temperature and high humidity environments. They speculated that salicylic acid (SA), jasmonic acid (JA) and ethylene (ET) regulate the expression of *CaWRKY40* leading to expression of downstream defence as well as thermotolerance-related genes. *CaWRKY40* may thus be used as a tool in genetic engineering of simultaneous disease and heat stress resistance of crops adapted to high temperature and high humidity environments.

Isbat *et al.* (2009) reported a pepper gene encoding BI-1 i.e., CaBI-1. CaBI-1 was isolated from a cDNA library of high temperature-stressed hot pepper, but its expression was inducible under various abiotic stresses like high temperature, cold, drought, submergence, high salintity and heavy metals.

Low Temperature

Low temperature (LT) is an environmental stress that affects crop production and quality and regulates expression of many genes, and the level of a number of proteins and metabolites. It has significant influence in plant growth affecting photosynthesis, uptake of water and nutrients, among others. Many studies showed that LT induced changes in expression of hundreds of genes followed by an increase in the levels of many metabolites, some of which have protective effects against the damaging effects of cold stress (Airaki *et al.,* 2012). Many economically significant crops like cotton, maize, pepper, rice, soybean, tomato, some tropical fruits (e.g. bananas, papayas) and subtropical fruits (e.g. grapes, oranges, mangoes) are LT sensitive, which affect their production and quality (Sharma *et al.,* 2005).

Results obtained in the work of Airaki *et al.* (2012) showed that in pepper plants exposed to LT during 24–48 h, a general imbalance of the reactive oxygen species (ROS) and reactive nitrogen species (RNS) metabolism was produced with the result of a rise in lipid peroxidation and protein tyrosine nitration, which strongly suggest the induction of oxidative and nitrosative stresses. Their work suggests that ascorbate, glutathione and the NADPH-generating dehydrogenases have a role in the process of cold acclimation through their effect on the redox state of the cell. The goal of their work was to study antioxidant metabolism and homeostasis of ROS and RNS in this plant species under LT conditions. The results showed that in pepper plants LT causes nitrosative and oxidative stress during the first 24 h but after this period, plants seem to recover by an acclimation of their metabolism, which involves important changes in their cellular antioxidant and redox state.

Light

Photo-oxidative stress is a common problem for all plants, particularly under unfavourable environmental conditions. Under such conditions, formation of active oxygen species is increased even at low light intensities (Asada, 1994; Foyer *et al.,* 1994). GraBes *et al.* (2001) examined significance of tocopherol content for protection of PSII against photoinactivation. They assumed that a reduced detoxification of ROS in the membrane due to reduced tocopherol content is responsible for the increased photoinactivation of PSII.

Plants are harmed when incident light fluxes exceed saturation point of photosynthesis. This initially causes a decline of photosynthetic efficiency and a reduction in photosynthetic yield (Montane *et al.,* 1999). Protection mechanism includes release of light-harvesting antennae from the PSII reaction centre by phosphorylation (Bennett, 1983), the activation of the xanthophyll cycle (Demmig-Adams, 1990; Eskling *et al.,* 1997), and a rise in ascorbate peroxidation (Schoner and Krause, 1990). Early light-inducible proteins (ELIPs) are produced as soon as the plants are transferred into high light (Adamska *et al.,* 1992). Genes specifically involved in the dynamic response to UV were determined by comparing expression profiles (Molinier *et al.,* 2005). High levels of UV-B can cause distress in plants causing elevated levels of ROS (Hideg *et al.,* 2013).

Heavy Metals

Soil contamination with heavy metals is a worldwide problem leading to losses in agricultural yield and hazardous health effects as they enter the food chain (Gill *et al.,* 2011). A high concentration of heavy metals like Cd, Pb, Hg, Cu, Zn and Ni affect growth, development and yield of plants (Pahlsson, 1989; Sresty and Rao, 1999). Heavy metal tolerance mechanisms vary from species to species, of which reduced transport across the cell membrane, metal binding to wall, chelation and sequestration of heavy metals by particular ligands such as phytochelatins and metallothioneins etc., are an important few (Tomsett and Thurman, 1988; Cobbet and Goldsbrough, 2002).

Nutrients

Apart from carbon, hydrogen, and oxygen derived from the atmosphere and soil water, plants require 13 more essential elements (nitrogen, phosphorus, potassium, calcium, magnesium, sulfur, iron, zinc, manganese, copper, boron, molybdenum, and chlorine). These are usually supplied either from soil minerals and soil organic matter or by organic or inorganic fertilizers (Silva and Uchida, 2000). Both deficiency and excess supply of nutrients affect plant growth and yield negatively (Lynch and Brown, 2001).

Plant nutrition studies gained immensely from complete sequencing of a number of genomes and the development of new genomic tools. If the expression of target genes is regulated by specific nutrient stresses, the corresponding miRNA may be also regulated under the same conditions. Regulation of miRNA expression by specific nutrient stress opens up a new additional dimension of sensing, signalling and adaptive regulation to nutrient stresses. Two miRNAs, miR395 and miR399, were identified to be important in nutrient stresses responses (Chiou *et al.,* 2007).

Biotic Factors

Plants are under constant assault by biotic agents, including viral, bacterial and fungal pathogens, parasitic plants and insect herbivores, with significant economic and ecological impact (Pimentel 1991, 2002). As soon as an attack is perceived, plant metabolism must balance potentially competing demands for resources to support defence versus requirements for cellular maintenance, growth and reproduction (Herms and Mattson, 1992; Zangerl and Berenbaum 1997, 2003; Berger *et al,* 2007). Exposure to biotic stress induces ROS production, which can cause irreversible oxidative damage to cells. Under ideal conditions, ROS are produced and detoxified in an orderly fashion and are used as signalling agents to initiate developmental and regulatory processes (Foyer and Noctor, 2005; Carol and Dolan, 2006; Wang and Song, 2008). About 80 defense-responsive genes which might participate in resistance gene-mediated defense against both viral and bacterial pathogens were identified (Marathe *et al.,* 2004).

The induction of specific defense genes in the response against certain pathogens depends on specific environmental conditions, suggesting existence of a complex signalling network that allows the plant to recognize and protect itself against pathogens and environmental stress (Fraire-Velázquez *et al.,* 2011). In the area of plant-pathogen interactions, transcript profiling has provided insight to the mechanisms underlying specific gene resistance and basal defense, host/ non-host resistance, pathogenicity of vascular/nonvascular pathogens, and many others (Wise *et al.*, 2007).

Plant viruses do not contain common nucleotide sequences and form a diverse group among plant pathogens. Micro-array technology provides the basis of specific testing for individual and even combination of viruses (Boonham *et al.*, 2007). Among the plant pathogenic bacteria, *Ralstonia solanacearum* is one of the most important having a wide host range including ginger (Ravindran and Nirmal Babu, 2005).

The defence responses to biotic factors are extremely variable, partly because of the unique co-evolutionary relationships between specific plant species and the particular damaging agents. Faced with this diversity, it is remarkable that biotic assault induces a uniform and apparently regulated reduction in the transcription of nuclear genes coding for major components of photosynthesis, regardless of the plant host or damage vector (Saibo, Lourenco and Oliveira, 2009). Common regulatory elements in the promoter region of photosynthetic genes and genes involved in sugar metabolism and ROS detoxification suggest that the transcriptional response to biotic stress is highly coordinated (Damla *et al.*, 2010).

Kong *et al.* (2001) reported that the pepper cyclophilin (CACÒP1) gene necessary to accelerate the protein folding may be involved in differential regulation of cellular metabolism in some tissues of diseased or stressed pepper plants. In particular, the levels of CACÒP1 mRNA expression were responsive to pathogen infection or abiotic elicitors and regulated by tissue specificity, indicating a possible function for this cyclophilin under conditions exposed by various stresses.

A WRKY transcription factor named CaWRKY30, which was rapidly induced upon inoculation with avirulent *Meloidogyne incognita* and several pathogens, was isolated from pepper plants by Jingyuan *et al.* (2011). They analyzed the structure of CaWRKY30 and studied the expression of the gene during the interaction of pepper plants with avirulent/virulent *M. incognita* and following infections with diverse pathogens and treatments with two phytohormones, SA and JA. The subcellular localization of CaWRKY30 protein was examined using green fluorescent protein (GFP). They suggested that CaWRKY30 is induced by pathogen infection and might play an important role during the pathogen stresses. They isolated and functionally characterized the pepper gene *CaPMEI1*, which encodes a pectin methylesterase inhibitor protein (PMEI), in pepper leaves infected by *Xanthomonas campestris* pv. *vesicatoria* (*Xcv*).

A novel pathogen-induced gene, designated CAPIP2, was isolated and functionally characterized by Lee *et al.* (2006) from pepper leaves infected with *Xanthomonas campestris* pv. *vesicatoria*. Their results suggested that pepper CAPIP2 gene may function as a defense-related gene against both biotic and abiotic stresses. To better understand the functions of the pepper CAPIP2 gene in plants, they generated CAPIP2-overexpressing transgenic *Arabidopsis* plants, and evaluated their resistance to bacterial infection, as well as their tolerance to both osmotic and oxidative stresses. In addition, they observed the subcellular localization of the CAPIP2 protein in the *Arabidopsis* protoplasts which expressed the GFP fusion protein. Early CAPIP2 expression was induced through bacterial infection. Ectopic expression of CAPIP2 in transgenic *Arabidopsis* plants resulted in enhanced resistance to Pst DC3000. Tolerance to osmotic stress generated by NaCl and mannitol were found increased in the CAPIP2-transgenic plants.

Several fungal and bacterial diseases create unmanageable hurdles in the production of ginger (*Zingiber officinale* Rosc.). The expression analysis of defense-related genes in *Zingiber* species provided an opportunity for intensive studies to manipulate wild genetic resources for genetically engineering soft-rot resistance into ginger (Kavitha and Thomas, 2008b). The study conducted by Kavitha and Thomas (2008a) revealed high genetic diversity and variability for pathogen resistance among *Zingiber zerumbet* accessions and confirmed the value of *Z. zerumbet* as a potential donor for genetically improving ginger. Aswati Nair *et al.* (2010) identified a member of the pathogenesis-related protein group 5 (PR-5) gene family in *Zingiber zerumbet* (L.) Smith (Zingiberaceae), that is expressed constitutively but upregulated in response to infection by *Pythium aphanidermatum*. They analyzed its expression pattern and suggested the likely function of the protein encoded by this gene in governing defense mechanisms in *Z. zerumbet*. In ginger Swetha and Subrahmanyan (2008) identified the presence of a Resistance (R) gene of CC-NBS-LRR (class of plant resistance genes) showed very high homology to R-genes. They found that these R-genes provide a new resource of molecular markers for rapid identification of fusarium yellows resistant ginger varieties.

An *et al.* (2008) determined that CaPMEI1 from pepper played a role as an antifungal protein and has an inhibitory effect on pectin methylesterase (PME). A PCR-based SSH was found to be suitable for the identification and cloning of pathogenesis related (PR) genes in *Piper colubrinum* (Dicto and Manjula, 2005). This exotic wild *Piper* showed a high degree of resistance to fungal pathogens and is a potential source of resistance genes which makes it a good donor plant in biotechnological approaches like molecular resistance breeding for improvement of cultivated species, *Piper nigrum*.

Molecular Breeding

Marker-assisted (MAS) is one of the powerful tools which will benefit biotechnological breeding and genomic programs, aiding researchers to select plant traits quickly and easily. By the development of DNA markers such as restriction fragment length polymorphisms (RFLPs), random amplified polymorphic DNAs (RAPDs), amplified fragment length polymorphisms (AFLPs) and simple sequence repeats (SSRs, microsatellites), the obstructions associated with the assessment of genetic diversity in plant species have been abducted (Garcia *et al*, 2004). Marker assisted selection (MAS) of resistant varieties is a reliable and faster method of selecting the right varieties for cultivation

Philip *et al.* (1999) reported genotypic difference among three female varieties of *P. longum* as revealed by RAPD analysis and described an efficient protocol for plant regeneration from their leaf explants. The RAPD markers used for the analysis of genetic diversity in black pepper and ginger have given prominent reports (Pradeepkumar *et al.*, 2003, Nayak *et al.*, 2005). Chavan *et al.*, 2008 studied DNA-based RAPD markers that can be applied for the identification and differentiation of the commercially important plant *Z. officinale* Roscoe from the closely related species *Zingiber zerumbet* (pinecone, bitter or 'shampoo' ginger) and *Zingiber cassumunar* [cassumunar or plai (Thai) ginger]and were further cloned and sequenced to develop SCAR(sequence-characterized amplified region) markers.

The development of a linkage map in *Capsicum* has been greatly aided by the use of tomato-derived restriction fragment length polymorphism (RFLP) probes (Kang *et al.*, 2001). Fingerprinting analysis with AFLP has proved to be an ideal tool for cultivar identification and phylogenetic studies of *P. nigrum*. AFLP shows the high level of polymorphism and the unique characterization of the major cultivars thereby providing valuable information for molecular breeding programmes (Joy *et al.*, 2007, Morales *et al.*, 2013). The usefulness of AFLP was assessed by Gordo *et al.* (2012) for cultivar identification and to determine genetic relationships of important cultivars of *P.nigrum*. Species-specific AFLP markers for identification of different Zingiber sps were developed which can also be used to authenticate Zingiber sp-derived drugs (Ghosh *et.al.*, 2011). From turmeric (*Curcuma longa* L.) Siju etal.(2010) developed and characterised polymorphic microsatellite markers which could be used for genetic diversity analysis and resolving the taxonomic confusion prevailing in the genus. Analysis of SSRs in the data generated has great potential for applications in studies of genetic diversity, plant breeding and reproduction. The Simple Sequence Repeat (SSR) or microsatellite based assessment of genetic diversity among different genotypes and species of black pepper in South India by Joy *et al.*, (2011) served as an initiative towards molecular profiling of unexploited indigenous accessions. The developed polymorphic informative markers could be utilized to identify genotypes, true hybrids and also for mapping, linkage analysis, association studies of candidate genes etc. An SSR-based genetic map of pepper (*Capsicum annuum* L.) was developed (Mimura *et.al.*, 2012) which served as an anchor for the alignment of major pepper maps. Kong *et al.*, 2012 also demonstrated the value of mining the redundant sequences in public sequence databases for the development of polymorphic SSR markers, which can be used for marker-assisted breeding in pepper. .

Transcriptome from high throughput sequencing-by-synthesis is a good resource of molecular markers. Next-generation sequencing (NGS) technologies offer a great opportunity for the rapid access to genetic information. NGS technologies generate large numbers of reads with high sampling rates of cDNA libraries, providing a deeper and more complete view of transcriptomes (Morozova *et al.*, 2009, Lu *et al.*, 2012). The first leaf transcriptome of black pepper (Joy *et al.*, 2013) was carried out the *de novo* sequencing using illumine HiSeq^{TM} 2000. RNA-Seq technology was also employed to analyse the root transcriptome of black pepper (Gordo *et al*, 2012) in order to understand the molecular interaction between the pathogens and the host's root region which is important for obtaining resistant cultivars by biotechnological breeding. The miRNA based phylogenetic analysis of plants belonging to the Piperaceae family which includes PCR-based detection and cloning strategy of miRNAs from tissues of black pepper was carried out by (Asha *et al.*, 2012). Due to their conservation in the secondary structure, deeply conserved miRNAs can be a potential resource in the evolutionary study of molecular phylogenetics.

Conclusion

The molecular stress biology of spices is rarely addressed and deserves considerable attention. It is apparent that genetic modification is necessary to enable these plants to resist stress, thereby paving way to crop improvement. When the knots are untied at the molecular level, chances of combating stress using biotechnological applications become more fruitful. Promising developments are possible if intensive research in this area is encouraged. More explorations regarding the stress response of spice crops combined with the application of beneficial tools like genetic markers would aid in the development of stress-tolerant varieties in the near future, thereby conserving a wealth of important spices.

References

Adamska I, Ohad I and Kloppstech K 1992. Synthesis of early light- inducible proteins is controlled by blue light and related to light stress. *Proc Natl Acad Sci USA* 89: 2610-2613.

Airaki M, Leterrier M, Mateos R M, Valderrama R, Chaki M, Barroso J B, Río L A D, Palma J M and Francisco J 2012. Corpas1Metabolism of reactive oxygen species and reactive nitrogen species in pepper (*Capsicum annuum L.*) plants under low temperature stress. *Plant, Cell and Environment* 35: 281–295.

Ramakrishna A and Ravishankar G A 2011. Plant Signaling and Behavior . Influence of abiotic stress signals on secondary metabolites in plants. 6:(11) : 1720-1731.

An S H , Sohn K H, Choi H W , Hwang I S , Lee S C and Hwang B K 2008. Pepper pectin methylesterase inhibitor protein CaPMEI1 is required for antifungal activity, basal disease resistance and abiotic stress tolerance. *Planta* 228:61–78.

Asada K 1994. Production and action of active oxygen in photosynthetic tissues. (In) Foyer CH and Mullineaux PM (Eds) Causes of photo-oxidative stress and amelioration of defence systems in plants. CRC Press, Boca Raton, p 77-104.

Asha S, Joy N and E V Soniya 2012. In silico Characterisation and Phylogenetic Analysis of Two Evolutionarily Conserved miRNAs (miR166 and miR171) from Black Pepper (*Piper nigrum* L.) *Plant Mol Biol Rep.* DOI 10.1007/s11105-012-0532-5.

Aswati Nair R, Kiran A G, Sivakumar K C and Thomas G 2010. Molecular characterization of an oomycete-responsive PR-5 protein gene from *Zingiber zerumbet. Plant Molecular Biology Reporter.* 28:128–135.

Bennett J 1983. Regulation of photosynthesis by reversible phosphorylation of the light-harvesting chlorophyll a/b protein. *Biochem J.* 212: 1-12.

Berger S, Sinha A K and Roitsch T 2007. Plant physiology meets phytopathology: plant primary metabolism and plant-pathogen interactions. *Journal of Experimental Botany* 58: 4019–4026.

Boonham N, Tomlinson J and Mumford R 2007. Microarrays for rapid identification of plant viruses. *Annual Review of Phytopathology* 45: 307-328.

Carol R J and Dolan L 2006. The role of reactive oxygen species in cell growth: lessons from root hairs. *Journal of Experimental Botany* 57: 1829–1834.

Chevallier A 1996. The encyclopedia of medicinal plants, Dorling Kindersley Publishers, London, P: 237.

Cobbet C and Goldsbrough P. 2002. Phytochelatins and metallothioneins: roles in heavy metal detoxification and homeostasis. *Annual. Rev. Plant. Biol.* 53: 159-182.

Czarnecka E, Edelman L, Schoffl F and Key J .1984. Comparative analysis of physical stress responses in soybean seedlings using cloned heat shock cDNAs. *Plant Mol Biol* 3:45–58.

Damla D B, Zavala J A, Zhu J, Clough S J, Ort D R and De Lucia E H. 2010. Biotic stress globally downregulates photosynthesis genes. *Plant, Cell and Environment* 33: 1597–1613.

Dang F F, Wang Y N, Yu L, Eulgem T, Lai Y, Liu Z Q, Wang X, Qiu, A L, Zhang, T X, Lin J, Chen Y S, Guan D Y *et al.*, 2013. CaWRKY40, a WRKY protein of pepper, plays an important role in the regulation of tolerance to heat stress and resistance to *Ralstonia solanacearum* infection. *Plant, Cell & Environment*, 36: 757–774.

Deeba F , Pandey A K, Ranjan S, Mishra A , Singh R, Sharma Y K , Pramod A S and Pandey V. 2012 . Physiological and proteomic responses of cotton (*Gossypium herbaceum* L.) to drought stress. *Plant Physiology and Biochemistry* 53:6-18.

Dehaghi M A and Filabi A M .2011. Evaluation of Some Drought Resistance Criteria in Cumin (*Cuminum cyminum* L.) Landraces. *Advances in Environmental Biology*, 5(2): 237-242.

Demmig-Adams B 1990. Carotenoids and photoprotection in plants: a role for the xanthophyll zeaxanthin. *Biochim Biophys Acta* 1020: 1-24.

Dicto J and Manjula S 2005. Identification of elicitor-induced PR5 gene homologue in *Piper colubrinum* Link by suppression subtractive hybridization. *Curr. Sci.* 88(4): 624-627.

Dipnarayan S, Vajinder K, Shripad R B and Ramamurthy S 2011. Characterization of upstream sequences of the LOJ gene leads to identification of a novel enhancer element conferring lateral organ junction-specific expression in Arabidopsis thaliana. *Plant Mol Biol Rep* 29:265–277.

Eskling M, Arvidsson P O and Akerlund H E 1997.The xanthophylls cycle, its regulation and components. *Physiol Plant* 100: 806-816.

Hideg E, Jansen M A K and Strid A 2013.UV-B exposure, ROS, and stress: inseparable companions or loosely linked associates? *Trends in Plant Science* 18(2):107-15.

Foyer C H, Lelandais M and Kunert K J 1994. Photo-oxidative stress in plants. *Physiol Plant* 92:696-717.

Foyer G and Noctor C H 2005. Oxidant and antioxidant signalling in plants: a re-evaluation of the concept of oxidative stress in a physiological context. *Plant, Cell and Environment* 28: 1056–1071.

Garcia A A F , Benchimol L L, Barbosa A M M, Geraldi I O, Souza C L Jr and de Souza A P 2004. Comparison of RAPD, RFLP, AFLP and SSR markers for diversity studies in tropical maize inbred lines. *Genetics and Molecular Biology*, 27, (4) : 579-588.

Ghamarnia H, Jalili Z and Daichin S 2012. The effects of saline irrigation water on different components of black cumin (*Nigella sativa* L.). *International Journal of AgriScience* 2 (10):915-922.

Gill S S, Nafees A K, Anjum N A and Tuteja N 2011. Amelioration of Cadmium Stress in Crop. Plants by Nutrients Management: Morphological, *Physiological and Biochemical Aspects* pp 1-23.

Gordo S M C, Pinheiro D G, Moreira E C O, Rodrigues S M, Poltronieri M C, *et al.*, 2012. High-throughput sequencing of black pepper root transcriptome. BMC Plant Biol 12: 168 doi:10.1186/1471-2229-12-168.

GraBes T, Grimm B, Koroleva O and Jahns P 2001. Loss of á-tocopherol in tobacco plants with decreased geranyl reductase activity does not modify photosynthesis in optimal growth conditions but increase sensitivity to high-light stress. *Planta* 213: 620-628.

Guidi L and Soldatinit G F 1997. Chlorophyll fluorescence and gas exchange in flooded soybean and sunflower plants, *Plant Physiol. Biochem.*, 35: 713.

Herms DA and MattsonWJ 1992. The dilemma of plants: to grow or defend. *Quarterly Review of Biology* 67: 283–335.

Hyeon K, Hee-Juhn P, Hyun-Ju J, Jong-Won C, Kyu-Seok C and Joohun H *et al.*, 2003. Cinnamaldehyde induces apoptosis by ROS mediated mitochondrial permeability transition in human promyelocytic leukemia HL- 60 cells. *Cancer Lett;* 196:143-152.

Isbat M, Zeba N, Kim S R and Hong C B 2009. A BAX inhibitor-1gene in Capsicum annuum is induced under various abiotic stresses and endows multi-tolerance in transgenic tobacco. *Journal of Plant Physiology* 166: 1685-1699.

Jingyuan Z , Xuexiao Z , Zhenchuan M and Bingyan X 2011. A Novel Pepper (*Capsicum annuum* L.) WRKY Gene, CaWRKY30, is involved in Pathogen Stress Responses. *J. Plant Biol.* 54:329–337.

Jones J D G and Dangl J L 2006. The plant immune system. *Nature* 444: 323– 329.

Joy N, Abraham Z and E V Soniya 2007. A preliminary assessment of genetic relationships among agronomically important cultivars of black pepper. *BMC Genetics* 8:42.

Joy N, Asha S, Mallika V and Soniya E V 2013. *De novo* Transcriptome Sequencing Reveals a Considerable Bias in the Incidence of Simple Sequence Repeats towards the Downstream of 'Pre-miRNAs' of Black Pepper. PLoS ONE 8(3): e56694. doi:10.1371/journal.pone.0056694

Kampinga H H, Brunsting J F, Stege G J, Burgman P W and Konings A W 1995. Thermal protein denaturation and protein aggregation in cells made thermotolerant by various chemicals: role of heat shock proteins. *Exp. Cell Res.* 219: 536–546.

Kang B C, Nahm S H, Huh J H, Yoo H S, Yu J W, Lee M H, and Kim B D 2001. An interspecific (Capsicum annuum ×C. chinese) F2 linkage map in pepper using RFLP and AFLP markers. *Theor Appl Genet.* 102: 531–539.

Kavitha P G and Thomas G 2008a. Population genetic structure of the clonal plant *Zingiber zerumbet* (L.) Smith (Zingiberaceae), a wild relative of cultivated ginger, and its response to Pythium aphanidermatum. *Euphytica* 160:89-100.

Kavitha P G and Thomas G 2008b. Expression analysis of defense-related genes in *Zingiber* (Zingiberaceae) species with different levels of compatibility to the soft pathogen Pythium aphanidermatum. *Plant Cell Reports* 27:1767-1776.

Kong H Y, Lee S C and Hwang B K 2001. Expression of pepper cyclophilin gene is differentially regulated during the pathogen infection and abiotic stress conditions. *Physiological and Molecular Plant Pathology* 59: 189-199.

Lee S C, Kim S H, An S H, Yi S Y and Hwang B K 2006. Identification and functional expression of the pepper pathogen induced gene, CAPIP2, involved in disease resistance and drought and salt stress tolerance. *Plant Mol Biol* 62:151–164.

Linkemer G, Board J E and Musgrave M E 1998. Waterlogging effects on growth and yield components in late-planted soybean, *Crop Sci.*, 38: 1576.

Lu H, Xue T, Zhang A, Sheng W, Zhu Y, Chang L, Song Y and Xue J 2013. Construction of an SSH Library of *Pinellia ternate* Under Heat Stress, and Expression Analysis of Four Transcripts *Plant Mol Biol Rep* 31; pp 185-194 .

Lynch J and Brown K M 2001. Topsoil foraging- an architectural adaptation of plants to low phosphorus availability. *Plant Soil* 237: 225-237.

Mahajan S and Tuteja N 2006. Cold, salinity and drought stresses: An overview; Arch. Biochem. *Biophys.* 444: 139–158.

Mangal J L, Yadava A and Singh G P 1986. Effects of different levels of soil salinity on germination, growth, yield and quality of coriander and fennel. *South Indian Hort.* 34:26–31.

Marathe R, Guan Z, Anandalakshmi R, Zhao H Y and Dinesh Kumar S P 2004. Study of *Arabidopsis thaliana* resistome in response to Cucumber mosaic virus infection using whole genome microarray. *Plant Molecular Biology* 55: 501-520.

McCue K F and Hanson A D 1990. Salt-inducible betaine aldehyde dehydrogenase from sugar beet: cDNA cloning and expression. *Trends Biotechnol.* 8: 358–362.

Molinier J, Oakeley E J, Niederhauser O, Kovalchuk I and Hohn B 2005. Dynamic response of plant genome to ultraviolet radiation and other genotoxic stresses. Mutation Research- Fundamental and Molecular Mechanisms of Mutagenesis 571: 235-247.

Montane M H, Petzold B and Kloppstech K 1999. Formation of early-light-inducible-protein complexes and status of xanthophyll levels under high light and cold stress in barley (*Hordeum vulgare* L.). *Planta* 208: 519-527.

Morozova O, Hirst M, Marra M A 2009. Applications of new sequencing technologies for transcriptome analysis. *Annu Rev Genomics Hum Genet.*, 10:135–151.

Mimura Y, Inoue T, Minamiyama Y, Kubo N 2012. An SSR-based genetic map of pepper (*Capsicum annuum* L.) serves as an anchor for the alignment of major pepper maps. *Breed Sci.* 62(1):93-8.

Morales R G, Resende J T, Resende F V, Delatorre C A, Figueiredo A S, Da-Silva P R 2013. Genetic divergence among Brazilian garlic cultivars based on morphological characters and AFLP markers. *Genet Mol Res.* 4:12(1):270-81.

Nayak S, Naik P K, Acharya L, Mukherjee A K, Panda P C, Das P 2005. Assessment of genetic diversity among 16 promising cultivars of ginger using cytological and molecular markers.

Nirmal Babu K, Minoo, Divakaran, Geetha, S P, Ravindran, P N and Peter, K V 2005. Advances in Biotechnology of Spices and Herbs. *Indian Journal of Botanical Research.* 1 (2): 55-214.

Nisha Joy, Prasanth V P and Soniya E V 2011. Microsatellite based analysis of genetic diversity of popular black pepper genotypes in South India, *Genetica:* 139(8) : 1033-1043.

Nishida S, Kikuichi S, Yoshioka S, Tsubaki M, Fujii Y, Matsuda H *et al* 2003. Induction of apoptosis in HL-60 cells treated with medicinal herbs. *Am. J. Chin. Med.* 31:551-562.

Nybe E V and Mini Raj N 2007.Peter,K V (Ed). Spices: Vol.05 .Horticulture Science Series. New India Publishing Agency, New Delhi p.328.

Lu F H, Cho M C and Park Y J 2012.Transcriptome profiling and molecular marker discovery in red pepper, *Capsicum annuum* L. TF68. *Mol Biol Rep.* 39(3):3327-3335.

Pahlsson A M B 1989. Toxicity of heavy metals (Zn, Cu, Cd, Pb) to vascular plants: a literature review. *Water, Air, Soil, Pollut.* 47: 287-319.

Perez D E, Hoyer J S, Johnson A I, Moody Z R, Lopez J and Kaplinsky N J 2009. BOBBER1 is a noncanonical Arabidopsis small heat shock protein required for both development and thermotolerance. *Plant Physiol* 151: 241–252.

Peter K V 2012. (Ed) Handbook of herbs and spices.Vol.1-2 Woodhead Publishing Company, UK p.1207.

Philip S, Banerjee N S and Das M R 2000. Genetic variation and micropropagation in three varieties of *Piper longum* l. *Current Science* 78: 2, 25.

Pimentel D 1991. Diversification of biological control strategies in agriculture. *Crop Protection* 10: 243–253.

Pimentel D 2002. Biological Invasions: Economic and Environmental Costs of Alien Plant, Animal and Microbe Species. CRC Press, Boca Raton, FL, USA, p. 384.

Prabu G, Kawar P G, Pagariya M C and Prasad D T 2011. Identification of water deficit stress upregulated genes in sugarcane. *Plant Mol Biol Rep* 29:291–304.

Pradeepkumar Y, Karihaloo J L, Archak S, Baldev A 2003. Analysis of genetic diversity in Piper nigrum L. using RAPD markers. *Genet Resour Crop Evol* 50(5):469–475.

Ravindran P N and Nirmal Babu K. 2005. Ginger: the genus Zingiber- (Medicinal and aromatic plants-industrial profiles; v 41, p.293). ISBN 0-415-32468-8.

Ravindran P N. 2001. Black pepper, *Piper nigrum*. Medicinal and Aromatic Plants - Industrial Profiles, *Phytochemistry*, 58 : 827-829.

Saibo N J, Lourenco T and Oliveira M M 2009. Transcription factors and regulation of photosynthetic and related metabolism under environmental stresses. *Annals of Botany* 103: 609–623.

Sairam R K and Tyagi A 2004. Physiology and molecular biology of salinity stress tolerance in plants. *Cur. Sci.*, 86: 407-421.

Sangwan V, Orvar B L, Beyerly J, Hirt H and Dhindsa R S 2002. Opposite changes in membrane fluidity mimic cold and heat stress activation of distinct plant MAP kinase pathways. *Plant J.* 31: 629–638.

Sauìl Fraire-Velaìzquez, Rauìl Rodriìguez-Guerra and Lenin Saìnchez-Calderoìn 2011. Abiotic and Biotic Stress Response Crosstalk in Plants, Abiotic Stress Response in Plants - Physiological, Biochemical and Genetic Perspectives, Prof. Arun Shanker (Ed.), Chapter 1, p.3-26. ISBN: 978-953-307-672-0.

Schoner S and Krause G H 1990. Protective systems against active oxygen species in spinach: response to cold acclimation in excess light. *Planta* 180: 383-389.

Sepehr M F, Ghorbanli M and Amini F 2012.The Effect of Water Stress on Nitrate Reductase Activity and Nitrogen and Phosphorus Contents in *Cuminum cyminum* L. *Pak. J. Bot.*, 44(3): 899-903.

Sharma P, Sharma N and Deswal R 2005. The molecular biology of the low-temperature response in plants. *Bioessays* 27: 1048–1059.

Siju S, Dhanya K, Syamkumar S, Sasikumar B, Sheeja T E, Bhat A I, and Parthasarathy V A 2010. Development, characterization and cross species amplification of polymorphic microsatellitemarkers from expressed sequence tags of turmeric (Curcuma longa L.). *Mol Biotechnol.* 44(2):140-147.

Silva J A and Uchida R 2000.Plant Nutrient Management in Hawaii's Soils, Approaches for Tropical and Subtropical Agriculture, Chapter 3. p. 31-55.

Sresty T V S and Rao K V Madhava 1999. Ultrastructural alterations in response to zinc and nickel stress in the root cells of pigeon pea. *Environ. Exp. Bot.* 41: 3-13.

Susheela R 2006. Handbook of spices, seasonings and flavourings 2nd ed.

Swetha Priya R and Subramanian R B 2008. Isolation and molecular analysis of R-gene in resistant Zingiber officinale (ginger) varieties against Fusarium oxysporum f.sp. zingiberi. *Bioresour Technol.* Jul;99 (11):4540-4543.

Tomsett A B and Thurman D A 1988. Molecular biology of metal tolerance of plants. *Plant Cell Environ.* 11: 383-394.

Tzyy-Jen Chiou 2007. The role of microRNAs in sensing nutrient stress. *Plant, Cell and Environment* 30: 323–332.

Wang P and Song C P 2008. Guard-cell signalling for hydrogen peroxide and abscisic acid. *The New Phytologist* 178: 703–718.

Wang W K, Liu C C, Chiang T Y, Chen M T, Chou C H and Yeh C H 2011. Characterization of expressed sequence tags from flower buds of alpine *Lilium formosanum* using a subtractive cDNA library. *Plant Mol Biol Rep* 29:88–97.

Wang X X, Chen J, Wang B, Liu L J, Jiang H, Tang D L and Peng D X 2012. Characterization by suppression subtractive hybridization of transcripts that are differentially expressed in leaves of anthracnose-resistant ramie cultivar. *Plant Mol Biol Rep* 30: 547–555.

Weis E and Berry J A 1988. Plants and high temperature stress. *Symp Soc Exp Biol* 42:329–346.

Wise R P, Moscou M J, Bogdanove A J and Whitham S A 2007. Transcript profiling in host-pathogen interactions. *Annual Review of Phytopathology* 45: 329-369.

Xu Z, Zhou G and Shimizu H 2010. Plant responses to drought and rewatering. *Plant Signal Behav* 5:649-54.

Zangerl A R and Berenbaum M R 1997. Cost of chemically defending seeds: furanocoumarins and Pastinaca sativa. *The American Naturalist* 150: 491–504.

Zangerl A R and Berenbaum M R 2003. Phenotype matching in wild parsnip and parsnip webworms: causes and consequences. *Evolution* 57:806–815.

Zhu J K 2001. Over expression of a delta-pyrroline-5-carboxylate synthetase gene and analysis of tolerance to water and salt stress in transgenic rice. *Trends Plant Sci.* 6: 66–72.

7

Mycorrhizal Symbiosis for Sustainable Horticulture

D. J. Bagyaraj

Sustainability refers to productive performance of a system over time. It implies use of natural resources to meet the present needs without jeopardizing the future potential. The Technical Advisory Committee of the Consultative Group of International Agricultural Research has defined sustainability as "Successful management of resources for agriculture to satisfy changing human needs while maintaining or enhancing the quality of the environment and conserving natural resources" (TAC, 1989). A sustainable system must be sustainable both ecologically and economically.

During the past several decades, production of horticultural crops has undergone a tremendous change in both developed and developing countries due to the development and use of fertilizers, pesticides, mechanization, soil fumigation, container production in soil-less media, breeding, micropropagation, pathogen elimination and a number of other innovations for attaining its future stability and sustainability. Recently, in modern horticulture, enhanced productivity depends heavily on inputs of fertilizers and pesticides to maintain an economic level of productivity. On the other hand, social and economic pressures are demanding that agricultural/horticultural practices, that have a damaging impact on the environment and pose risk to health of mankind, be curtailed. Thus throughout the world, agriculture including horticulture is at crossroads in terms of its future stability and sustainability. The solution for the above lies in developing new technology with reduced use of fertilizer and pesticides besides maintaining a sustainable level of productivity. The increased concern about the sustainability and undesirable side-effects of chemistry-based agriculture is prompting many researchers to more seriously evaluate alternative approaches. One such effort was put forth in the management of microbes that live in the rhizosphere soil around the roots of nearly all terrestrial plants. It has been believed that the main orchestrators of microbial activity in the rhizosphere are mycorrhizal fungi.

Frank (1885), the German botanist first coined the term "mycorrhiza" which literally means "fungus root" to describe the mutualistic association between roots of higher plants and certain fungi. These associations are grouped, based on morphological and anatomical characters, as ectomycorrhizae and endomycorrhizae. Ectomycorrhizae are common among temperate forest tree species and can be cultured on laboratory media. Endomycorrhizae, include arbutoid, monotropoid, ericoid, orchid and arbuscular mycorrhizal (AM) forms. Arbuscular mycorrhizae are the most common and widely occurring of all the mycorrhizal associations and have great economic significance. They cannot be cultured on laboratory media.

AM associations are the most common type of fungal-root symbiosis, abundant in soils from all continents (Harley, 1991). Evidence from fossil record (Stubblefield *et al.,* 1987), morphological studies (Morton, 1990) and r-DNA nucleotide sequence analysis (Simon *et al.,* 1992) suggest that these associations co-evolved with plants. AM fungi have been reported from almost all environments including aquatic (Bagyaraj *et al.,* 1979) to xerophytic (Williams and Aldon, 1976), from lowland tropical rainforests (Janos, 1987) to high altitudes (Allen *et al.,* 1987), and even in the canopy of epiphytes (Nadkarni, 1985).

AM fungi have the widest host range and distribution of all the mycorrhizal associations. It is estimated that about 90% of vascular plants normally establish mutualistic relationships with AM fungi. AM fungi have been observed in 1000 genera of plants representing some 200 families. There are at least 300,000 receptive hosts in the world flora and there are about 220 species of AM fungi. If the hosts are divided up evenly among the fungi, with no overlap in host range, each fungus would have more than 1,360 potential partners. We know that the host range overlaps extensively, suggesting that some individual AM fungi may well have access to thousands of host.

Arbuscular Mycorrhizal Fungi: Classification, Importance and Occurrence

Arbuscular mycorrhizal fungi are obligate symbionts and cannot be cultured on synthetic media. Their penetration takes place through root hairs or epidermal cells and then grows intercellularly or intracellularly in the root cortex, ultimately developing short haustoria like structures called arbuscules within the cortical cells. These arbuscules function as sites of nutrient exchange between the fungus and host roots. Vesicles are formed in the cortical cells, which are thin walled structures of various sizes and shapes and function as storage organs. The presence of vesicles and arbuscules is the criteria for identifying AM fungus in the roots.

AM fungi belong to the phylum Glomeromycota, which has a single class Glomeromycetes with 4 orders Glomerales, Diversisporales, Paraglomerales and Archaeosporales. There are 11 families, 17 genera and 228 species of AM fungi (Schüßler and Walker, 2010). The commonly occurring genera of AM fungi are *Glomus, Gigaspora, Scutellospora, Acaulospora* and *Entrophospora*.

Improved plant growth due to inoculation of soil with AM fungi has been demonstrated especially under P deficient conditions (Mosse, 1977). The growth improvement is mainly because of enhanced P uptake. AM fungi can also enhance tolerance or resistance to root pathogens (Borowicz, 2001) and abiotic stresses such as drought and metal toxicity (Meharg and Cairney, 2000). AM fungi play a role in the formation of stable soil aggregates, build up a macroporous structure of soil that allows penetration of water and air and prevents erosion (Miller and Jastrow, 1992). There is well documented evidence that AM fungi have important effects on plant P uptake. Greater soil exploration by mycorrhizal roots as a means of increasing phosphate uptake is well established. In phosphate deficient soils immobile phosphate ions develop a phosphate depletion zone around the roots. The hyphae spread beyond this zone and directly translocate nutrients from the soil to the root cortex (Hayman, 1983). Experiments with ^{32}P labeled phosphate indicate that AM fungal hyphae obtain their extra phosphate from the labile pool rather than by accessing insoluble phosphate by solubilizing it (Raj *et al.*, 1981). Sparingly soluble rock phosphate is better utilized by the hyphae by closer physical contact with the ions dissociating at the particle surface.

The increased growth of plants inoculated with AM fungi is not only attributed to improved phosphate uptake but also to better availability of other elements like Zn, Cu, K, Al, Mn, Fe etc. AM fungi also affect the levels of plant hormones. Allen (1991) measured levels of plant hormones like cytokinins and gibberlin-like substances. AM fungi can tolerate a wide range of soil water regimes and also improve water relationships of many plants. It is still unclear whether the observed effects are directly due to the fungus itself or indirectly due to some alteration in host physiology as a result of improved P nutrition. Anatomical and other physiological studies have brought out that mycorrhizal plants have increased rates of respiration, photosynthesis and increased amounts of sugars, amino acids, RNA etc. and larger and/ or more number of chloroplasts, mitochondria, xylem vessels, motor cells, etc. Changes in the root exudations and altered rhizosphere microorganisms (which also affect plant growth) may result because of colonization of roots by AM fungi (Caroline Machado and Bagyaraj, 1995).

Mycorrhizal colonization may also allow introduced populations of beneficial soil organisms like *Azotobacter, Azospirillum* and phosphate solubilizing bacteria to be maintained in high numbers than around non-mycorrhizal plants and to exert synergistic effects on plant growth. It is apparent from the investigations on AM fungi-plant pathogen interaction that AM fungi can usually (though not always) deter or reduce the severity of disease caused by soil-borne pathogens. All these studies bring out that AM fungi help the host plant in more than one way and that AM fungal inoculation helps plants' growth.

AM fungi, in addition to their widespread distribution throughout the plant kingdom, are also geographically ubiquitous and occur in plants growing in arctic, temperate and tropical regions (Mosse, 1981). As an explanation for their remarkably widespread distribution, Trappe (1977) proposed that AM fungi were disseminated intercontinentally prior to the continental drift. The super continent Gondwanaland is thought to have begun to break apart and drift north about 125 million years ago. Fossil records of plants containing AM like structures have occurred as Trappe suggested. In general, AM fungal population is more in cultivated soil and their numbers decrease markedly below the top 15cm (Redhead, 1977). They are normally not found in depths beyond the normal root range of plants (Mosse, 1981). Although AM fungi are ubiquitous in soils, the patterns or distribution of individual species have not been fully understood. Studies on the distribution of species have either covered large geographical areas or smaller regions. The distribution of species of AM fungi varies with climatic and edaphic environment as well as with land use. For example, *Acaulospora laevis* is common in Western Australia but occurs less frequently in soils of eastern Australia. *Glomus* spp. appear to have the widest distribution. *Gigaspora* and *Sclerocystis* spp. are more common in tropical soils. *Acaulospora* seems to be better adapted to soils with pH < 5.0. Infact, certain AM fungi have been linked to particular kind of soil: *Glomus mosseae* with fine textured, fertile high pH soils; *Acaulospora laevis* with coarse textured, acid soils; and *Gigaspora* species with sand dune soils (Bagyaraj, 1991).

Need for AM Fungal Inoculation

Once the importance of AM symbiosis is recognized, a decision must be made as to whether the native population of AM fungi suffices as starting material or if there is a need to supplement the native species by external inoculation. If the native fungal population is adequate, then its efficacy in terms of plant growth improvement should be assessed. Generally spore counts are employed for assessing the native fungal population. This does not clearly indicate the population of AM fungi as sporulation depends on host and environmental factors. Alternatively, young fresh roots of native plants growing in the soil may be collected and stained for AM colonization, but this is not always possible. Variations in extent of root colonization by a particular AM fungal species can depend on the host, fungal and environmental factors (Dodd and Jeffries, 1986) and an absence of infection at a particular time cannot indicate that the soil is devoid of AM fungi. A best indicator would be to grow the intended crop in the soil to which it would be transplanted and to evaluate the benefits for plant growth derived from the native AM fungi.

Some of these fungi are reported to cause no growth improvement in spite of maximum root colonization (Lovato *et al.*, 1992). Hence, inoculation with a fungus with high levels of colonization alone is not sufficient to harness maximum plant growth benefits. Further, AM fungi are known to have host preference, though not host specificity (Mosse, 1981). A fungal species and even isolates of the same species are known to vary markedly in the extent to which they colonize roots and in the benefits they confer to host plants (Rea and Tullio, 2005). This problem is further complicated by the fact that the host plant species (Plenchette *et al.*, 1983) and even cultivars of the same species (Graham and Syversten, 1985) respond differently to a specific fungus. This has led to selection of an efficient fungus for a particular host species (Reddy *et al.*, 1996). In all these cases it was observed that native AM fungal inoculum or native species were not efficient in enhancing the growth and nutrition of hosts to a great extent as compared to introduced AM inoculum. Hence, the need arises to screen and select an efficient fungus for each crop and to inoculate the substrate with such a fungus to get maximum benefits from this association.

Mycorrhizal Dependency of Plants

Plants differ greatly in their mycorrhizal dependence. Gerdemann (1975) defined Relative Mycorrhizal Dependency (RMD) as the degree to which a plant is dependent on the mycorrhizal condition to produce maximum growth or yield at a given level of soil fertility. The objective of this was to determine the extent of growth increase attributed to mycorrhizal condition. Plenchette *et al.* (1983) proposed a formula to calculate RMD of crop plants under field conditions by comparing plants in fumigated and unfumigated soils. This measures the extent of growth increase due to native endophytes and the calculated values are presented on a scale of 0 to 100%. There have been many reports that the introduction of mycorrhizal fungi can improve plant growth under unsterile conditions in spite of the presence of native endophytes (Jarstfer and Sylvia, 1992). Hence, Bagyaraj *et al.* (1988) proposed another formula which enables calculation of the Mycorrhizal Inoculation Effect (MIE) to assess the growth improvement brought about by inoculation with a mycorrhizal fungus in unsterile soil with indigenous AM fungi. MIE is very useful for the assessment of the extent to which introduced fungi compete with native endophytes to bring about a plant growth response. This information is of great value in practical agriculture, especially in developing countries, where farmers do not fumigate or sterilize either the nursery or the field.

Selection of Efficient AM Fungi

Where inoculation with selected AM fungi is required, it is essential to characterize the native AM fungal population first, as well as the soil characteristics, particularly soil P status which influences the efficacy of the fungus to a great extent. The selection process may then be carried out with host plants and inocula of various strains/species of AM fungi. Such experiments should be carried out under conditions which are nearest possible to the field conditions. Such pot experiments/micro plot experiments will be realistic and effective only when the exact host plant is used, as it is well-known that host plant species and even cultivars of the same species respond differently to a particular fungus. Further, the soil or substrate used must be the same or similar to the field soil conditions. This is also true with respect to the management practices like water, fertilizers and pesticide levels, as used in a sustainable agricultural system.

In the preliminary screening, generally where different species of AM fungi obtained from different sources or a culture collection are screened, performance of each species must be based on the plant growth and nutritional parameters apart from colonization levels. In the secondary screening, three or four fungal species which have performed well in the preliminary screening are compared with two to three predominant fungal isolates selected from the root zone soil of the host species under study. These fungi are again evaluated for their efficacy in terms of host plant growth and nutrition, alleviation of stress factors and ability to colonize and proliferate in the rhizosphere soil. This is essential because a fungus may heavily colonize the roots without considerably promoting plant growth. In such cases, where a plant species fails to significantly curtail the extent of root colonization by the mycorrhizal fungus, a depression in plant growth has been observed (Lovato *et al.*, 1992) and it has been attributed to the utilization of photosynthates by the fungus without compensation for better mineral nutrition of the host plant. The technique developed by Giovannetti and Mosse (1990) is most commonly used to evaluate the level of AM fungal colonization after staining the roots according to the procedure described by Kormanik *et al.* (1980). In this technique, arbuscules, the structures where exchanges between plant and fungus occur, cannot be evaluated. Trouvelot *et al.* (1986) developed another technique based on visual evaluation of the volume of the cortex occupied by AM fungi with a simple notation for richness in arbuscules. Similarly several other techniques, such as estimating total fungal biomass in the roots using non-vital staining (Koske and Gemma, 1989) and other staining techniques (Kormanik *et al.*, 1980), do not suffice as these procedures fail to differentiate between dead and alive fungal material. To combat these limitations, staining procedures based on physiological activities of fungi have been developed (Tisserant *et al.*, 1992). Hence, a combination of non-vital, vital and functional staining procedures can be reliable in an AM fungi selection program.

Appropriate Technology for Nursery Raised Crops

AM inoculum of suitably selected strains can be used for inoculation in the nursery bed. Growers only need to incorporate inoculum in the nursery beds or containers at the appropriate rate by hand. Seedlings thus raised will be colonized by the introduced fungus and then can be planted out in the field. In developed countries, seedlings are usually raised in fumigated soil or potting mix. Inoculation of such substrate with selected AM fungi for the crop species is an easy and appropriate technology. Commercial AM inoculation by this method in the nursery production of citrus is applied in the USA (Palazzo *et al.*, 1994). Inoculation of the nursery with efficient AM fungi has been successful in the large-scale production of many horticultural plants in France (Gianinazzi *et al.*, 1990).

In many developing countries and sometimes in developed countries, seedlings are raised in unsterile soil or substrate. Inoculation with appropriate fungi in containers containing unsterilized potting mix was found to give vigorously growing seedlings of asparagus that performed well when planted out in the field (Powell *et al.*, 1985). The inoculation of unsterile nursery beds with efficient AM fungi enhances the growth of rootstocks of citrus and mango (Balakrishna Reddy and Bagyaraj, 1994). Further AM inoculation increased yield of chillies, tomato, capsicum and other vegetables (Bagyaraj, 2007). Increased growth of the seedlings of plantation crops like cashew was observed when inoculated with AM fungi (Lakshmipathy *et al.*, 2000). It was also observed that in AM inoculated plants there was an increased number of flowers, and the vase life of cut flowers of chrysanthemum and China asters was enhanced (Bagyaraj and Mallesha, 2000).

Use of AM Fungi in Rooting of Cuttings, Air Layering and in Overcoming Transplant Shock

Many horticultural plants are propagated through cuttings. Tropical root crops, such as cassava and sweet potato, are also propagated through cuttings. In such cases, rooting of the cuttings is very important. Enhanced rooting of cuttings through inoculation with AM fungi has been reported in apple (Plenchette *et al.*, 1981), cassava and sweet potato (Potty, 2004). More rooting in tamarind and cashew plants propagated through air layering was reported (Bagyaraj and Mallesha, 2000). Such plants not only withstood transplant shock but also established better when planted in field sites. AM-inoculated avocado plants withstood transplant shock better than uninoculated stock (Menge *et al.*, 1980).

Application Methods and Time of Application

It is important to standardize the dose and time of application, apart from the method of application, in order to harness the maximum benefit from the AM-host association. In a potential inoculum, every propagule is capable of colonizing the host root. However, in order to ensure a threshold level of colonization and quicken the process of colonization, a higher density of propagules should be present in the inoculum. Our experience has shown that about 150-200 infective propagules (determined by four-fold dilution) per g of substrate (sand: soil 1:1) will result in good colonization and establishment of the host. However, the number varies with other factors, such as their ability to tolerate soil (pH, nutrient status) and environmental factors (light intensity, temperature) and to proliferate. In horticultural species, where planting is carried out at comparatively larger distances, high inoculum density is preferred, unlike in other agricultural crops where seedling density is high, hence roots have a greater probability of encountering the infective propagules.

Considering the time of application, it has been well demonstrated that earlier the inoculation, the greater the benefits to the host. Our experience reveals that inoculation at the time of sowing/planting will result in better colonization of the roots as they emerge, hence help in better establishment of the host. Experiments conducted with micropropagated banana and *Ficus benzamina* have revealed that the best time for inoculation would be just before planting out hardened plantlets (Shashikala *et al.*, 1999a, b; Ravolanirina *et al.*, 1989). However, Vidal *et al.* (1992) observed that inoculation of micropropagated avocado plantlets four weeks after acclimatization resulted in better establishment of these plantlets. Such varying views indicate that apart from identifying an efficient fungus for each host plant, there is a need to determine the optimum time of application for that particular host in order to maximize the benefits from AM fungi.

Methods of application of AM fungi generally include hand placement, placing below the seed material or in case of pot experiments, uniformly mixing the inoculum with the substrate. The importance of method of application of inoculum arises with the need to initiate colonization in the early stages of plant growth. Hence, from this view point, it is essential to place the inoculum close to the seed material such that the roots will come into contact with the fungal material as soon as they emerge. An experiment was conducted to standardize the method of inoculum application for fruit tree species raised in polybags. Polybags were filled with sand:soil mix in the ratio of 1:1. *Glomus mosseae* inoculum (15g) was inoculated into the polybags by different methods, like thoroughly mixing with the substrate, or placing as a layer at different depths, or placing as a band at different depths, or placing at a point at different depths. Papaya was the host used in this study. Based on the plant parameters, it was concluded that placing AM inoculum

8cm below the surface of the soil (7cm below the seed), at a point or as a layer, was the best method of inoculum placement for seedlings raised in polybags (Mamatha and Bagyaraj, 2000a). For seedlings raised on the nursery beds, placing the inoculum 2 cm below the seed was found to be the best approach (Mamatha and Bagyaraj, 2000b).

Production of Seedlings Inoculated with AM Fungi

Plants multiplied by seedlings such as vegetables, fruits and ornamental plants can be inoculated with AM fungi due to the possibility of using a small amount of inoculum and to the easy handling, disinfecting and adjustment of the fertility of the substrate. The growth of the seedling in semi-controlled conditions promotes the establishment of inoculated AM fungi. Furthermore, the use of subsoil and organic material in the composition of the substrate, which have few or no AM fungal propagules, decrease the competition of the native AM fungi with the inoculated fungus. In many cases, there is also a disinfecting of the substrate to avoid pathogens and weeds killing the native AM fungi. This makes inoculation of the seedlings viable and necessary when it is related to plants that present high mycorrhizal dependency or to seedlings that will be planted in areas that have few AM fungal propagules, such as those that are degraded or where there was a longtime culture of a non-AM species, such as *Pinus*.

There are many reports on nutritional improvement after inoculation of AM fungi and on the increase in the growth of seedlings, accelerating their formation and reducing the time and cost of nursery. There are, however, other less observed benefits promoted by the inoculation of seedlings. In the production of seedlings, both by conventional and cloning techniques, a difficult factor to control is the non-uniformity of plant growth. Even plants cloned *in vitro* show variation among the individuals in the formation of seedlings. This variability can be diminished with the inoculation of AM fungi. However, few works evaluate the uniformity of the seedlings, which should be considered to evaluate mycorrhizal seedlings, since it is an important characteristic that affects the production costs and the commercial value of the seedlings. Another benefit that seedling inoculation can bring is the increase in tolerance and even in protection against attacks by soil pathogens (Saggin Junior and da Silva, 2006). This happens due to the promotion of a better nutrition of plants with roots damaged by fungi or nematodes, and by direct or indirect antagonistic effects on the pathogens (Sylvia and Chellemi, 2001).

Several technical aspects should be considered in the inoculation of seedlings with AM fungi, from the choice of the inoculum to the care that should be taken in the handling of the inoculated seedlings. Technology should be adapted to the local conditions of seedling production areas. The cost should be low to make the inoculation attractive to producers and nursery owners since, most times, the aim should not be the benefit of inoculation in seedling production, and one should search for the valuation of the produced seedling and the post-transplantation benefit.

Phosphorus Response Curves

Using the most efficient fungus for inoculating a particular crop, the amount of P fertilizer that can be saved may be worked out. It should then be possible to select the fertilizer level at which the response to inoculation can be optimized. These kinds of experiments suggest the amount of P fertilizer that could be saved with mycorrhizal inoculation to produce the same amount of growth and yield. Mycorrhizal inoculation will only be useful in horticulture if it enables the farmer to reach an optimum yield with a significantly lower level of P fertilizer. The savings from the reduced fertilizer requirement will be dependent on the shape of the P response curves, and then can be compared with the costs of supply and introduction of the mycorrhizal inoculant.

Examples of Growth Response to AM Fungi

Several workers have studied the influence of AM fungi on the growth and nutrition of a variety of horticultural crops. Most of these studies have included the plant growth benefits and nutrient content, particularly phosphorus. Some have studied their response in terms of flowering and fruiting. However, a majority of these studies are greenhouse experiments.

There is great potential for utilizing AM fungi in horticultural crops raised in fumigated nursery and/or field soil. Fumigation not only destroys indigenous AM fungi but also other microorganisms. Inoculation with AM fungi in such situations resulted in substantially increased plant growth and yield. Such growth increases in fumigated nursery has been demonstrated in citrus (Timmer and Leyden, 1978), peach (Lambert *et al.,* 1979) and avocado (Menge *et al.,* 1980). Similarly, large acreages of strawberry, tomato and rose are produced in fumigated soil to control pathogens in several countries. Inoculation of the field or transplanting inoculated seedlings in the field markedly increased yield of these crops (Maronek *et al.,* 1980).

Soil-less media like vermiculite, perlite, potting mixes, etc. are used for raising several horticultural crops. Such media are usually fumigated or heat sterilized. Inoculation of these soil-less media, enhancing seedling growth and yield, has been reported in woody ornamentals (Crews *et al.*, 1978), asparagus (Powell *et al.*, 1985), citrus (Yao and Li, 1999), papaya (Manjunath *et al.*, 2001), passion fruit (Wamocho *et al.*, 2001), potato (Hernandez-Zardon and Bertolil-Herrera, 2001), eggplant (Yang *et al.*, 2001), onion (Sharma and Adholeya, 2000), maize (Kabir and Koide, 2000) and cucumber (Yun *et al.*, 2002). AM fungi have also been reported to survive in a commercial nursery container production system and enhance plant productivity (Davies *et al.*, 2000).

Significant increase in plant growth and yield of several horticultural crops, due to AM inoculation in unsterile soils containing less or inefficient indigenous endophytes, have been reported by several workers in crops like Easter lily (Ames and Linderman, 1978), poinsettia (Barrows and Rancadori, 1977), tomato and capsicum (Mallesha and Bagyaraj, 1997), chillies (Sreeramula and Bagyaraj, 1986), chrysanthemum and China aster (Mamatha and Bagyaraj, 2001).

Further, working with citrus rootstocks, it has been observed that mycorrhizal inoculation resulted in plants ready for budding four to five months in advance compared to uninoculated plants (Vinayak and Bagyaraj, 1990). A similar observation was also made in mango rootstocks, where inoculated plants were ready for grafting two months early (Balakrishna Reddy and Bagyaraj, 1994). Early grafting and faster establishment after mycorrhizal inoculation has been reported in cashew rootstocks as well (Lakshmipathy *et al.*, 2000). Again, inoculation of chrysanthemum with *Glomus mosseae* not only increased the flower yield per plant, but also the vase life of cut flowers (Mamatha and Bagyaraj, 2001).

In recent years, many high value crops are being raised through tissue culture. Micropropagated plantlets, cultivated in agar rooting medium, are usually delicate, hence hardened before planting in the field. During the process of hardening, micropropagated plantlets get acclimatized to the outside environment. During the process usually 20-40% mortality is observed. Inoculation of micropropagated plantlets with AM fungi after hardening, also improved plantlet vigour and growth in cassava, coffee, grapevine, apple, avocado, pineapple, kiwifruit, strawberry, raspberry, asparagus and banana (Varma and Scheupp, 1995; Shashikala *et al.*, 1999a; Bressan *et al.*, 2000; Calderon *et al.*, 2001; Mathews *et al.*, 2003). It was also observed that AM inoculation significantly reduced *Rhizoctonia solani* infection in micropropagated potato (Yao *et al.*, 2002).

AM fungal inoculation, enhancing the growth and yield, has been reported in a number of tropical and many subtropical fruit plants

(Chundawat and Singh, 2001), coriander (Prabhu *et al.*, 2002a), capsicum (Olsen *et al.*, 1999), okra (Prabhu *et al.*, 2002b), gaillardia (Rathod *et al.*, 2002) and snapdragon (Besmer and Koide, 1999). In snapdragon flowers, there was decreased ethylene production after AM inoculation. Thus, AM inoculation is a viable alternative to toxic ethylene inhibitors like silver thiosulfate, which are used to delay ripening (Besmer and Koide, 1999). Drought tolerance was observed in *Phaseolus vulgaris* when inoculated with AM fungi. Inoculated plants showed recovery from water stress as well (Tohamy *et al.*, 1999).

Performance of Mycorrhizal Inoculation of Seedlings In the Nursery when Planted in the field site

The application of AM fungi in the production of seedlings is promising as the beneficial effects have become evident in the initial stages of seedling production and transplantation to the fields, being able to increase productivity. Inoculated seedlings have superior growth and survival in the field over seedlings not inoculated with AM fungi. Many studies have shown that inoculation of nursery grown seedlings with selected mycorrhizal fungi enhances plant survival, growth, and establishment following out planting.

In annuals, seedlings raised in nursery beds or containers supplied with selected AM fungi and transplanted to the field have produced economic growth responses. This has been reported in agronomically important crops like chilli, finger millet, tobacco etc. (Bagyaraj, 2003) and horticultural crops such as tomato, capsicum, marigold etc. (Bagyaraj, 2007; Bagyaraj and Kehri, 2012). Further, these studies also brought out that application of phosphatic fertilizer can be reduced by nearly 50%. In medicinal and aromatic plants AM fungal inoculation in the nursery not only increased the crop yield in the field but also that of the active ingredient (Arpana *et al.*, 2010). In *Solanum quitoense* inoculated seedlings established and grew vigorously where as uninoculated seedlings remained very small (Collazos *et al.*, 1986).

There are only a few reports on the post-transplant effect of AM fungi inoculated seedlings of perennial plants like plantation crops and fruit trees. Coffee seedlings inoculated with AM fungi in the nursery exhibited higher survival rate and grew more vigorously in the field compared to uninoculated seedlings (Sieverding, 1991). The author felt that it may not only be due to improved plant nutrition but can be because of other positive effects like enhanced water uptake. In cashew we observed that seedlings raised with AM fungi in the nursery, exhibited higher level of grafting success and higher survival rate when planted in the field compared to seedlings which received no AM fungi in the nursery (Lakshmipathy *et al.*, 2004)

Inoculation of AM fungi along with Other Beneficial Soil Microorganisms

There are several reports on the interaction between AM fungi and other beneficial soil microorganisms like nitrogen fixers, phosphate solubilizers and PGPRs. These studies suggest that the interaction is synergistic with consequential benefit on plant growth. Recent studies have shown that inoculation with microbial consortia consisting of an efficient AM fungus together with a nitrogen fixer, P solubilizer and PGPR carefully screened and selected for a particular crop plant is more beneficial than AM fungus alone in improving the growth, biomass and yield. The fundamental means of assessing the effectiveness of inoculating the nursery seedlings are their survival and growth in the field. Such studies have been done with ectomycorrhizal fungi mostly in the USA and Canada. The success of ectomycorrhizal fungal inoculation in improving seedling growth and survival after out planting is well recognized. However such studies with AM fungi alone or as microbial consortia are megre.

Bagyaraj and Menge (1978) studied the interaction between *Azotobacter chroococcum* and the AM fungus *Glomus fasciculatum* in tomato and found a synergistic effect on plant growth. Mycorrhizal infection increased the *A. chroococcum* population in the rhizosphere, which was maintained at a high level for a longer period and *A. chroococcum* enhanced colonization and spore production by the mycorrhizal fungus. Synergistic interaction between AM fungi and associative nitrogen fixing bacteria *Azospirillum* spp. and *Acetobacter diazotrophicus* have been reported (Tilak *et al.*, 2010). A tripartite interaction study between the free-living nitrogen fixing bacterium *Beijerinckia mobilis*, phosphate solubilizing fungus *Aspergillus niger* and the mycorrhizal fungus *G. fasciculatum* was conducted. A synergistic beneficial effect on the growth of onion with all 3 organisms was observed. These were attributed to hormone production rather than, or in addition to nitrogen fixation. The hormones produced by these bacteria could exert some synergistic effect on plant growth or mycorrhizal efficiency. The studies conducted so far reveal a definite positive interaction between free-living nitrogen fixing bacteria and AM fungi in the rhizosphere with consequential improvement in plant growth (Manjunath *et al.*, 1981; Arpana *et al.*, 2010). Microbial consortia consisting of *Glomus mosseae* together with *Tricoderma harzianum* reducing the severity of wilt disease of *Coleus forskohlii* better compared to *G. mosseae* inoculation alone has also been reported (Boby and Bagyaraj, 2003).

Interactions between AM fungi and Mycorrhiza Helper Organisms

It has been found that some microorganisms present in soil, can modify the establishment of the mycorrhizal symbiosis. The organisms associated with the rhizosphere of mycorrhiza which, stimulate the growth of the fungus and mycorrhizal colonization are called mycorrhiza helper organisms (MHO). Therefore, MHO selectively helps AM fungal species to form the symbiotic association by increasing its competitive ability against other potential symbionts. It was suggested that it could be due to either the production of enzymes or growth promoting substances which increase the cell wall plasticity and thereby increase fungal colonization.

Several workers have reported that co-inoculation of AM fungi with MHO enhanced mycorrhizal colonization and in turn plant growth and yield. In high-tech agriculture some crops are raised in disinfected soil/nursery. Disinfection kills several groups of organisms. In such soils, mycorrhizal inoculation is recommended. If MHO are also killed then colonization by AM fungi will not be proper. It is suggested that milder disinfection treatments which selectively kill indigenous AM fungi and root pathogens but conserving MHO should be developed. It is probable that, in the near future, commercial mycorrhizal inoculum will contain MHO, improving the efficiency of inoculation in a wider range of conditions and reducing the quantity of inoculum needed. Recent studies have shown that inoculation with microbial consortia consisting of an efficient AM fungus together with a nitrogen fixer, P solubilizer and PGPR carefully screened and selected for a particular crop plant is more beneficial than AM fungus alone in improving the growth, biomass and yield.

The fundamental means of assessing the effectiveness of inoculating the nursery seedlings are their survival and growth in the field. Such studies have been done with ectomycorrhizal fungi mostly in the USA and Canada. The success of ectomycorrhizal fungal inoculation in improving seedling growth and survival after out planting is well recognized. However such studies with AM fungi alone or as microbial consortia are megre. It has been reported recently that use of microbial consortia not only reduced that application of chemical fertilizers by nearly 50% but also enhanced the yield of the crop and the concentration of secondary metabolites in aromatic plants like patchouli and sweet basil (Bagyaraj, 2011). Further use of microbial consortia alleviating the severity of disease more effectively compared to only AM fungus has been shown through recent studies (Boby and Bagyaraj, 2003; Gupta *et al.*, 2007; Bagyaraj, 2011). In turmeric, combined application of FYM, *Azospirillum*, phosphobacteria and AM fungi exhibited higher rhizome yield (33313.00 kg ha^{-1}) as compared to the application of 50 per cent recommended dose of fertilizer + phosphobacteria (18636.00 kg ha^{-1}). The highest curing percentage (21.09 per cent), cured rhizome yield and quality parameters like curcumin and oleoresin content were also regimed supreme by application of FYM, *Azospirillum*, phosphobacteria and AM fungi (Bagyaraj, 2011; Hemalatha *et al.*, 2010).

Conclusion

Many horticultural plants, e.g. tomato, onion and capsicum, some agronomic crops, tobacco and most of the tropical fruit trees are first established in seedbeds or maintained during early development in nurseries before transplanting to the field. Ornamentals which are often important export products of tropical countries are raised in nursery containers.

The special circumstances of the pre-establishment of plants under "controlled" conditions allow a wide range of methods of incorporating the AM fungi inoculation technology into these production systems. Often, the seedbeds or potting substrates are disinfected (either with steam or with general biocides) to eradicate or to prevent soil-borne plant diseases and pests; furthermore, general biocides are often applied to avoid weed problems in nursery soil substrates. As said earlier, general biocides greatly reduce propagule density of indigenous AM fungi. Sub-soils which have low pressure of soil-borne pathogens are also often used as potting material in nurseries; these soils contain little in the way of AM fungal populations. Some artificial potting media, like vermiculite, perlite or composted organic materials, do not have any AM fungi. There are many reports in literature that selected AM fungi (within all sources of inocula) have been successfully introduced to seedbeds and nursery potting materials.

The inoculation of seedlings with AM fungi is considered a safe and eco-friendly biotechnology, since these fungi occur naturally in the soils of most land ecosystems. Moreover, it is a low-cost technology, which makes its application very viable. Further, it is essential to develop microbial consortia consisting of suitable AM fungi + PGPRs to inoculate a crop plant in the nursery. This simple nursery technology will not only help to produce healthy, vigorously growing seedlings in the nursery but will also reduce transplant shock and ensure better survival, growth and productivity when out planted in the field. Finally, there is still a gap in the research and development framework, and technology transfer in this area. It is likely that inoculation with microbial consortia consisting of effective AM fungi + PGPRs will become an integral part of the nursery technology in the future. This simple technology will not only improve plant growth and productivity but will also reduce the usage of fertilizers and pesticides thus minimizing environmental pollution.

Acknowledgement

The author wish to thank Mr. R. Ashwin and Ms. Anshu Beulah Ram for their help in bringing out this chapter.

References

Allen, M.F. 1991. *The Ecology of Mycorrhizae*. Cambridge University Press, Cambridge, U.K.

Allen, E. B., Chambers, J.C., Conner, K. F., Allen, M. F. and Brown, R. W. 1987. Natural establishment of mycorrhiza in disturbed alpine ecosystems. *Arch. Alpine Res.*19: 11 - 20

Ames, R.C. and Linderman, R.G. 1978. The growth of Easter lily (*Lilium longiflorum*) as influenced by vesicular arbuscular mycorrhizal fungi, *Fusarium oxysporum* and fertility level. *Can. J. Bot.* 56:2773-2780

Arpana, J., Bagyaraj, D.J., Prakasa Rao, E.V.S., Parameswaran, T.N. and Rahiman, B.A. 2010. Evaluation of growth, nutrition and essential oil content with microbial consortia. *J. Soil Biol. Ecol.* 30: 56-71

Bagyaraj, D. J. 1991. Ecology of vesicular-arbuscular mycorrhizae. In: Arora, D.K., Rai, B., Mukerji, K.G. and Knudsen, G.R. (eds.) *Handbook of Applied Mycology*. Marcel Dekker, Inc., New York, USA. pp. 3-34

Bagyaraj, D.J. 2003. Mycorrhizal biotechnology for improved crop productivity. In: Marwaha, S.S. and Arora, J.K. (eds.) *Biotechnological stratagies in agro-processing*. Asiatech. Pub. Inc., New Delhi. pp. 80-84

Bagyaraj, D.J. 2007. Arbuscular mycorrhizal fungi and their role in horticulture. In: Keshva Chandran, R., Nazeem, P.A., Girija, D., John, P.S. and Peter, K.V. (eds.) *Recent Trends in Horticultural Biotechnology*. New India Pub. Agency, New Delhi. pp. 53-58

Bagyaraj, D.J. 2011. Microbial biotechnology for sustainable agriculture, horticulture and forestry. New India Pub. Agency, New Delhi, India.

Bagyaraj, D.J. and Menge, J.A. 1978. Interaction between VA mycorrhiza and *Azotobacter* and their effect on rhizosphere microflora and plant growth. *New Phytol.* 80:567-573

Bagyaraj, D.J. and Mallesh, B.C. 2000. Potentials for the use of VA mycorrhiza in horticulture. In: Aneja, K.R., Charaya, M.D., Aggarwal, A., Hans, D.K. and Khan, S.A. (eds.) *Glimpse in Plant Sciences.* Pragathi Prakashan, Meerut, India. pp. 37-41

Bagyaraj, D.J. and Kehri, H.K. 2012. AM fungi: Importance, nursery inoculation and performance after out planting. In: Bagyaraj, D.J., Tilak, K.V.B.R. and Kehri, H.K. (eds.) *Microbial diversity and functions.* New India Pub. Agency, New Delhi. pp. 641-667

Bagyaraj, D. J., Manjunath, A. and Patil, R. B. 1979. Occurrence of vesicular arbuscular infection in some tropical aquatic plants. *Trans. Br. Mycol. Soc.* 72:164-177

Bagyaraj, D.J., Manjunath, A and Govinda Roa, Y.S. 1988. Mycorrhizal inoculation effect on marigold, eggplant and citrus in an Indian soil. *J. Soil Biol. Ecol.* 8: 98-103

Balakrishna Reddy and Bagyaraj, D.J. 1994. Selection of efficient vesicular arbuscular mycorrhizal fungi for inoculating the mango root stock cultivar 'Nekkare'. *Scientia Hort.* 59: 39-73

Barrows, J.B. and Rancadori, R.W. 1977. Endomycorrhizal synthesis in *Gigaspora margarita* in poinsettia. *Mycologia.* 69:1173-1184

Besmer, Y.L. and Koide, R.T. 1999. Effect of mycorrhizal colonization and phosphorus on ethylene production by snapdragon (*Antirrhinum majus* L.) flowers. *Mycorrhiza.* 9:161-166

Boby, V.U. and Bagyaraj, D.J. 2003. Biological control of root-rot of *Coleus forskohlii* Briq. using microbial inoculants. *World J. Microbiol. Biotechnol.* 19:175-180

Borowicz, V. A. 2001. Do arbuscular mycorrhizal fungi alter plant-pathogen relations? *Ecology.* 82: 3057-3068

Bressan, W., Decarvalho, C.H.S. and Sylvia, D.M. 2000. Inoculation of somatic embryos of sweetpotato with an arbuscular mycorrhizal fungus improves embryo survival and plantlet formation. *Can. J. Microbiol.* 46:741-743

Calderon, L.D., Gomez, L., Blanco, F. and Uribe, L. 2001. Inoculation with *Glomus manihotis* on growth and development of *in vitro* propagated cassava plants during acclimatization. *Agronomica-Costarricense.* 24:25-29

Caroline Machado and Bagyaraj, D.J. 1995. Mycorrhization bacteria and its influence on growth of cowpea. In: Adholeya, A. and Sujan Singh (eds.) *Mycorrhizae, Biofertilizers for the Future.* Tata Energy Research Institute Pub. New Delhi. pp. 192-196

Chundawat, B.S and Singh, R. 2001. Integrated nutrient management in tropical and sub tropical fruits. *Indian J. Hort.* 58:59-69

Collazos, S.H.H., Figueroa, B.P., Jordan, S.A., Patino, C.H. and Sieverding, E. 1986. Estudio preliminar sorbe micorriza vesiculo-arbuscular (MVA) en lulo (*Solanum quitoense* Lam.). *Acta Agronomica.* 36: 34-36

Crews, C.L., Johnson, C.R and Joiner, J.N. 1978. Benefits of growth and development of three woody ornamentals. *Hort. Sci.*13:429-430.

Davies, F.T. Jr., Grossi, J.A.S, Carpio, L. and Estrada-Luna, A.A. 2000. Coloniaztion and growth effects of the mycorrhizal fungus *Glomus intraradices* in a commercial nursery container production system. *J. Environ. Hort.* 18:247-251

Dodd, J.C. and Jeffries, P. 1986. Early development of vesicular arbuscular mycorrhizas in autumn-sown cereals. *Soil. Biol. Biochem.* 18: 149-154

Frank, A. B. 1885. Uber die out Wurzetsymbiose berohende ernahausing gewisser Baume durch utnerimdinche. *Berdent Bot. Gessel.* 3: 128-145

Gerdemann, J.W. 1975. Vesicular arbuscular mycorrhizae. In: Torrey, J.C and Clarkson D.T. (eds.) *The Development and Function of Roots.* Academic Press, New York, USA. pp.575-591

Gianinazzi, S., Gianinazzi-Pearson, V. and Trouvelot, A. 1990. Potentialities and procedures for the use of endomycorrhizas with emphasis on high value crops. In: Whips, J.M. and Lumsden, B. (eds.) *Biotechnology of Fungi for Improving Plant Growth.* Cambridge University Press, Cambridge, U.K. pp.41-54

Giovannetti, M and Mosse, B. 1990. An elevation of techniques for measuring vesicular arbuscular mycorrhizal infection in roots. *New Phytol.* 84: 489-500

Graham, J.H. and Syversten, J.P. 1985. Host determinants of mycorrhizal dependency of citrus root stock seedlings. *New Phytol.* 101: 667-676

Gupta, R.P., Kalia, A. and Kapoor, S. 2007. Bioinoculants: a step towards sustainable agriculture. New India Pub. Agency, New Delhi, India.

Harley, J. L. 1991. Ectomycorrhiza: the state of art. *Methods in Microbiol.* 24: 1-24

Hayman, D.S. 1983. The physiology of vesicular arbuscular endomycorrhizal symbiosis. *Can. J. Bot.* 61: 944-963

Hemalatha, P., Velmurugan, M., Harisudan, C. and Davamani, V. 2010. Importance of mycorrhizae for horticultural crops. In: Thangadurai, D., Busso, C.A. and Hijri, M. (eds.) *Mycorrhizal Biotechnology.* Capital Pub. Co., New Delhi, India. pp. 128-139

Hernandez-Zardon, A.F. and Bertolil-Herrera, M.P. 2001. Agronomic management of substrates, biofertilizers and sowing methods in the sustainable production of seed potato tubers. *Proc. Inter-American Soc. Trop. Hort.* 43:42-49

Janos, D.P. 1987. VA mycorrhizas in humid tropical ecosystems. In: Safir, G. R. (ed.) *Eco physiology of VA mycorrhizal plants.* CRC Press, Boca Raton, FL., USA, pp. 107-134

Jarstfer, A.G. and Sylvia, D.M. 1992. Inoculum production and inoculation strategies for vesicular arbuscular mycorrhizal fungi. In: Metting, B.(ed.) *Soil Microbial Technologies: Application in Agriculture, Forestry and Environmental Management.* Marcel Decker, New York, U.S.A. pp. 349-377

Kabir, Z. and Koide, R.T. 2000. The effect of dandelion or a cover crop on mycorrhiza inoculum potential, soil aggregation and yield of maize. *Agric. Ecosyst. Environ.* 78:167-174

Kormanik, P.P., Bryan, W.C. and Schultz. R.C. 1980. Procedure and equipment for staining large number of plant root samples for endomycorrhizal assay. *Cand. J. Microbiol.* 26: 536-538

Koske, R.E. and Gemma, J.N. 1989. A modified procedure for staining roots to detect VA mycorrhiza. *Mycol. Res.* 92: 486-505

Lakshmipathy, R., Balakrishna, A.N., Bagyaraj, D.J. and Kumar, D.P. 2000. Symbiotic response of cashew root stocks to different VA mycorrhizal fungi. *The Cashew.*14: 20-24

Lakshmipathy, R., Balakrishna, A.N., Bagyaraj, D.J., Sumana, D.A. and Kumar, D.P. 2004. Evaluation, grafting success and field establishment of cashew rootstock as influenced by VAM fungi. *Indian J. Exp. Biol.* 42: 1132-1135

Lambert, D.H., Stouffer, R.F. and Cole, J.R. 1979. Stunting of peach seedlings following soil fumigation. *J. Am. Soc. Hort. Sci.* 104:433-435

Lovato, P., Guillemin, J.P. and Gianinazzi, S. 1992. Application of commercial arbuscular endomycorrhizal fungi inoculants to the establishment of micropropagated grapevine rootstock and pineapple plants. *Agron. J.* 12: 873-880

Mallesha, B.C. and Bagyaraj, D.J. 1997. Field response of tomato and capsicum to inoculation with arbuscular mycorrhizal fungi. *J. Soil Biol. Ecol.* 17:34-38

Mamatha, G. and Bagyaraj, D.J. 2000a. Effect of different methods of VAM inoculum application on plant growth and nutrient uptake of papaya raised in polybags. *J. Soil. Biol. Ecol.* 20: 51-59

Mamatha, G. and Bagyaraj, D.J. 2000b. Effect of different methods of VAM inoculum application on plant growth and nutrient uptake of tomato seedlings grown in raised nursery beds. *J. Soil. Biol. Ecol.* 20: 71-77

Mamatha, K.B. and Bagyaraj, D.J. 2001. Effect of soil inoculation with *Glomus mosseae* at different P levels on flowering response of chrysanthemum. *Proc. Nat. Acad. Sci.* India. 71(B):157-163

Manjunath, A., Mohan, R. and Bagyaraj, D.J. 1981. Interaction between *Beijerinckia mobilis, Aspergillus niger* and *Glomus fasciculatus* and their effects on growth of onion. *New Phytol.* 87:723-727

Manjunath, V.G., Patil, C.P, Swamy, G.S.K. and Patil, P.B. 2001. Effect of different VAM fungi on growth parameters of papaya cv. Sunset Solo. *J. Maharashtra Agric. Univ.* 26:269-271

Maronek, D.J., Hendrix, J.W and Kiernan, J. 1980. Different growth response to the mycorrhizal growth fungus *Glomus fasciculatus* of Southern magnolia and Bar Harbor Juniper grown in containers in composted hardwood bark shale. *J. Am. Soc. Hort. Sci.* 105:206-208

Mathews, D., Hegde, R.V. and Sreenivasa, M.N. 2003. Influence of arbuscular mycorrhizae on the vigor and growth of micropropagated banana plantlets during acclimatization. *Karnataka J. Agric. Sci.* 16:438-442

Meharg, A.A. and Cairney, J.W.G. 2000. Co-evaluation of mycorrhizal symbionts and their hosts to metal-contaminated environments. *Adv. Ecol. Res.* 30: 69-112

Menge, J.A., Larue, J., Labanauskas, C.K. and Johnson, E. 1980. The effect of two mycorrhizal fungi upon growth and nutrition of avocado seedlings grown with six fertilizer treatments. *J. Amer. Soc. Hort. Sci.* 105: 400-404

Miller, R.M. and Jastrow, J.D. 1992. The application of VA mycorrhizae to ecosystem restoration. In: Allen, M.F. (ed.). *Mycorrhizal Functioning.* Chapman and Hall Ltd., London, England. pp. 438-467

Morton, J.B. 1990. Evolutionary relationships among arbuscular mycorrhizal fungi in the Endogonaceae. *Mycologia.* 82:192-207

Mosse, B. 1977. Plant growth responses of vesicular-arbuscular mycorrhiza-IV. Soil given additional phosphorus. *New Phytol.* 72: 127-136

Mosse, B. 1981. *Vesicular Arbuscular Mycorrhizal Research for Tropical Agriculture.* Honolulu University of Haiwaii Press, HI, U.S.A. 54p

Nadkarni, N.M. 1985. Roots that go out on a limb. *Natural History.* 94:42 – 48

Palazzo, D., Pommerening, B. and Vanadia, S. 1994. Effects of soil sterilization and vesicular arbuscular mycorrhiza on growth of sour orange (*Citrus aurantium* L.) seedlings. *Proc. Int. Soc. Citriculture.* 2: 621-623

Plenchette, C., Furlan, V. and Fortin, J.A. 1981. Growth stimulation of apple trees in unsterilized soil under field conditions with mycorrhizal inoculation. *Can. J. Bot.* 59: 2003-2008

Plenchette, C., Fortin, J.A. and Furlan, V. 1983. Growth responses of several plant species to mycorrhiza in a soil of moderate P fertility. *Plant Soil.* 70: 199-209

Potty, V.P. 2004. Arbuscular mycorrhizal fungi - a biofertilizer. Abstract of National Seminar on Horticulture Based Sustainable Income and Environmental Protection, Feb. 24-26, Kohima, Nagaland, India. p. 12

Powell, C.L., Bagyaraj, D.J., Clark, G.E and Cladwell, K.I. 1985. Inoculation with arbuscular mycorrhizal fungi in greenhouse production of asparagus seedlings. *N.Z. J. Agric. Res.* 28: 293-298

Prabhu, T., Narwadkar, P.R., Sajindranath, A.K. and Rathod, N.G. 2002a. Effect of integrated nutrient management on growth and yield of coriander (*Coriandrum sativum* L.). *South Indian Hort.* 50:680-684

Prabhu, T., Syed-Ismail, Sajindranath, A.K. Savithiri, R. and Ismail, S. 2002b. Effect of integrated nutrient management on yield, dry matter and nutrient content in okra. *Orissa J. Hort.* 30:52-56

Raj, J., Bagyaraj, D.J and Manjunath, A. 1981. Influence of soil inoculation with arbuscular mycorrhiza and a phosphate dissolving bacterium on plant growth and ^{32}P-uptake. *Soil Biol. Biochem.* 13:105-108

Rathod, N.G., Narwadkar, P.R., Sajeendranath, A.K. and Prabhu, T. 2002. Effect of integrated nutrient management on growth and yield of gaillardia. *J. Maharashtra Agric. Univ.* 27:318-319

Ravolanirina, F., Gianinazzi, S., Trouvlet, A. and Carre, M. 1989. Production of endomycorrhizal explants of micropropagated grapevine rootstocks. *Agric. Ecosyst. Environ.* 29:323-327

Rea, E. and Tullio, M. 2005. The agro-industrial production valorization: The possible role of arbuscular mycorrhizal fungi. In: Ray. R.C.(ed.) *Microbial Biotechnology in Agriculture and Aquaculture.* Science Publishers, Inc., Enfield, NH, USA. pp. 101-124

Reddy, B., Bagyaraj, D.J. and Mallesha, B.C. 1996. Selection of efficient VA mycorrhizal fungi for acid lime. *Indian J. Microbiol.* 36:13-16

Redhead, J.F. 1977. Endotrophic mycorrhizas in Nigeria: Species of the Endogonaceae and their distribution. *Trans. Br. Mycol. Soc.* 69:275-280

Saggin Junior, O.S. and da Silva, E.M.R. 2006. Production of seedlings inoculated with arbuscular mycorrhizal fungi and their performance after outplanting. In: Rai, M.K. (ed.) *Handbook of Microbial Biofertilizers.* International Book Distributing Co., Lucknow, India. pp. 354-386

Schüßler, A. and Walker, C. 2010. http:// www. lrz.de/~schuessler/amphylo/.

Sharma, M.P. and Adholeya, A. 2000. Enhanced growth and productivity following inoculation with indigenous AM fungi in four varieties of onion (*Allium cepa* L.) in an alfisol. *Biol. Agri. Hort.* 18:1-14

Shashikala, B.N., Reddy, B.J.D. and Bagyaraj, D.J. 1999a. Influence of *Glomus mosseae* on the growth of micropropagated *Syngonium* and *Spathiphyllum* at varied levels of fertilizer phosphorus. *Crop. Res.* 18:454-460

Shashikala, B.N., Reddy, B.J.D. and Bagyaraj, D.J.1999b. Response of micropropagated banana plantlets to *Glomus mosseae* at various levels of fertilizer phosphorus. *Indian. J. Exptl. Biol.* 37:499-502

Sieverding, E. 1991. Vesicular-arbuscular mycorrhiza management in tropical agricultural systems. TZ-Verlagsgesellscaft Rossdorf, Germany.

Simon, L., Lalonde, M. and Bruns, T. D. 1992. Specific amplification of 18S fungal ribosomal genes from vesicular-arbucular endomycorrhizal fungi colonizing roots. *Appl. Environ. Microbiol.* 58:291-295

Sreeramula, K.R. and Bagyaraj, D.J. 1986. Field response of chilly to VA mycorrhiza on black clayey soil. *Plant Soil.* 93:299-302

Stubblefield, S. P., Taylor, T. N. and Trappe, J. M. 1987. Fossil mycorrhizas: a case for symbiosis. *Science.* 237:59-60

Sylvia D.M. and Chellemi, D.O. 2001. Interactions among root-inhabiting fungi and their implications for biological control of root pathogens. *Adv. Agron.* 73:1-33

TAC Secretariat. 1989. *Sustainable Agricultural Production: Implication for International Agricultural Research,* Consultative Group on International Agricultural Research, Washington, D. C.

Tilak, K.V.B.R., Pal, K.K. and Dey, R. 2010. Microbes for sustainable agriculture. I. K. International Publishing House Pvt. Ltd., New Delhi.

Timmer, L.W. and Leyden, R.F. 1978. Stunting citrus seedlings in Texas and its correction by phosphorus fertilization and inoculation with mycorrhizal fungi. *J. Am. Soc. Hort. Sci.* 103:533-537

Tisserant, B., Gianinazzi-Pearson, V., Gianinazzi, S. and Gollute, A. 1992. Plant histochemical staining of fungal alkaline phosphatase activity for analysis of efficient endomycorrhizal infections. *Mycol. Res.* 97:245-250

Tohamy, E.W., Schnitzler, W.H., Behairy, E.U. and Beltagy, E. 1999. MS effect of VA mycorrhiza on improving drought and chilling tolerance of bean plants. (*Phaseolus vulgaris* L.). *Angewandte Botanik.* 73:178-183

Trappe, J.M. 1977. Biogeography of hypogeous fungi: Trees, mammals and continental drift. In: Bigelow, H.E. and Simmons, E.G. (eds.) Abstr. 2nd Int. Mycol. Congr. University of South Florida, Tampa, pp. 675

Trouvelet, A., Kough, J.L. and Gianinazzi-Pearson, V. 1986. Measure du taux de mycorrhization VA d'un systeme radiculaire. Recherche de methods ayant une significant functionelle. In: Gianinazzi-Pearson, V. and Gianinazzi, S. (eds.) *Physiological and Genetic Aspects of Mycorrhizae.* Proceedings 1st Eur. Symposium on Mycorrhizae. Instiut National de la Researche Agronomique. Paris, France. pp. 217-221

Varma, A. and Scheupp, H. 1995. Mycorrhization of micropropagated plantlets. In: Adholeya, A. and Singh, S. (eds.) *Mycorrhizae: biofertilizers for the future.* Tata Energy Research Institute Pub., New Delhi, India. pp. 322-328

Vidal, M.T., Azcon-Aguilar, C.K., Barea, J.M. and Pliego-Alfaro, F. 1992. Mycorrhizal inoculation enhances growth and development of micropropagated plants of avacado. *Hort. Sci.* 27:785-787

Vinayak, K. and Bagyaraj, D.J. 1990. Selection of efficient VA mycorrhizal fungi for trifoliate orange. *Biol. Agri. Hort.* 6:305-311

Wamocho, L., Michieka, R., Yamashita, K., Ishii, T. and Byrne, D. 2001. Charcoal application and VAM inoculation promote vine growth, yield and quality of passion fruits in Kenya. In: Perez-Gonzalez, S., Dennis, F. Jr., Mondragon, C., Queretarom and Michoacan, P. (eds.) *Proceedings of 6th International Symposium on Temperate Fruit Growing in the Tropics and Subtropics*, June 12-16, 2000, Mexico, pp. 115-119

Williams, S. E. and Aldon, E. F. 1976. Endomycorrhizal (vesicular- arbuscular) associations of some arid zone shrubs. *South West Nature.* 20:437-444

Yang, K.J., Chang, S.N, Yang, K.J. and Chang, S.N.D. 2001. Influence of arbuscular mycorrhiza on the growth of eggplant seedling in plug system and their growth response in the field. *J. Chinese Soc. Hort. Sci.* 47:175-184

Yao, M.K., Tweddell, R.J. and Desilets, H. 2002. Effect of two vesicular-arbuscular mycorrhizal fungi on the growth of micropropagated potato plantlets and on the extent of disease caused by *Rhizoctonia solani. Mycorrhiza.* 12:235-242

Yao, Q. and Li, D. 1999. Influence of VA mycorrhiza on juice components and rind colour of citrus fruit. *J. Fruit Sci.* Guoshu Kexue, China. 16:38-42

Yun, L.G., Lin, C.G., Hui, Q.G., Lu, G.Y., Chen, G.L. and Qi, G.H. 2002. Study on substrate with mycorrhiza for cucumber (*Cucumis sativas*) seedling. *China Vegatables.* 5:9-12

8

Carbohydrate Metabolism During Fruit Spoilage

Neelima Garg and V.K.Singh

Consumption of fruit and vegetables has dramatically increased globally during the past a few decades. These are living perishable biological entities which perform a number of metabolic functions and have a short life span. Once the fruit is harvested, respiration and transpiration continue, but only for as long as the fruit can draw on its own food reserves and moisture. It is this limited ability to continue vital metabolic functions which define fruit as perishable. It is also estimated that about 20% of all fruits and vegetables produced are lost each year due to spoilage. The process of the fruit spoilage is just the decomposition of the fruit itself. Once harvested, this begins to deteriorate until eventually it becomes unfit for consumption. This deterioration is known as decay and that leads to spoilage. Latter may be defined as "any sensory change (physical, visual, olfactory or flavour) which the consumer considers to be unacceptable. The gradual changes which cause deterioration and decay in fruits are due to certain organisms and chemicals present in the food and outside the food. Carbohydrates play a major role in the fruit spoilage process (Prasanna *et al.*, 2007) by way of depolymerization leading to decreased molecular size with concomitant increase in the levels of ripening inducing specific enzymes, whose target differ from fruit to fruit. The major classes of cell wall polysaccharides that undergo modifications during ripening are pectins, cellulose, starch, and hemicelluloses (Hallet *et al.,* 2005). The focus of this chapter is to provide information about carbohydrate metabolism during fruit spoilage.

Before understanding the carbohydrate metabolism, it is important to know about fruit cell wall structure. The cell wall is a complex structure, composed of polysaccharides, imposing cell shape and rigidity. These are carbohydrate polymers consisting of tens to hundreds to several thousand monosaccharide units (Hallett *et al.,* 2005) . All of the common polysaccharides contain glucose as the monosaccharide unit. Primary cell walls of plants are considered a network of cellulose microfibrils embedded in a matrix of hemicellulose and pectin (Rose and Bennett 1999; Cosgrove 2001).The cell walls of fruits do not have secondary walls and contain very low amount of lignins. Instead, they contain a high level of pectins, which can be modified more easily at different stages of plant development (Rose *et al.,* 2003; Harris, 2005), therefore allowing a fine-tuning of cell wall properties during fruit maturation.

Cellulose

The major component in the rigid cell walls in fruits is cellulose(Updegraff,1969). Cellulose is a linear polysaccharide polymer with many glucose monosaccharide units with beta-1,4 linkage between glucose molecules (Fig 1).

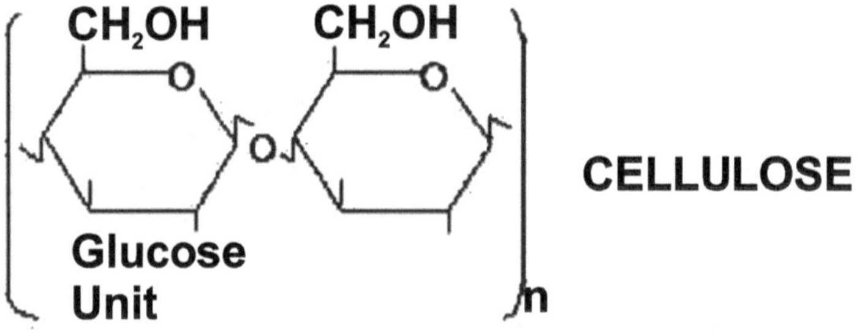

Fig. 1 : Cellulose structure

Cellulose fibers usually consist of over 500,000 cellulose molecules. It is the H-bonding that is the basis of the high tensile strength of cellulose(Fig 2).

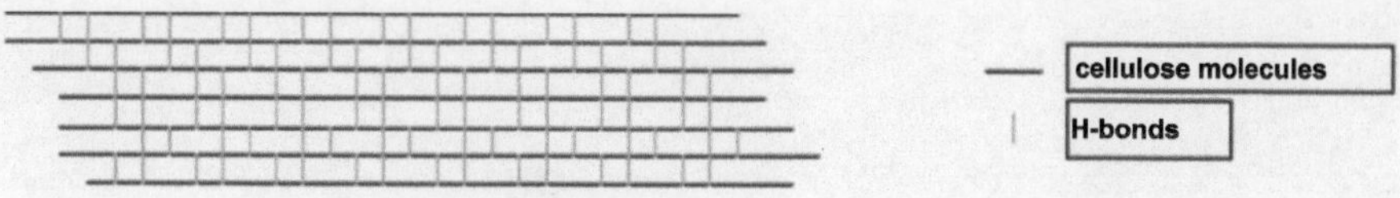

Fig. 2 : Arrangement of Cellulose molecules in fibre

Pectin

Pectins are the common and major components of primary cell wall and middle lamella, contributing to the texture and quality of fruits. Their degradation during ripening seems to be responsible for tissue softening of a number of fruits (Barras *et al.*, 1994). Structurally, pectins are a diverse group of heteropolysaccharides containing partially methylated D-galacturonic acid residues with side chain appendages of several neutral polysaccharides. The degree of polymerization/ esterification and the proportion of neutral sugar residues/side chains are the principal factors contributing to their (micro-) heterogeneity.

Pectin substances are the negatively charged and acidic complex polysaccharide in which the primary chain is composed with α-(1,4) linked molecules of α-D-galacturonic acid(Fig 3). The side chains also contain molecules of L-rhamnose, arabinose, galactose and xylose that are connected to the main chain through their first and the second carbon atom. The carboxyl groups of galacturonic acid are partially esterified by methyl groups and partially or completely neutralized by calcium, potassium,magnesium, iron, ammonium or other ions. Some of the hydroxyl groups on the second and the third carbon atom can be acetylated.

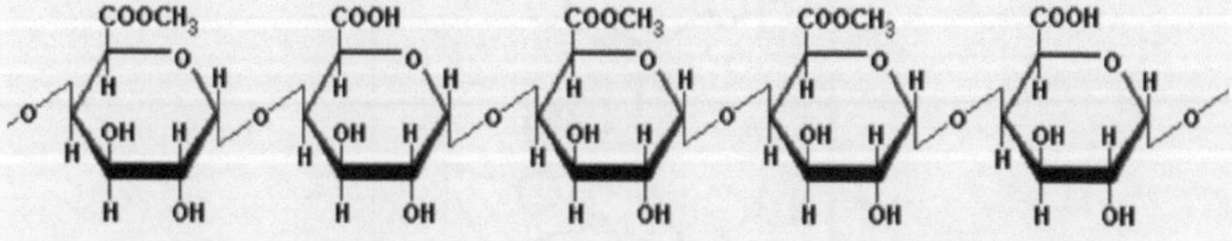

Fig. 3 : Structural formula of the primary chain of pectin- the polygalacturonic acid

Pectic substances may be divided into four types:

Protopectin: It is the water insoluble pectic substance present in intact tissue. Protopectin on restricted hydrolysis yields pectin or pectic acids. The structure of protopectin is shown in Fig. 4.

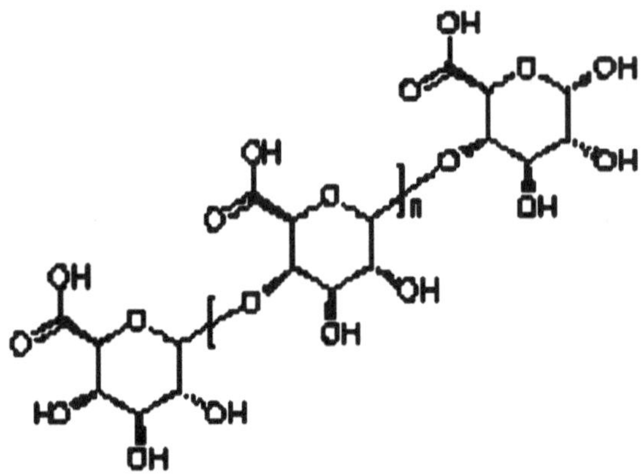

Protopectin structure

Fig.4 : Protopectin structure

Pectic Acid

It is a major component of middle lamella but also found in primary walls is the soluble polymer of galacturonans that contains negligible amount of methoxyl groups (Fig 5). It forms salts and salt bridges with Ca++ and Mg++ that are insoluble gels.

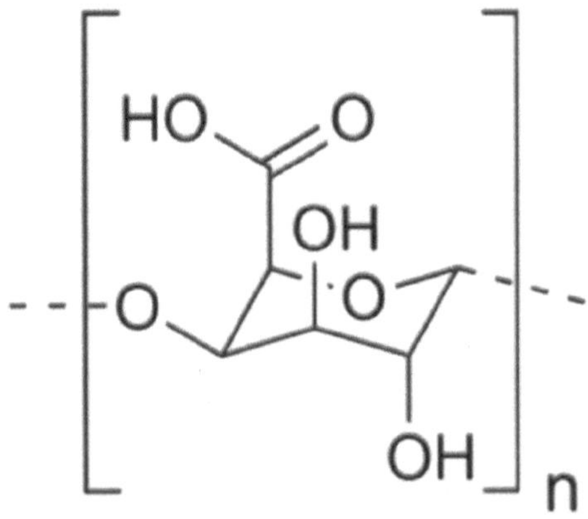

Fig 5 : Pectic acid structure

Pectinic acids: is the polygalacturonan chain that contains >0 and

<75% methylated galacturonate units. Normal or acid salts of pectinic acid are referred to as pectinates. The structure of protopectin is shown in Fig 6.

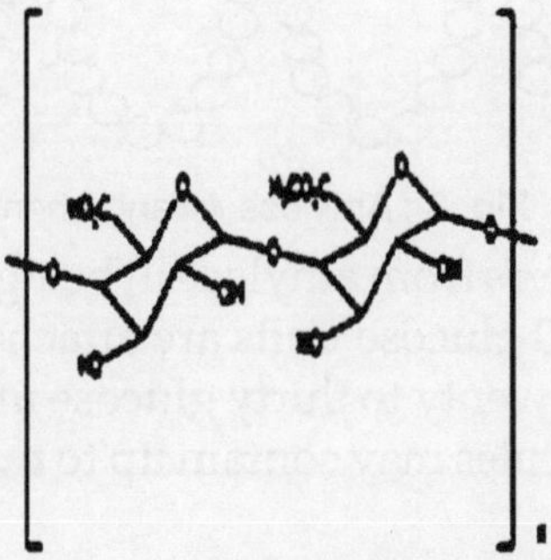

Fig.6 : Pectinic acid structure

Starch

Plants store glucose as the polysaccharide starch. However, instead of a beta-1,4 linkage between glucose molecules, starch uses an alpha-1,4 linkage(Fig. 7). The difference is due to the conformation of the ring structure. The alpha-1,4 linkage causes the polymer to take on a twisted configuration instead of the linear shape of cellulose. Thus, starch forms globular structures. Starch molecules are often branched, which also prevents from forming linear arrays (Kojim,1994). Natural starches are mixtures of amylose (10-20%) and amylopectin (80-90%). Amylose forms a colloidal dispersion in hot water whereas amylopectin is completely insoluble.The structure of amylose consists of long polymer chains of glucose units connected by an alpha acetal linkage.

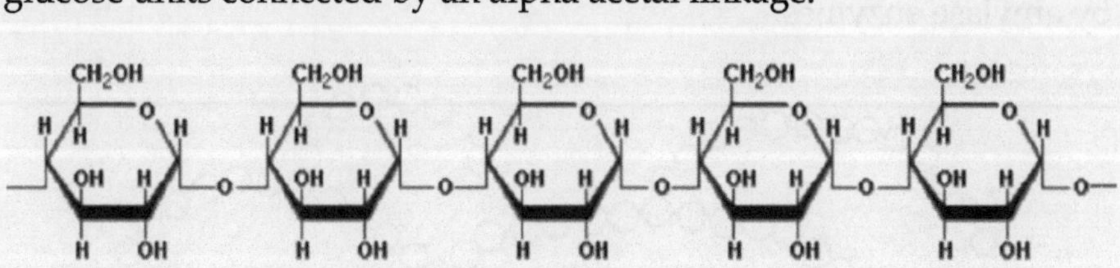

Fig. 7 : Amylose structure

The amylose chain is floppy, and it tends to coil up into a helix (Fig 8).

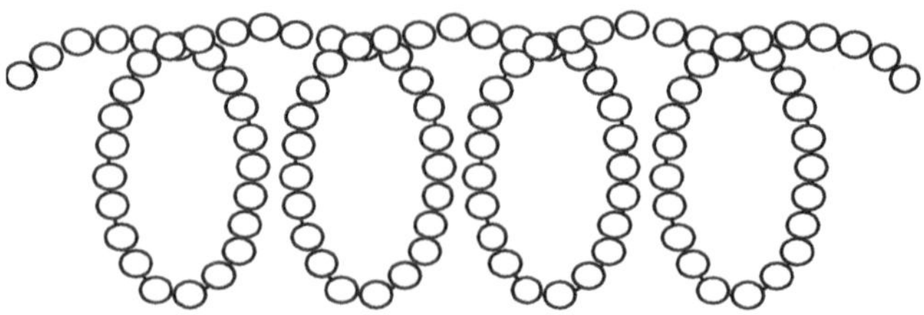

Fig. 8 : Amylose Arrangement

Amylopectin differs from amylose in being highly branched. Short side chains of about 30 glucose units are attached with 1α →6 linkages approximately every twenty to thirty glucose units along the chain (Fig 9). Amylopectin molecules may contain up to two million glucose units.

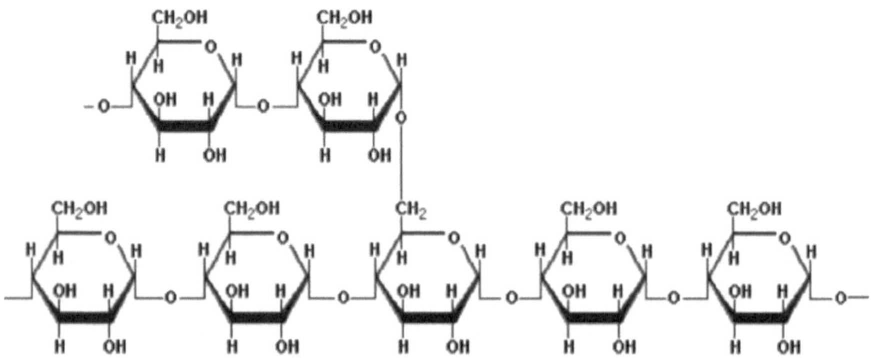

Fig. 9 : Amylopectin structure

This gives it a more open molecular structure than amylase (Fig. 10). Because it has more ends, it can be broken more quickly than amylose by amylase enzymes.

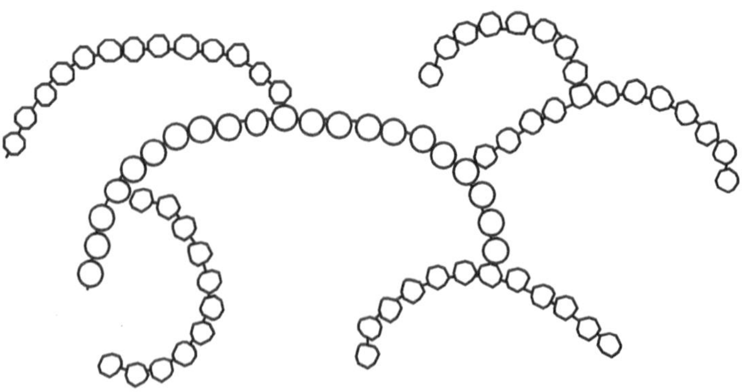

Fig. 10 : Amylopectin arrangement

Hemicellulose is a collective term for various components of cell walls in plants (glucans, galactans, mannans, pentosans, xylans). These are polysaccharides – long chains of sugars viz. glucans, galactans, mannans, pentosans, xylans (with the exception of cellulose) – that are part of the plant cell wall. Hemicellulose is a cross-linking agent and interconnects cellulose microfibrils. It can also link the cellulose to other cell wall components.

Causes of fruit spoilage

Fruit spoilage is caused by two main factors, namely;

- Natural decay
- Microbial contamination

A. Natural Decay in Food

This comes about as a result of moisture loss and the action of enzymes.

I. Moisture loss

Fruits and vegetables generally contain large amount of water. After harvesting these continue to respire and lose water through their leaves and skin. As a result these shrinks in size, becomes limp and the skin becomes wrinkled and leathery.

II. Enzyme action in the food

Most fruits first become ripened because of the release of a gaseous hormone (ethylene, $H_2C{=}CH_2$) within the fruit and the plant . Ethylene causes the breakdown which in itself causes a production of enzymes which break down the structures of the fruit (eg.: amylase, pectinases present inside the cell itself). Enzymes are mainly composed of protein which speed up chemical changes that result in loss of flavour, colour and texture. They are active at room temperature. As, they are sensitive to heat, their activity retard or slow down at low temperature. The activity of these enzymes stops when they are heated above 70 °C. Once activated, enzymes speed up the process of decay by breaking down the tissues and components of the fruits in various ways. Examples involving endogenous enzymes include a) the post-harvest senescence and spoilage of fruit and vegetables; b) oxidation of phenolic substances in plant tissues by phenolase (leading to browning); c) sugar - starch conversion in plant tissues by amylases; d) post-harvest demethylation of pectic substances in plant tissues (leading to softening of plant tissues during ripening, and firming of plant tissues during processing).

B. Microbial Spoilage of Fruits

The activity of enzymes in food makes it easier for micro-organisms to enter inside. Fruits and vegetables present nearly ideal conditions for survival and growth of many types of microorganisms. The internal tissues are nutrient rich and the structure is comprised mainly of the polysaccharides cellulose, hemicellulose, and pectin. The principal storage polymer is starch. Spoilage microorganisms exploit the host using extracellular lytic enzymes that degrade these polymers to release water and the plant's other intracellular constituents for use as nutrients for their growth.

Microbiological spoilage defects of fresh fruits and vegetables include microbial colony formation or visible microbial growth mainly due to microorganism proliferation, off-aroma and off-flavor formation mainly due to fermentation of sugar, soft-rot/water soak and sliminess due to enzymatic pectolyzation, and discoloration. Most microorganisms that are initially observed on whole fruit or vegetable surfaces are soil inhabitants, members of a very large and diverse community of microbes that collectively are responsible for maintaining a dynamic ecological balance within most agricultural systems.

Vectors for disseminating these microbes include soil particles, airborne spores, and irrigation water (Janisiewicz and Korsten, 2002, Andrews and Harris, 2000). Spoilage microorganisms can also be introduced to the crop on the seed itself, during crop growth in the field, during harvesting and postharvest handling, or during storage and distribution. Those same types of soil-borne spoilage microbes that occur on produce are the same spoilage microorganisms that are present on harvesting equipment, on handling equipment in the packinghouse, in the storage facility, and on food contact surfaces throughout the distribution chain.

The main micro-organisms responsible for the fruit spoilage are moulds, yeasts and bacteria.

Moulds

Moulds are able to grow on all kinds of food: cereals, meat, milk. fruit, vegetables. nuts, fats and products of these. Because of the acidic and sugar rich composition, fruits are suitable substrate for moulds to grow. These develop as a multicellular structure visible to the naked eye. Most molds grow at a pH range of 3 to 8 and some can grow at very low water activity levels (0.7–0.8) on dried fruit. The fungal spores present in the air, settle and multiply on fruit surface under high humidity conditions, at mesophilic temperatures. Spores can tolerate harsh environmental conditions but most are sensitive to heat treatment. An exception is Byssochlammys, whose spores have a D value of 1–12 minutes at 90°C. Different mold species have different optimal growth temperatures, with some able to grow in refrigerators. They have a diverse secondary metabolism producing a number of toxic and carcinogenic mycotoxins (Pitt and Hocking,1977). Some spoilage molds are toxigenic while others are not. Spoilage molds can be categorized into four main groups:

a. Zygomycetes are considered relatively primitive fungi but are widespread in nature, growing rapidly on simple carbon sources in soil and plant debris, and their spores are commonly present in indoor air. Generally they require high water activities for growth and are notorious for causing rots in a variety of stored fruits and vegetables, including strawberries and sweet potatoes. The most common spoilage species are *mucor* and *rhizopus*.

b. *Penicillium* and related genera are present in soils and plant debris and are distinguished by their reproductive structures that produce chains of conidia. Many species are important spoilage organisms, and some produce potent mycotoxins (patulin, ochratoxin, citreoviridin, penitrem). *Penicillium spp.* cause visible rots on citrus, pear, and apple fruits and cause enormous losses in these crops. They also spoil other fruits and vegetables.A related genus, Byssochlamys, is the most important organism causing spoilage of pasteurized juices because of the high heat resistance of its spores.

c. *Aspergillus* and related molds generally grow faster and are more resistant to high temperatures and low water activity than Penicillium spp. and tend to dominate spoilage in warmer climates. Many aspergilli produce mycotoxins such as aflatoxins, ochratoxin, territrems, cyclopiazonic acid. Aspergilli spoil a wide variety of fruits and vegetables.

d. Other molds, belonging to several genera, have been isolated from spoiled fruits and vegetables. These generally are not major causes of spoilage but can be a problem for some foods.

The mould growth may result in several kinds of food-spoilage: off-flavours. Toxins, discolouration, rotting and formation of pathogenic or allergenic propagules . The deterioration of sensorial properties is often due to the production of an extensive set of carbohydrate-active enzymes specifically dedicated to degrade plant polysaccharides during growth (Brink and Vries ,2011; Bigelis, 1992). However, these sets differ between fungal species. For instance, *Trichoderma reesei* has a highly efficient set of enzymes involved in cellulose degradation (Martinez *et al.*, 2008; Kubicek *et al.*, 2011), while *Aspergillus* species produce many enzymes to degrade pectin (Martens and Schaap 2009).

Once inside the food these enzymes may continue their activities independent of destruction or removal of the mycelium. The enzymatic activities may give rise to different types of off flavours. Some times mycotoxins are formed during growth of moulds on fruits (Swanson,1987). Some mycotoxins are only present in the mould, while most of them are excreted in the foods. In fruits like peaches, pears and tomatoes the diffusion of mycotoxins can be very fast, leaving no part of the product uncontaminated (Reddy *et al.*, 2010). In solid fruits like apples the diffusion is slow leaving the major part of the product uncontaminated.

Yeasts

Yeasts are microscopic fungi, found in the air and soil, and on fruit surface. Some are able to tolerate fairly high acidic, salt and sugar concentrations and can grow without the presence of oxygen. They can cause spoilage in syrups, fruits, fruits juices and jam especially as they can survive without air (Tournas and Katsoudas,2005). Yeast multiply by budding or fission under conditions similar to those required by other fungi. They need oxygen, warmth, food and moisture in order to grow successfully. However, they may grow under oxygen stress conditions also. Yeast grow best at temperatures between 25°C and 30°C. Extreme heat destroys all yeasts and most are destroyed at temperatures above 60°C.

There are five main groups of spoilage yeasts:

a. Zygosaccharomyces and related genera tolerate high sugar and high salt concentrations and are the usual spoilage organisms in foods such as honey, dried fruit, jams and soy sauce (Martorell *et al.*, 2007). They usually grow slowly, producing off-odors and flavors and carbon dioxide that may cause food containers to swell and burst.

b. *Debaryomyces hansenii* can grow at salt concentrations as high as 24%, accounting for its frequent isolation from salt brines used for cured meats, cheeses, and olives. This group also includes the most important spoilage organisms in salad dressings (Martorell ,2005).

c. *Saccharomyces* spp. are best known for their role in production of bread and wine but some strains also spoil wines and other alcoholic beverages by producing gassiness, turbidity and off-flavors associated with hydrogen sulfide and acetic acid. Some species grow on fruits, including yogurt containing fruit, and some are resistant to heat processing (Loureiro and Ferreira , 2003.)

d. Candida and related genera are a heterogeneous group of yeasts, some of which also cause human infections. They are involved in spoilage of fruits, some vegetables and dairy products (Casey and Dobson, 2003).

e. Dekkera/Brettanomyces are principally involved in spoilage of fermented foods, including alcoholic beverages and some dairy products. They can produce volatile phenolic compounds responsible for off-flavors (Phister and Mills, 2003).

Bacteria

Bacteria are the most widespread of the micro-organisms found in food. They are minute single cells of various shapes. Under ideal conditions, they divide into two every 20 minutes, consequently, millions of them may develop in contaminated food in a short time. They are more dangerous than moulds and yeast because food may be severely infected but not smell, taste or look bad. As all bacteria thrive in similar conditions, it is important to avoid conditions which favour their growth in order to prevent infection. They are active over a wide range of temperatures. Some like warmth and are active at 75ºC. Others like cold conditions and grow at temperatures as low as 5ºC. Although normal bacteria are destroyed during heat treatment by boiling, some spores survive boiling for hours. They can resume normal activity when conditions become more favourable and contaminate some preserved foods.

The natural acidity of most fruits also serves as a barrier to many spoilage microbes, especially bacteria Bacteria are killed in an acid medium, and therefore they are not a problem in preserving fruits and making jams. However some bacteria such as *Erwinia carotovora* subsp. Carotovora, species of *Xanthomonas, Pseudomonas, Clostridium, Cytophaga,* and *Bacillus* are effective spoilage microbe that causes soft rot across a broad host range of vegetables and some fruits (Lund 2000).

Enzymatic Breakdown of Fruit Carbohydrates during Spoilage of Fruits due to plant or Microbial Enzymes

Cellulases

These are inducible enzymes synthesized by a large diversity of microorganisms including both fungi and bacteria during their growth on cellulosic materials . These microorganisms can be aerobic, anaerobic, mesophilic or thermophilic. Cellulase is a family of at least 3 groups of enzymes (Morana *et al.,* 2011)

1. endo-(1,4)-β-D-glucanase (EC 3.2.1.4)
2. exo-(1,4)-β-D-glucanase (EC 3.2.1.91
3. β-glucosidases(EC 3.2.1.21).

The exoglucanase (CBH) acts on the ends of the cellulose chain and releases β-cellobiose as the end product; endoglucanase (EG) randomly attacks the internal *O*-glycosidic bonds, resulting in glucan chains of different lengths; and the β-glycosidases act specifically on the β-cellobiose disaccharides and produce glucose (Fig.11).

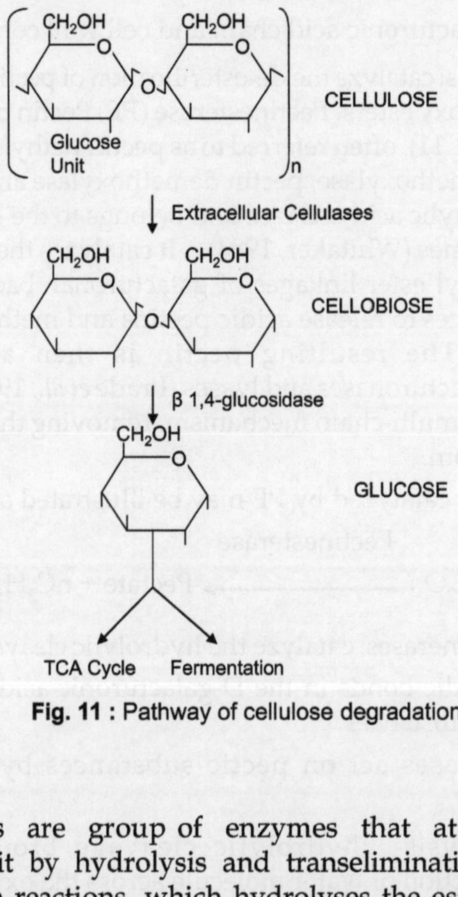

Fig. 11 : Pathway of cellulose degradation

Pectinases

Pectinases are group of enzymes that attack pectin and depolymerise it by hydrolysis and transelimination as well as by deesterification reactions, which hydrolyses the ester bond between carboxyl and methyl groups of pectin (Ceci and Loranzo, 1998; Wong, 2008).

The pectinolytic enzymes may be divided in three broader groups:

(I) Protopectinases (PPase): degrade the insoluble protopectin and give rise to highly polymerized soluble pectin by following equation.

$$\text{Protopectin (insoluble)} + H_2O \xrightarrow{\text{PPase}} \text{Pectin (soluble)}$$

PPases are classiûed into two types, on the basis of their reaction mechanism

- A-type -react with the polygalacturonic acid region of protopectin
- B-type - react on the polysaccharide chains connecting the

polygalacturonic acid chain and cell wall constituents

(II) Esterases: catalyze the de-esterification of pectin by the removal of methoxy esters. Pectinesterase (PE, Pectin pectlylhydrolase, E.C.3.1.1.11), often referred to as pectinmethylesterase, pectase, pectin methoxylase, pectin demethoxylase and pectolipase, is a carboxylic acid esterase and belongs to the hydrolase group of enzymes (Whitaker, 1984). It catalyzes the deesteriûcation of methyl ester linkages of galacturonan backbone of pectic substances to release acidic pectins and methanol (Cosgrove, 1997). The resulting pectin is then acted upon by polygalacturonases and lyases (Prade *et al.*, 1999). Fungal PEs act by a multi-chain mechanism, removing the methyl groups at random.

The reaction catalyzed by PE may be illustrated as follows:

$$\text{Pectin} + n\,H_2O \xrightarrow{\text{Pectinesterase}} \text{Pectate} + nC_2H_5OH$$

(III) Depolymerases: catalyze the hydrolytic cleavage of the u-(1,4) -glycosidic bonds in the D-galacturonic acid moieties of the pectic substances.

Depolymerases act on pectic substances by two different mechanisms:

- Hydrolysis- hydrolytic cleavage brought about by introduction of water molecule across the oxygen bridge .
- Trans-elimination lysis- The glycosidic bond is broken by a trans-elimination reaction without any participation of water molecule.

Depolymerases are subdivided into four different categories based on the preference of enzyme for the substrate, the mechanism of breakdown of the glycosidic bonds.

Enzyme	substrate	mechanism
Polygalacturonase	pectate	hydrolysis
Polymethylgalacturonase	pectin	hydrolysis
Polygalacturonate lyase	pectate	β elimination
Polymethylgalacturonate lyase	pectin	β elimination

Polygalacturonases

Polygalacturonases (PGases) are the pectinolytic enzymes that catalyze the hydrolytic cleavage of the polygalacturonic acid chain with the introduction of water across the oxygen bridge. They are the most extensively studied among the family of pectinolytic enzymes (Herron *et al.,* 2000). The polygalacturonase involved in the hydrolysis of pectic substances are endo-PGase (E.C. 3.2.1.15) and exo-PGase(E.C. 3.2.1.67)

Lyases

Lyases (or transeliminases) perform non-hydrolytic breakdown of pectates or pectinates, characterized by a trans-eliminative split of the pectic polymer. The lyases break the glycosidic linkages at C-4 and simultaneously eliminate H from C-5, producing a " 4:5 unsaturated product .

Based on the pattern of action and the substrate acted upon Lyases may be classified into four types .

(I) endopolygalacturonate lyase (Endo PGL, E.C. 4.2.2.2);

(II) exopolygalacturonate lyase (Exo PGL, E.C. 4.2.2.9);

(III) endopolymethylgalacturonate lyase (Endo PMGL, E.C. 4.2.2.10);

(IV) exopolymethylgalacturonate lyase (ExoPMGL).

The reactions catalyzed by lyases is as follows:

$$\text{Polygalacturonate} \xrightarrow{\text{polygalacturonate lyase}} \Delta 4:5 \text{ unsaturated galacturonates}$$

$$\text{Polymethylgalacturonate} \xrightarrow{\text{polymethylgalacturonate lyase}} \Delta 4:5 \text{ unsaturated methylgalacturonates}$$

Mode of action of pectinases is depicted in figure 12.

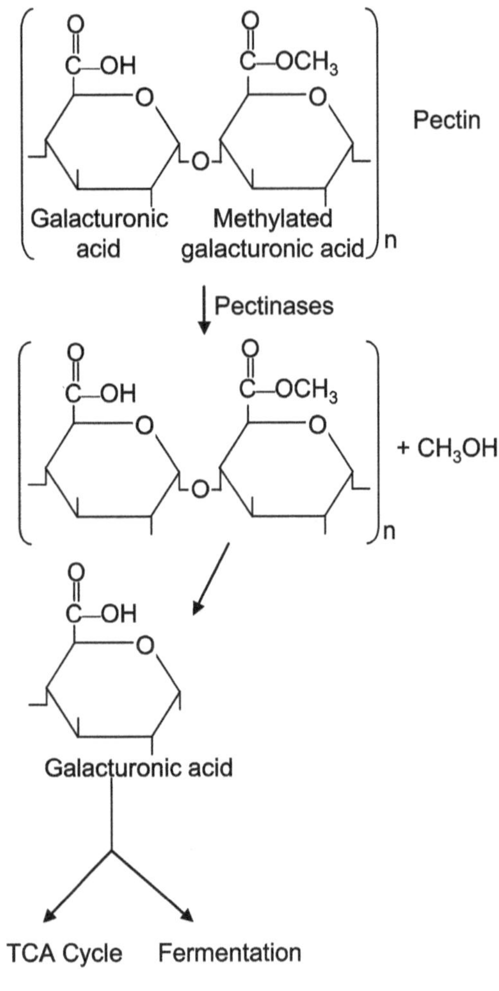

Fig.12 : Pathway of Pectin degradation

Amylases

These are glycoside hydrolases and act on á-1,4-glycosidic bonds and catalyse the breakdown of starch into sugars. Both amylose and amylopectin are broken down by the enzyme amylase into maltose, though at different rates.

The enzyme can only break down the 1-4 bond not the 1-6 so starch is broken down into glucose and a disaccharide (Fig. 13).

α-amylases

The α-amylases (EC 3.2.1.1) (also known as : 1,4-α-D-glucan glucanohydrolase) act at random locations along the starch chain and breaks down long-chain carbohydrates, ultimately yielding maltotriose and maltose from amylose, or maltose, glucose and "limit dextrin" from amylopectin.

β-amylase

β-amylase (EC 3.2.1.2) (also known as: 1,4-α-D-glucan maltohydrolase) act from the non-reducing end and catalyzes the hydrolysis of the second α-1,4 glycosidic bond, cleaving off two glucose units (maltose) at a time. During the ripening of fruit, β-amylase breaks starch into maltose, resulting in the sweet flavour of ripe fruit.

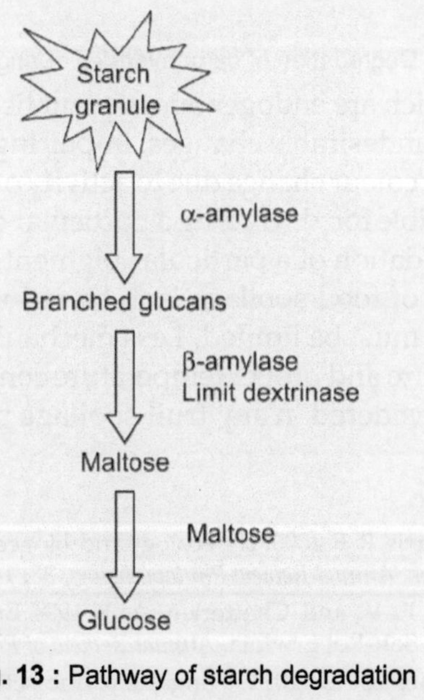

Fig. 13 : Pathway of starch degradation

The over all degradation of carbohydrates is depicted in following figure (Fig.14).

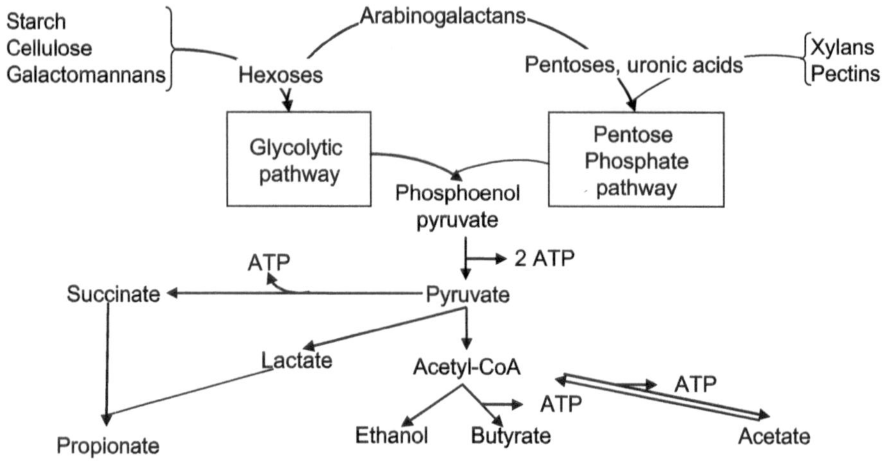

Fig 14 : Degradation of carbohydrates during fruit spoilage

Enzymes which are endogenous to plant tissues or from microbial origin, can bring undesirable changes, impairing its physical character such as softness or colour change due to activity of one or more enzymes. Latter are responsible for degrading a particular component of the plant cell wall, or degradation of a particular pigment leading to a change in colour. If the risk of food spoilage is to be reduced, conditions which promote spoilage must be limited. Less mechanical handling, reduced exposure to moisture and proper temperature control are three strategies that should be considered in any fruit spoilage prevention program.

References

Andrews, J. H. and Harris, R. F. 2000. The ecology and biogeography of microorganisms on plant surfaces. *Annual Review Phytopathology* 38: 145–180.

Barras, F, Gijsegem, F. V. and Chatterjee, A. K.1994.Extracellular Enzymes and Pathogenesis of Soft-Rot Erwinia. *Annual Review of Phytopathology*.32: 201-234.

Brink J V.D., and Vries R.P.D. 2011. Fungal enzyme sets for plant polysaccharide degradation. *Appl Microbiol Biotechnol.* 91(6):1477-92.

Casey, G.D. and Dobson, A.D.W. 2003. Molecular detection of *Candida krusei* contamination in fruit juice using the citrate synthase gene cs1 and a potential role for this gene in the adaptive response to acetic acid. *J Appl Microbiol* 95:13–22.

Ceci, L. and Lozano, J. 1998. Determination of enzymatic activities of commercial pectinases for the clarification of apple juice. *Food Chem.* 61 : 237- 241.

Cosgrove, D.J. 1997. Assembly and enlargement of the primary cell wall in plants. *Annu Rev Cell Dev Biol*;13:171–201.

Cosgrove, D.J., 2001. Wall structure and wall loosening. A look backwards and forwards. *Plant Physiol.*, 125: 131–134.

Hallett C. , P. W. Sutherland , F. R Harker , and E. A. MacRae .2005. Fruit Cell Walls, Texture and Convenience. Microsc. *Micro. Anal.* 11(Suppl 2):1144-45.

Harris, P.J. 2005. Diversity in plant cell walls. In: Plant Diversity and Evolution: Genotypic and Phenotypic Variation in Higher Plants (ed R.J. Henry). Pp. 201-227. Wallingford: CAB International Publishing

Herron, S.R., Benen, J. A.E., Scavetta, R. D. , Visser, J. and Jurnak, F. 2000. Structure and function of pectic enzymes: Virulence factors of plant pathogens. *Proc Natl Acad Sci U S A.* 97(16): 8762–8769.

Janisiewicz, W. J. and Korsten, L. 2002. Biological control of postharvest diseases of fruits. *Annual Review of Phytopathology* 40: 411–441.

Kojim K., Sakurai N., and Kuraishi S. 1994. Fruit softening in banana: correlation among stress-relaxation parameters, cell wall components and starch during ripening. *Physiologia Plantarum.* 90(4): 772–778.

Kubicek CP, Herrera-Estrella A, Seidl V, Le Crom S, MartinezDA, Druzhinina IS, Zeilinger S, Casas-Flores S, Horwitz BA,Mukherjee PK, *et al.*, 2011. Comparative genome sequence analysis underscores mycoparasitism as the ancestral lifestyle of Trichoderma. *Genome Biol.* 12(4):R40..

Loureiro ,V. and Ferreira, M. 2003. Spoilage yeasts in the wine industry. *Int J Food Microbiol* 86:23–50.

Lund, B.M. and Snowdon, A.L. 2000.Fresh and processed fruits. In: Lund BM, Baird-Parker AC, Gould GW. (Eds) The Microbiological Safety and Quality of Foods. Gaithersburg, MD: ASPEAN, 738-58.

Martens-Uzunova, E.S., and Schaap, P.J.2009. Assessment of the pectin degrading enzyme network of Aspergillus niger by functional genomics. *Fungal Genet Biol.* 46 Suppl 1:S170-S179

Martinez, D., Berka, R.M., Henrissat, B., Saloheimo, M., Arvas, M., Baker, S.E., Chapman, J., Chertkov, O., Coutinho, P.M., Cullen, D. 2008. Genome sequencing and analysis of the biomassdegrading fungus *Trichoderma reesei* (syn. *Hypocrea jecorina*). *Nat Biotechnol.* 26:553–560.

Martorell, P., Fernandez-Espinar, M.T. and Querol, A. 2005. Molecular monitoring of spoilage yeasts during the production of candied fruit nougats to determine food contamination sources. *Int J Food Microbiol* 101:293–302.

Martorell, P., Stratford, M., Steels, H., FernandezEspinar, M.T. and Querol, A. 2007. Physiological characterization of spoilage strains of Zygosaccharomyces bailii and Zygosaccharomyces rouxii isolated from high sugar environments. *Int J Food Microbiol* 114:234–242.

Miedes, E. and Lorences, E. P. 2004. Apple (*Malus domestica*) and tomato (*lycopersicum*) fruits cell wall hemicelluloses and xyloglucan degradation during penicillium expansum infection. *Journal of Agricultural and Food Chemistry* 52: 7957–7963.

Morana, A., Maurelli, L., Ionata, E., Cara, F. La,. and Rossi, M . 2011. Cellulases from Fungi and Bacteria and their Biotechnological Applications. *(In)* Adam E. Golan (ed.) Cellulase: Types and Action, Mechanism, and Uses. Nova Publishers, N.Y :1-80

Phister, T.G. and Mills, D.A. 2003. Real-time PCR assay for detection and enumeration of *Dekkera bruxellensis* in wine. *Appl Environ Microbiol* 69:7430–7434.

Pitt JI and Hocking AD. 1997. Fungi and Food Spoilage. Blackie Academic and Professional, New York

Prade, R.A., Zohan, D., Ayoubi, P. and Mort ,A.J. 1999. Pectins, pectinases and plant-microbe interactions. *Biotechnol Genetic Eng* Rev;16: 361–91.

Prasanna, V. Prabha, T.N. Tharanathan RN. 2007.Fruit ripening phenomena—an overview. *Crit Rev Food Sci Nutr.* 47(1):1-19.

9

Processing Potatoes in Sub-Tropical Climates

Dinesh Kumar, Parveen Kumar, SK Pandey and SV Singh

The potato (*Solanum tuberosum* L.) is one of the four most important food crops of the world, the other three being rice, wheat and maize. It is a high yielding and nutritive crop and produces highest dry matter, carbohydrates, edible protein, minerals and vitamin C and B per unit of area and time. In view of the shrinking cultivable land area and increasing population of India, the potato has a high potential to solve the problems of hunger and malnutrition in our country. But, potato is a perishable commodity and its harvest time (February/March) in Indo-Gangetic plains (which account for nearly 87% of the country's potato production) coincides with steep rise in ambient temperatures. From April onwards, temperatures start shooting up and the produce has either to be consumed within a short period of time or is required to be preserved in the cold stores. Due to inadequate, expensive and unevenly distributed refrigerated storage facilities in the country, there have been frequent gluts in the market causing substantial economic losses to farmers and wastage of precious food. Such gluts occur every 3 to 5 years. The glut years are followed by a reduction in potato area and production leading to high potato prices in alternate years which again encourage farmers to increase potato area in subsequent years resulting again in a glut situation and price crashes. This boom and bust cycle has thus continued for long (Pandey *et al.*, 2009).

One of the solutions to this perpetual cyclic problem appears to be diversification of potato utilization. The strategy has potential for sustaining continued increase in potato production, thereby, meeting targets of production for future, besides stabilizing potato prices in the country. Potato processing is a potentially viable media for increasing potato utilization in the current socio-economic scenario (Marwaha 1997). Processing leads to value addition and preservation of the semi-perishable potatoes. Captive potato production by processing industry could be an added advantage to ensure assured returns to the farmers. However, availability of raw material for processing is a major bottleneck for the growth of industry in the country (Rana and Pandey 2007). The raw material requirement of Indian potato processing industry during 2007-08 was estimated equal to 1.68 million tonnes (Rana and Pandey 2007). The current requirement of processing industry is expected to be near 10% of national potato production. However, the composition of raw material utilization of Indian potato processing industry is drastically different than that of developed nations. On an average developed nations use more than 60% processing quality potatoes for French fries making, this proportion is nearly 1% in India. In India nearly 90% of processing quality potatoes is consumed by the potato chips industry and rest by other potato products (Rana and Pandey 2009). It is a peculiar situation with gluts on one hand and shortage of trade potatoes for processing on the other in some of the years. The reason being that the industry had special quality requirements of potatoes which include high tuber dry matter for getting high quantity of finished products and low reducing sugars to ensure high quality light coloured products acceptable in the market (Marwaha 1997). In the early 90s indigenous varieties were not found suitable for quality processed products, therefore several multinational companies (MNC) insisted on bulk import of potato varieties from Europe and America (Pandey *et al.*, 2009). However, in view of quarantine problems related to bulk import, involving risk of entry of new pests and diseases in the country, the Govt. of India allowed import of limited quantity of seed of a few processing varieties like Atlantic and Frito Lay hybrids. These varieties when grown under Indian conditions produced low yields (Singh *et al.*, 1999; Singh *et al.*, 2003; Marwaha *et al.*, 2005) and were also susceptible to late blight, the most devastating disease of potato. Thus, availability of processing quality potatoes became a major bottleneck for growth of potato processing industry in India. For making superior quality processed products, the potatoes must possess certain minimum quality characters mentioned in Table 1 and discussed below:

1. **Shape, size and depth of eyes:** For making chips of uniform size, the tubers should be round to oval in shape. For French fries, oblong to long tubers are preferred, while for flakes, though round to oval shape is desirable, the requirement of shape is not very strict. For canning, small tubers of round to oval shape are suitable. Medium to large size tubers with shallow eyes are preferred for most of the processed products as these result in lower peeling losses.
2. **Tuber dry matter content:** The dry matter content of tubers is the most important character determining the quality and yield of fried and dehydrated products. The higher dry matter or solids content results in higher recovery of processed products, lower oil absorption, lesser energy consumption and imparts a crispy texture to the product. The lower oil content in the fried products results in longer shelf life of the product. Conversely, for canning the tubers with low dry matter are preferred as higher dry matter content results in sloughing of tubers. Specific gravity of potatoes is commonly used by the potato processing industry as a tool for quick estimation of dry matter content, as both of them are positively correlated.
3. **Reducing sugar content:** The reducing sugars play a critical role in determining the colour of fried products which develops during frying at high temperatures due to the 'Maillard reaction' between reducing sugars and free amino acids present in the tubers (Schallenberger *et al.*, 1959). Presence of excessive amounts of reducing sugars in potato tubers result in unacceptably dark colour and bitter taste in fried products also known as non-enzymatic browning. Besides affecting the colour and flavour of fried products, Maillard reaction has also been related to the formation of acrylamide, which is considered a potentially toxic compound.

Table 1 : Tuber quality requirements for different processed products

Parameters	Chips	French fries	Flakes	Canned Potatoes
Shape	Round	Oblong	Round to oval	Round
Size	45-88 mm	>88 mm	>35 mm	18-38 mm
Eyes	Shallow	Shallow	Shallow	Shallow
Dry matter (%)	>20	>20	>20	<18%
Specific gravity	> 1.080	> 1.080	> 1.080	<1.070
Reducing sugars (%)	< 0.1	<0.15	<0.15	<0.5
After cooking discoloration	Slight	Slight	Nil	Nil
Enzymic discoloration	Slight	Nil	Nil	Nil
Texture	Fairly firm to mealy	Fairly firm	Fairly firm to mealy	Waxy

Development of Processing Varieties Suitable for Indian Conditions

The conditions under which potato is grown in India are very different from those in Europe and North America. Potato is a short duration crop in sub-tropical Indian plains as compared to the long duration crop in temperate climates of western countries (Shekhawat *et al.*, 1999, Table 2).

Table 2 : Potato growing conditions in sub-tropical and temperate regions

Parameters	Sub-tropical	Temperate
Day/night temperatures	25-30°C / 2-15°C	25°C / 15°C
Photoperiod	10 hr/day	14 hr/day
Frosting	Common	Absent
Crop duration	90-100 days	120-180 days
Post-harvest conditions	Harvesting followed by hot summer and rains	Harvesting followed by severe winter
Result	Low yields, low dry matter, high reducing sugars, short dormancy and poor keeping quality	High yields, high dry matter, low sugars, long dormancy and good keeping quality

India's first two processing potato varieties were released in 1998. Till date a total of five varieties have been developed for processing. A brief description of these varieties has been given in following lines.

(i) **Kufri Chipsona-1**: This is the most popular indigenous processing variety as on date. The variety produces oblong tubers having high dry matter content (> 20%) and very low reducing sugar content (< 0.1 % on fresh weight basis) (Gaur *et al.*, 1998). The variety has become well adapted in Punjab, Uttar Pradesh and Madhya Pradesh. The chemical maturity of the variety is 90-110 days depending upon the location. The variety has been found highly suitable for making chips and flakes and can be used up to some extent for making French fries also. The major drawbacks observed in the variety over the years are lower percentage chipping grade potatoes in the total produce (around 60%), knobbiness in larger sized tubers and oblong shape (not desirable for making chips) (Pandey *et al.*, 2006).

(ii) **Kufri Chipsona-2:** This variety has very high dry matter content (> 22%) and highly desirable round shape for making chips (Gaur *et al.*, 1999). But this variety has not become popular because of relatively longer duration of crop (110-120 days) and relatively poor keeping quality during storage.

(iii) **Kufri Chipsona-3**: This variety was released in 2005 and was an improvement over the variety Kufri Chipsona-1 (Pandey *et al.*, 2006). The variety has round to oblong tubers with reasonably good dry matter content (21%). The variety produces 11-15% more total and processing grade tuber yields when compared with the best control variety Kufri Chipsona-1. The variety achieves chemical maturity in 100-110 days depending upon the location (Pandey *et al.*, 2008). Development of this variety has offered a greater choice to the farmers as well as processors to choose among the indigenous processing varieties.

(iv) **Kufri Himsona:** Presently, the potato variety Kufri Chipsona-1 is the major source of raw material for processing industries in India. From April onwards, the industries utilize potatoes of this variety stored at intermediate temperatures (10-12°C) with the application of CIPC (Isopropyl-N-(3 (chlorophenyl) carbamate. However, during July end or August, tubers of Kufri Chipsona-1 start accumulating higher amounts of reducing sugars and become unsuitable for processing. This is a very crucial period for the industry, since fresh potatoes are not available in near by areas in northern Indian plains for use as raw material use. Therefore, processing industries located in the north have to procure potatoes from Hassan in Karnataka spending huge expenses on transportation thus leading to increased cost of production. The chipping variety 'Kufri Himsona' for the north western hills was developed in 2007 (Pandey *et al.*, 2008), which can be harvested from August onwards to provide suitable processing potatoes to the industry in this lean period.

(v) **Kufri Frysona:** The international market for various processed potato products, particularly frozen potatoes, is growing rapidly. World over western-style cuisine is gaining popularity where frozen French fries and other frozen products are generating billions of dollars in sale. It is estimated that nearly 60-70% of total world out put of French-fries is shared among U.S, Canada and the Netherlands. However off late there is increasing demand of frozen potato products in developing Asian and Latin American countries. China and India have the most potential for growth of frozen potato products where economic boom is likely to generate strong demand of convenience foods. This is mainly because of socio-economic change, rise in number of working couples where women are increasingly spending less time in kitchen and more in office,

alternative career, need to spend quality time with children and greater media exposure. Processing of potatoes into French fries, however, requires certain quality attributes and mainly includes oblong to long-oval tubers of preferably more than 85mm size with shallow eyes to yield desirable long sticks of international standard with minimum peeling losses. The tubers need to have 20% or more dry matter with reducing sugars content below 200 mg/100 g. fresh tuber weight so as to yield crisp and light coloured French fries. One of the major problems for supply chain of frozen French fries was the non-availability of specially bred varieties for French fry making. The varieties used by the industry namely Kufri Chipsona-1 and Kufri Surya have certain deficiencies. As already mentioned Kufri Chipsona-1 suffers from lesser percentage of processing grade tubers hence sufficient French fry grade tubers are not available in this variety. In case of variety Kufri Surya, the dry matter content is not consistent and varies from 19-21%. Therefore, the variety 'Kufri Frysona' was bred for making French fries (Singh *et al.*, 2009).

Agro Technology for Growing Processing Grade Potatoes

Nutrient management, water management, crop geometry, population density, etc. are widely recognized as important manageable agronomic factors influencing potato tuber growth, development, productivity and quality. Agronomic practices have been traditionally standardized for growing seed and table potatoes. However, different package of practices are required for production of processing grade potatoes meant for processing. In following sections these agronomical aspects have been mentioned:

i) Nutrient management

NPK requirements of potato being grown for seed and table purposes have been worked out for different agro-climatic zones in India. Chipsona varieties (including Kufri Frysona) are relatively late maturing (100-110 days) with high tuber dry matter and it can be anticipated that input requirements of the Chipsona varieties would be higher. It has been shown that 270 kg N/ha was optimum dose to achieve maximum proportion of processing-grade tubers and total tuber yield, highest benefit and cost ratio (B: C), high dry matter/specific gravity and permissible level of low reducing sugars and superior chip colour both at the time of harvest and after storage at 10-12°C with CIPC (Kumar *et al.,* 2007f). It was also observed that application of N in two split dose *i.e.* half at planting and rest at time of eathing up produces higher yields and results in higher N recovery than applying entire dose at planting. However, the foliar spray of urea (2%) on potato processing cultivars Kufri Chipsona-1 and Kufri Chipsona-2 at 60 or 80 days after planting resulted in deterioration of processing quality traits at harvest and during storage (Kumar *et al.,* 2011b). Studies conducted with different levels of phosphorus (0, 40, 80, 120 and 160 kg P_2O_5/ha) for optimizing the phosphorus requirements of Chipsona varieties with respect to processing-grade tuber yield, revealed that 80 kg P_2O_5/ha was the optimum dose to obtain maximum yield/yield attributing characters and processing quality traits with net returns (Rs. 59,503/ha) and highest B: C ratio (2.43) (Kumar *et al.,* 2007c). Similarly experiments conducted to study the effect of graded levels of potassium (0, 50, 100, 150 and 200 kg K_2O/ha), it has been observed that 150 kg K_2O/ha produced maximum yield of processing-grade tubers and total yield with highest B:C ratio (2.45). Potassium level, however, did not influence the quality parameters (Kumar *et al.,* 2004a). Besides the NPK requirements, the source of potassium for processing varieties can also influence the quality. Potassium is required by plants for translocation of sugars and synthesis of starch. Since the Chipsona varieties are rich in starch hence their requirement of potassium is relatively higher (150 kg K_2O ha^{-1}). It is reported that tuber quality parameters such as dry matter, specific gravity, sugar concentration, flesh colour are affected by K fertilization, specific gravity in particular is affected by the source of K fertilizer. The most widely used form of K fertilizer is potassium chloride. Generally K application above the recommended dose decreases the dry matter with K-chloride having a higher effect than K-sulphate. The decrease in dry matter is largely due to chloride ion rather than K ion. K-sulphate usually results in a higher dry matter of potato tubers than equivalent quantities of K in the chloride form. Specific gravity and tuber dry matter are of prime importance to the processors as the recovery of the finished product is directly linked to them. It has been found that for chipping potatoes K-sulphate should be used instead of K-chloride, since it increases the tuber dry matter percentage and chips recovery and decreases the oil percentage of chips without affecting tuber yield (Kumar *et al.,* 2007d).

ii) Water management

It has been reported that for the cultivars Kufri Chipsona -1 and -2 a continual decrease in processing-grade and total tuber yield with increase in irrigation levels from 15 mm CPE (cumulative pan evaporation) to 35 mm CPE. However, the decline in tuber yield was not significant between 15 and 20 mm CPE. It is recommended that in west-central plains Chipsona cultivars should be irrigated at 20 mm CPE (Kumar *et al.*, 2007e).

iii) Crop geometry/plant density

Inter-/intra-row spacing and plant density directly affects the unit area available to the plant for its foliage and tuber development. Variation in inter-/intra-row spacing affects the tuber size distribution and generally wider spacing results in increased proportion of large-size tubers and vice-versa. The tuber-size requirements of the processing potato are very precise, i.e. for chips (45-85 mm) and for French fries (>75 mm). For the indigenous potato varieties 67.5 × 20 cm crop geometry (74, 000 plants/ha) has been recommended since at this spacing the proportion of chip-grade was higher. Problem of green tubers is also gets reduced to a great extent at 67.5 × 20 cm spacing as compared to to standard spacing's of 60 × 20cm (Kumar *et al.*, 2004b). To realize the higher proportion of French fry grade tuber yield (56%) from potato variety Kufri Frysona crop geometry of 67.5 × 25 cm (59,300 plants/ha) was found optimum (Kumar *et al.*, 2012)

Storage Technology for Processing Potatoes for Off-Season Availability of Quality Potatoes to the Industry

About 90% of the potato crop is grown from October to March and fresh potatoes are available from February to April; thereafter potatoes need to be stored. Potatoes are semi-perishable since they contain around 80% water and get spoiled due to rottage, water loss and sprouting if kept under ambient conditions for long time. To protect potatoes from high temperatures and rains, tubers are stored at low temperature. Refrigerated storage or low temperature storage is used for long term (> 3 months) storage and it protects tuber from microbial spoilage, water loss (evaporation) and sprouting. Traditional methods of storage include storage by heap or pit methods. The heap or pit storage is used for short-term storage of 2 to 3 months. But, majority of potatoes are stored by refrigerated storage at 2-4°C. The major disadvantage of storing potatoes storage at 2-4°C is development of sweetening in tubers (Kumar, 2011). The phenomenon known as 'low temperature sweetening' (LTS) results from conversion of starch into free sugars at low storage temperatures. The sweetened potatoes become unsuitable for making processed products and are not liked by majority of the people for table consumption also. The low temperature storage of 2-4°C is therefore, suitable for storage of seed potatoes only. Therefore, different storage practices are followed for processing potatoes and discussed in following section.

(i) On-farm storage by traditional methods

The non-refrigerated storage of potatoes in heaps and pits can provide a viable option for short-term storage. Heaps and pits are popular in several parts of the country for the following reasons (i) These are cheap and easy to prepare (ii) Locally available materials are used in their construction (iii) They enable the farmer to hold back his produce to sell it later for a better price (Ezekiel *et al.*, 2002). The heaps and pits maintain lower temperatures and higher relative humidity as compared to ambient conditions (Mehta *et al.*, 2007). High temperature in traditional non-refrigerated stores helps in maintaining low reducing sugar levels in potatoes. Kumar *et al.* (2005) have shown that during storage by traditional methods there was a significant reduction (33.84%) in reducing sugar content of potatoes and improvement (31.63 %) in chip colour. Sugar levels in stored potatoes were within the acceptable limit for processing and table purposes indicating that potatoes stored upto 90 days on-farm in heap and pits are suitable for these two purposes. The potatoes meant for ware market or cottage level processing (like Alu Bhujia, Tikkis, Samosas) *etc.* can be stored on-farm in heaps at a low cost and supplied to the local markets and unorganised processing industries at premium price as these contain low sugars. However, sprouting of tubers is still a major concern in heap storage. The dormancy period of all the Indian potato varieties is around six to eight weeks, therefore tubers start sprouting under ambient conditions after one month of storage. The sprouts are 100 times more permeable as compared to potato skin and thus evaporation of water is very high. Due to evaporative losses, potatoes become shriveled and fetch lower prices in ware and processing market. Secondly sprouts themselves consume dry matter content of potatoes for their growth and extra labour is required for de-sprouting of tubers before sending them to market. Therefore, control of sprouting in tubers is a must for potatoes stored in heap, to further minimize the losses to farmers and to keep potatoes marketable for table and processing market. As already mentioned, heap stored potatoes have superior processing quality with respect to chips making. But excessive shriveling leads to increased peeling losses making heap stored potatoes unsuitable for processing.

For control of tuber sprouting, CIPC is the most popular chemical used worldwide. But its efficacy comes down with the increase in storage temperatures, and is generally used with temperature range of 8-12°C. In heap the average temperatures are above 25°C after around one month of storage. Both CIPC fog and CIPC spray have been tried in heap stored potatoes. Both fogging and direct spray considerably reduced the quantitative and qualitative losses in the heap after three months of storage. However, due to practical convenience of application, CIPC spray has been recommended for control of sprouting in heap stored potatoes (Mehta *et al.*, 2007).

(ii) Potato storage at intermediate temperatures (10-12°C) with sprout suppressant CIPC (Isopropyl N-(3-chlorophenyl) carbamate)

Potatoes stored with traditional methods remain available for only 2-3 months (March-May) or till the rains start, while the raw material for industry needs to be stored for 6-8 months till the arrival of the next potato crop. Therefore, the potatoes need to be stored at a temperature where the accumulation of reducing sugars is minimum and they produce light coloured chips throughout the storage period. Storage at temperatures of 10-12°C is recommended for best storage (Ezekiel *et al.*, 2005). Indigenous processing varieties like Kufri Chipsona-1, Kufri Chipsona-2 and Kufri Chipsona-3, if harvested at right chemical maturity and stored at an appropriate time at 10-12°C with CIPC (a chemical that acts as sprout suppressant), accumulate low reducing sugars up to 6 months in comparison to cold store (2-4°C) and provide processing potatoes to the chipping industry and can as well be sold as low sugar table potatoes at a premium price in the market (Kumar *et al.*, 2007b).

Crop from Different Seasons for Increased Availability of Fresh Potatoes to the Industry

(i) **Availability of fresh potatoes in November/December:** Early potato crop which is planted in north western plains in the middle of September and harvested by end of November has high potential of providing fresh potatoes to processing industries, when fresh potatoes are not available anywhere (Pandey *et al.*, 2003). The produce of *Kharif* crop from Karnataka and high hills is exhausted by August/September. Kufri Surya is most suitable for growing as an early planted crop, providing not only very good yield (100.8 q/ha) within 75 days, but also producing satisfactory dry matter of 18-21% and light coloured chips of acceptable grade (3 or <3) on frying thus meeting the parameters for processing. It has been shown that the cultivation of variety Kufri Surya, a heat-tolerant variety as an early crop in the states of Punjab and western Uttar Pradesh can provide much needed raw material to the potato processing industries located in north-western plains, during the lean period from November to December till fresh potatoes from main crop season become available.

(ii) **Availability of fresh potatoes in April/May through Spring crop:** Temperature starts rising from March onwards in northern plains and potatoes are cold stored to save them from rotting. These cold stored potatoes after few months accumulate large amounts of reducing sugars and are not preferred for table consumption. To attract better prices for fresh potatoes, a spring crop is planted in a very limited area in Punjab where prolonged winter helps in tuberization (Pandey *et al.,* 2003). Kufri Chipsona-2 is aptly suitable for growing processing varieties in spring season in Punjab and can provide suitable raw material for industries in the months of April-June (Marwaha *et al.,* 2007).

Possible Future Research Requirements

(i) **Development of short duration/early maturing varieties for Indo-Gangetic plains:** With changing climatic conditions, biotic and abiotic stresses are increasing and season for potato growing is constricting in the Indo-Gangetic plains. Therefore, varieties are required which can attain physical and chemical maturity in 75-90 days for different areas in the Indo-Gangetic plains. This will not only help in increasing the availability of raw material to the industry in the coming times, but will provide raw material right from the month of January in the vicinity areas of the industry.

(ii) **Development of short duration and late blight resistant varieties for plateau region and West Bengal:** Late blight disease is causing extensive damage to potato crop being grown in Kharif season of Karnataka and in the Rabi season of West Bengal. Secondly, potato growing season is of 80-90 days in both these states. Therefore, processing varieties which have maturity period of 85-90 days coupled with late blight resistance are need of the hour for these states (Singh *et al.,* 2009). Karnataka serves as very important source of raw material during the months of September/October, when stored crop becomes unfit for processing.

(iii) **Development of heat tolerant processing varieties:** Varieties having heat tolerant trait are required for planting in early and spring seasons in Indo-Gangetic plains and to source raw material from non-traditional areas of the country (Kumar *et al.,* 2007a). Development of such a variety will increase the availability of freshly harvested raw material to the industry.

(iv) **Development of varieties with increased length of storage at 10-12°C with CIPC:** Industry is storing processing grade potatoes at the temperatures of 10-12°C with the use of sprout suppressant CIPC to make available raw material to the industry during off-season. However, most of the varieties develop unacceptably higher levels of reducing sugars after 4-6 months of storage. Therefore, in the months of September to November, there is acute shortage of raw material forcing the industry to source fresh potatoes from different parts of the country by spending a lot on transportation charges. In the coming time, even the limited area supplying material during this period may not be able to fulfil the growing demand of upcoming processing industry. Therefore, there is a need to develop varieties which can maintain low reducing sugars upto 8-10 months at 10-12°C with CIPC treatment. In this regard there is need to identify biochemical and molecular factors regulating sugar accumulation at 10-12°C and this information needs to be passed on to breeders for proper identification of parents.

(v) **Development of varieties with coloured flesh:** To bring diversification in the product range, several companies in the West have introduced pigmented chips. In India also preliminary work has been initiated to screen the germplasm with coloured flesh. In the coming years varieties will be required to have certain special characters like pigmented flesh, higher anti-oxidant content, higher protein content etc. to attract consumers.

(vi) **Development of varieties resistant to cold induced sweetening:** As already discussed storage of potatoes at low temperatures (< 8.0°C) lead to excessive accumulation of free sugars making potatoes unsuitable for processing. But storage at lower temperatures offer several advantages, which include lesser losses because of disease and sprouting, better maintenance of relative humidity, delayed senescence sweetening and avoidance of the use of sprout suppressant chemicals. Therefore varieties with this trait needs to be developed.

References

Ezekiel R, Dahiya PS and Shekhawat GS 2002. Tradational methods of potato storage in the Malwa region of Madhya Pradesh. Technical Bulletin No. 57, Central Potato Research Institute, Shimla. 39p

Ezekiel R, Mehta A, Singh B, Kumar D, Kumar NR, Paul V and Das M 2005. CIPC (Isopropyl N-(3-chlorophenyl) carbamate) For Sprout Suppression in Potatoes during Storage. Technical Bulletin No. 69, Central Potato Research Institute, Shimla. 50p

Gaur PC, Pandey SK, Singh SV, Kumar D, Marwaha RS and Kumar D 1998. Kufri Chipsona-1: A potato variety for processing. *J Indian Potato Assoc* 25: 113-118

Gaur PC, Singh SV, Pandey SK, Marwaha RS, Kumar D and Kumar D 1999. Kufri Chipsona-2 A new high dry matter potato variety for chipping. *Curr Sci* 76: 722-724

Khurana SMP, Singh BP, Kumar NR, Rawal S and Kumar D 2003. Production, processing and export potential of Indian potatoes: An overview. Conference on processing and export potential of potatoes within Asia, March 10, 2003, CPRIC, Modipuram, India.

Kumar D and Ezekiel R 2003. Sugar accumulation in potato tubers at two different temperatures of storage. Proceedings of the International Conference on Vegetables (Nov 11-14, 2002 Bangalore) "Vegetables for sustainable food and nutritional security in the new millennium" Pp 413-415

Kumar D, Singh SV and Pandey SK 2007a. Processing quality of heat tolerant potato (*Solanum tuberosum* L.) genotypes. Proceedings-NSRTC 2, 2005 pp 42-47, CTCRI, Thiruvananthapuram, Kerela, India

Kumar D, Singh SV, Pandey SK, Marwaha RS, Singh BP and Kumar P 2007b. Processing potential of advanced potato hybrids stored at intermediate temperatures with CIPC. *Indian J Hort* 64: 324-327

Kumar D 2011a. Cold induced sweetening development in Indian potato (Solanum tuberosum L.) varieties. *Indian J Biochem Biophys* 48: 123-127.

Kumar P, Pandey SK, Singh SV, Rawal S and Kumar D 2004a. Effect of potassium fertilization on processing grade tuber yield and quality parameters in potato (*Solanum tuberosum*). *Indian J Agri Sci* 74: 177-179.

Kumar P, Pandey SK, Rawal S, Singh SV and Kumar D 2004b. Optimizing crop geometry of potato cultivars for chipping. *Potato J* 31: 141-146.

Kumar P, Pandey SK, Singh BP, Singh SV and Kumar D 2007c. Optimizing phosphorus requirements of Chipsona varieties for west-central plains of India. *Potato J* 34: 199-202.

Kumar P, Pandey SK, Singh BP, Singh SV and Kumar D 2007d. Influence of Source and Time of Potassium Application on Potato Growth, Yield, Economics and Crisp Quality. *Potato Res* 50: 1-13.

Kumar P, Pandey SK, Singh SV and Kumar D 2007e. Irrigation requirements of chipping potato cultivars under west-central Indian plains. *Potato J* 34: 193-198.

Kumar P, Pandey SK, Singh BP, Singh SV and Kumar D 2007f. Effect of nitrogen rate on growth, yield, economics and crisp quality of Indian potato processing cultivars. *Potato Res* 50: 143-155.

Kumar P, Pandey SK, Singh SV, Singh BP, Singh K, Kumar D, Rawal S and Singh S 2011b. Effect of growth duration, N application and row spacing on productivity, profitability and processing quality of potato. *Potato Journal* 38 : 137-142.

Kumar P, Singh BP, Kumar R, Kumar D, Singh SV, Pandey SK, Singh S and Rawal S 2012. Standardization of intra-row spacing of cultivar Kufri Frysona for higher French fry grade tuber yield with good fry quality. *Potato Journal* 39 : 88-91.

Marwaha RS, Pandey SK, Singh SV and Khurana SMP 2005. Processing and nutritional qualities of Indian and exotic potato cultivars as influenced by harvest date, tuber curing, pre-storage holding period, storage and reconditioning under short days. *Adv Hort Sci* 19:1-11

Marwaha RS 1997. Processing of potatoes: Current status, need, future potential and suitability of Indian varieties - A critical appraisal. *J Food Sci Technol* 34:457-471

Marwaha RS, Pandey SK, Singh SV and Kumar D 2007. Yield, chipping and nutritive qualities of spring grown potatoes in north-western plains. *Potato J* 34: 61-62

Mehta A, Singh B, Kumar D and Ezekiel R 2007. Evaluation of CIPC spray for sprout inhibition in potatoes under traditional storage methods. *Potato J* 34: 69-70

Pandey SK, Kumar D, Singh SV and Tomar TPS 2003. Staggered planting-A solution for producing chipping potatoes for northwestern plains. *J Indian Potato Assoc* 30: 89-90

Pandey SK, Singh SV, Kumar D, Marwaha RS, Manivel P and Kumar P 2008. Performance of newly released 'Kufri Chipsona 3' Indian potato variety during different crop seasons in West central plains. *Indian J Agri Sci* 78: 116-121

Pandey SK, Marwaha RS, Kumar D and Singh SV 2009. Indian potato processing story: Industrial limitations, challenges ahead and vision for the future. *Potato J* 36: 1-13

Pandey SK, Singh SV, Kumar D, Manivel P, Marwaha RS, Kumar P, Singh BP and Gupta VK 2008. Kufri Himsona: A chipping cultivar for hill regions of India. *Potato J* 35: 1-8.

Pandey SK, Singh SV, Kumar D, Manivel P, Marwaha RS, Kumar P and Singh, BP 2006. Kufri Chipsona-3: A high yielding potato variety for chipping with defect free tubers. *Potato J* 33: 26-34

Rana RK and Pandey SK 2007. Processing quality potatoes in India: An analysis of indigenous demand. *Processed Food Industry* 10: 26-35

Rana RK, and Pandey SK 2009. Indian potato processing industry: Strengths, weaknesses and business opportunities. Presented in 15th Triennial Symposium of ISTRC-CIP, Lima-Peru during November 2-6, 2009

Rana RK, Pandey NK, Pandit A and Pandey SK 2009. Profitability Analysis Of Kufri Chipsona-1 Cultivation in Uttar Pradesh. *Potato J* 36: 166-172

Shekhawat GS, Lal SS, Pandey SK, Singh BP, Ezekiel R and Sukumaran, NP, Naik PS 1999. Potato Research in India: A success story of fifty years, Central Potato Research Institute, Shimla, India, p2

Singh SV, Kumar D, Gaur PC, Pandey SK and Singh JP 1999. Performance of Indian and Dutch processing varieties for chipping in north western plains. *J Indian Potato Assoc* 26: 65-66

Singh SV, Kumar D and Pandey SK 2003. Yield and processing attributes of indigenous and exotic potato varieties for processing in late planted crop in north western plains. *Potato J* 30: 253-258

Singh SV, Marwaha RS, Kumar D, Kumar P and Pandey SK 2009. Suitability of potato varieties grown in north-eastern Indian plains for processing. *Potato J* 36: 25-34

Singh SV, Pandey SK, Kumar D, Marwaha RS, Manivel P, Kumar P, Singh BP and Bhardwaj V. 2010. Kufri Frysona: First high yielding potato variety for French fries in India. *Potato J* 37: 103-109

10

Postharvest Physiology and Technology of Pomegranate Fruit

Sunil Pareek

Pomegranate is one of the oldest known edible fruits and is associated with ancient civilizations of the Middle East. From its origin in the area now occupied by Iran and Afghanistan, the pomegranate spread east to India and China and west to Mediterranean countries (Turkey, Egypt, Tunisia, Morocco, and Spain). Spanish missionaries brought the pomegranate to the Americas in the 1500s (Hodgson, 1917; LaRue, 1980). The primary commercial pomegranate growing regions of the world are the Near East, India and surrounding countries, and southern Europe. The fruit is consumed fresh or processed into juice, syrup, jams, or wine (Fig. 1) (LaRue, 1980; Adsule and Patil, 1995; Kumar, 1990; Patil and Karale, 2001; Roy and Wasker, 1997).

This chapter provides an overview of nutritional composition, health benefits, postharvest physiology and technology of pomegranates in relation to maintaining their quality between harvest and fresh consumption or processing.

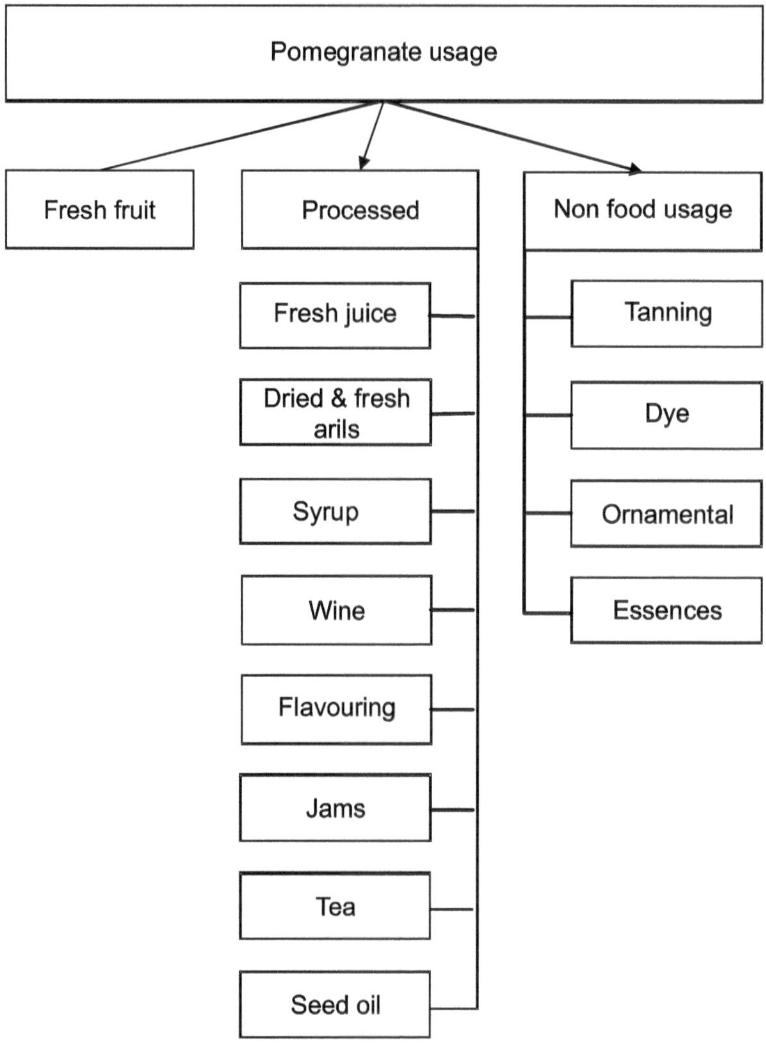

Fig. 1 : Usage of pomegranate fruit

1. Production Statistics

Total world production of pomegranate is estimated currently at 2 million tons/year. The main producing countries are given in Table 1. India is the largest producer followed by Iran and USA. Fresh fruit are exported from India, Spain, Turkey and Iran whereas juice is exports by USA. Israel exports minimally processed arils.

Maharashtra is the leading pomegranate producing state (Fig. 2) which has acreage of around 91 thousand ha followed by Karnataka state which has an area of 12.4 thousand ha, then comes Andhra Pradesh (area 7.7 thousand ha), followed by Gujarat and Tamil Nadu i.e. 3.8 and 0.4 thousand ha, respectively.

Table 1 : Top pomegranate producing countries in 2007

Countries	Production (in thousand tones)
India	900
Iran	800
USA	110
Turkey	80
Spain	80
Iraq	80
Afghanistan	75
Azerbaijan	65
Egypt	43
Uzbekistan	35
Israel	25

Source : USDA Foreign Agriculture Services

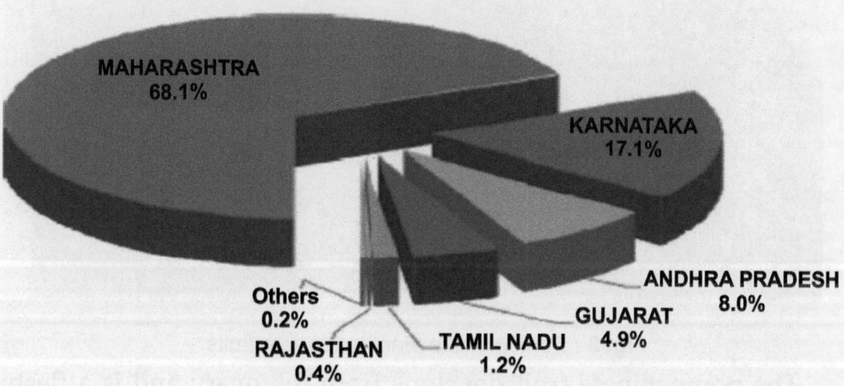

Fig. 2 : Leading pomegranate producing states

2. Botany and Morphology

Pomegranates (*Punica granatum* L.) have been variously placed in the Lythraceae family, or Punicacea, depending on the taxonomist and whether they are considering morphological or molecular data. The former is a large family comprised of 31 genera and 620 species, whereas Punicacea contains only two species, *Punica granatum* and *Punica protopunica* Balf. f., with the latter endemic to the small island of Socotra off the coast of Yemen. When defined in the broadest sense, the Lythraceae family comprises four subfamilies, with one, Punicoideae, sometimes regarded as a satellite family of Lythraceae *sensu stricto*. The generalized morphology of the family, together with the very distinctive genera, suggests that the family is of great age (Graham *et al.*, 2000).

Pomegranate grows as a shrub or small tree up to 5 m; its leaves are simple and opposite, with entire margins and a deciduous habit (Fig. 3). The flowers are bisexual, with radial symmetry, and the fruit is classified as a berry that is divided into irregular sections, each containing many edible seeds, with the seeds surrounded by a fleshy pulp, called an aril.

Fig. 3 : Pomegranate plant laden with fruits

The pomegranate fruit develops from the ovary and is a fleshy berry 6.25 to 12.5 cm wide, and weighs 200 to 650 g (Fig. 4). The nearly round fruit is crowned by a prominent calyx and connected to the tree by a short stalk. The pomegranate has a leathery smooth skin and is divided by thin inedible membranes into a number of cells, each packed full of angular seeds contained in a juicy pulp sac. Fruits may be white or green, pink or red with a yellow background on the exterior; the pulp color follows that of the exterior. The arils contain a juicy edible layer that develops entirely from outer epidermal cells of the seed which elongate in a radial direction (Holland *et al.*, 2009).

Fig. 4 : Pomegranate fruit

3. Cultivars

The major pomegranate cultivars in main producing countries are given in Table 2. 'Wonderful' is by far the most widely planted pomegranate cultivar in USA (Fig. 5) since it offers the best combination of yield and quality. Mature trees of this cultivar discovered about 1896 in Porterville in a quantity of cuttings received from Florida, can yield more than 6000 kg ha^{-1}. The fruit is dark and deep purple red with a glossy appearance. The arils are tender, deep crimson with good flavor, and the skin of medium thickness making the fruit well adapted for both fresh consumption and processing for whole arils or juice (LaRue, 1980).

The prominent cultivars in India are Bhagwan, Ganesh, Jodhpur Red, Jalore Seedless. Ruby, Arkta and Mridula. The information on salient characteristics (Fig. 6) of the Indian cultivars is given in Table 3.

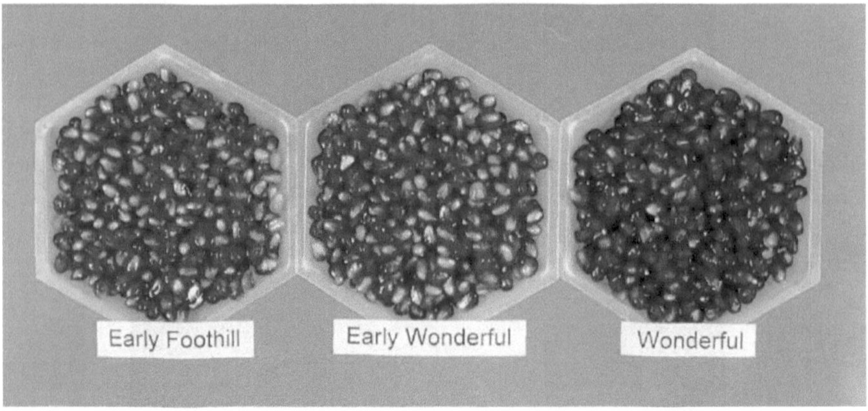

Fig. 5 : Intercultivar variation in pomegranate arils

Fig. 6 : Leading cultivars of pomegranate in India

Table 2 : List of important pomegranate cultivars in major growing countries

Country	Cultivar
China	Hongmanaozi, Taihanghong, Yushiliu 4, Hongyushizi, Mengliaihong, Zaoxuan 018, Taishan Dahongshiliou, Huashu Dahongshiliou
India	G-137, Mridula, Bhagwa, Ruby, Bedana, Kandhari,'Alandi, Dholka, Kabul, Muscat Red, Paper Shell, Poona, Spanish Ruby, Vellodu, Muscat White
Israel	Shani-Yonay, Wonderful, Red Loufani, Malissi, Ras el Baghl
USA	Wonderful, Ambrosia, Eversweet, Grenada, Kashmir, Red Sweet, Sweet pomegranate
Tunisia	Andaloussi, Beyounsi, Garoussi, Zehri
Azerbaijan	Nasimi, Vurgun, Aleko, Azerbaijan
Russian Republic	Desertnyi
Turkey	'Fellahyemez, Eksilik, Ernar, Hicaznar, Katirbasi, Asinar
Spain	Mollar de Elche, Roja, Valenciana

Contd.

Iran	Alak-E-Shirin, Bihasteh, Syah, Agha Mohammad Ali Shirin, Malase Shirin, Malase Torsh, Pust Sefeede Shirin
Italy	Ragana, Racalmuto, Profeta, Selinunte, Neirana, Dente di cavallo
Iraq	Ahmar, Aswad, Halwa
Saudi Arabia	Mangulati

Table 3 : Characteristics of commercially grown pomegranate cultivars

Cultivars	Characteristics
Bhagwa	The fruit is glossy red in colour with soft seeds and high TSS.
Ganesh	This cultivar has pinkish yellow to reddish yellow rind colour, having light pink arils and soft seeds. Fruit weighs between 225-250 g with medium TSS.
Arkta	Fruits are dark red in color with soft seeds and with high TSS.
Mridula	Fruit has bright red skin colour and weighs between 230-270 g. The arils are soft, having high TSS.
Ruby	The fruit skin is red in colour and weighs between 225-275 g. The seeds are soft, having high TSS.

4. Nutritional Composition, Health Benefits and Uses

The fruit can be divided into three parts – the seeds, juice, and the peels, that also include the interior network of membranes (Lansky and Newman, 2007). The edible portion of the pomegranates is about 55 to 60% of the total fruit weight and contains 80% juice and 20% seeds. The fresh juice contains 85% water and 15% sugars, pectins, ascorbic acid, polyphenolic flavanoids, anthocyanins and amino acids (Roy and Waskar, 1997; Kader, 2006). Pomegranate fruits are the good source of carbohydrates and minerals such as Ca, Fe and S and a moderate source of pectin. The nutritional composition of pomegranates is given in Table 4. Citric acid is the predominant organic acid in pomegranate, amounting to about 7-30 meq $100g^{-1}$ fresh weight (Malhotra *et al.*, 1983; Lee *et al.*, 1974). Glucose and sucrose are the main sugars in pomegranate. Juice contains fructose and glucose in similar quantities, Ca is 50% of its ash content, while the principal amino acids are glutamic and aspartic acids (Cemeroglu *et al.*, 1992; Aviram *et al.*, 2000). With respect to nutritive compounds in 'Mollar de Elche' cultivar the main sugars were glucose and fructose, while sucrose was found as a minor (Al-Maiman and Ahmad, 2002; Melgarejo *et al.*, 2000; Mirdehghan *et al.*, 2006). In 'Mollar de Alche' pomegranate the predominant acid was malic acid, with citric acid being the next most abundant, while oxalic, succininc and fumaric were found at much lower concentrations (Mirdehghan *et al.,* 2006). Different patterns for the organic acid profile have been reported depending on cultivar, with the main organic acids being citric and malic acid (Melgarejo *et al.,* 2000; Poyrazoglu *et al.,* 2002), although oxalic and tartaric acid were found as main acids in 'Assaria' cultivar (Miguel *et al.,* 2004a). During pomegranate growth and maturation, increases of anthocyanins and decreases of total phenolics and ascorbic acid are reported in arils (Kulkarni and Aradhya, 2005).

The chemical composition of the fruits (Table 5) differs depending on the cultivar, growing region, climate, maturity, cultivation practice, and storage conditions (Poyrazoglu *et al.,* 2002; Barzegar *et al.,* 2004; Fadavi *et al.,* 2005). Significant variations in organic acids, phenolic compounds, sugars, water soluble vitamins, and minerals of pomegranates have been reported over the years by various researchers (Aviram *et al.,* 2000; Mirdehghan and Rahemi, 2007; Cam *et al.,* 2009; Davidson *et al.,* 2009; Tezcan *et al.,* 2009). The edible part of the pomegranate fruit (50%) consists of 40% arils and 10% seeds. Arils contain 85% water, 10% total sugars, and 1.5% pectin, organic acid such as ascorbic acid, citric acid, and malic acid, and bioactive compounds such as phenolics and flavonoids, principally anthocyanins (Aviram *et al.,* 2000; Tezcan *et al.,* 2009). The seeds are a rich source of total lipids, pomegranate seed oil comprises 12% to 20% of total seed weight. The seeds also contain protein, crude fibres, vitamins, minerals, pectin, sugars, polyphenols, isoflavones, phytoestrogen coumestrol, and the sex steroid estrone (El-Nemr *et al.,* 2006; Syed *et al.,* 2007).

Pomegranates are rich in antioxidants and phytochemicals. The major class of pomegranate phytochemicals is the polyphenols that predominate in the fruit. Pomegranate polyphenols include flavonoids (flavonols, flavanols, and anthocyanins), condensed tannins (proanthocyanidins), and hydrolyzable tannins (ellagitannins and gallotannins). Other phytochemicals identified from the pomegranate are organic and phenolic acids, sterols and triterpenoids, fatty acids, triglycerides, and alkaloids.

Commercial pomegranate juice is obtained by a hydrostatic pressing process of whole fruits whereby two predominant types of polyphenolic compounds are extracted into juice: flavonoids and hydrolyzable tannins (Gil *et al.,* 2000). The flavonoids include flavonols such as luteolin, quercetin, and kaempferol found in the peel extract (Van Elswijk, 2004) and anthocyanins found in the arils (Hernandez *et al.,* 1999; Santagati *et al.,* 1984). Anthocyanins are the water-soluble pigments responsible for the bright red color of juice. Pomegranate anthocyanins include pelargonidin-3-glucoside, cyanidin-3-glucoside, delphinidin-3-glucoside, pelargonidin 3,5-diglucoside, cyanidin 3,5-diglucoside, and delphinidin 3,5-diglucoside (Hernandez *et al.*, 1999; Santagati *et al.*, 1984). Pomegranate seed oil contains levels of >60% of punicic acid, a cis-9, trans-11, cis-13 conjugated linolenic acid. Other fatty acids found in pomegranate seed oil include linoleic, oleic, palmitic, stearic, arachidic, and palmitoleic acid (Melgarejo *et al.*, 1995). In addition, there have been reports of sex hormones such as estrone and coumestrol in pomegranate seeds (Moneam *et al.*, 1988; Melgarejo *et al.*, 1995).

The antioxidant activities of the whole pomegranate fruit and its anthocyanidins (delphinidin, cyanidin, and pelargonidin) extracts were studied (Noda *et al.*, 2002). Pomegranate fruit extract exhibited scavenging activity against hydroxyl radicals and superoxide anions, and this effect could be related to the fruit anthocyanidins. The IC_{50} (the concentration needed to get 50% inhibition) values for superoxide anion scavenging capacity of delphinidin, cyanidin, and pelargonidin were 2, 22, and 456 μmol L^{-1}, respectively. The above anthocyanidins did not effectively scavenge NO radical, but inhibited H_2O_2 -induced lipid peroxidation in rat brain homogenates, with IC_{50} values of 0.7, 3.5, and 85 μmol L^{-1} for delphinidin, cyanidin, and pelargonidin, respectively (Noda *et al.*, 2002).

The soluble polyphenol content in pomegranate juice varies between 0.2 and 1.0%. It has been shown that the antioxidant activity of pomegranate juice is higher than that of red wine and green tea (Gil *et al.*, 2000). Pomegranate juice is an important source of phenolic compounds, with anthocyanins being one of the most important, especially the 3-glucosides and 3,5-diglucosides of delphinidin, cyaniding and palargonidin (Du *et al.*, 1975). These components alongwith gallagyl-type tannins, ellagic acid derivatives, and other hydrolysable tannins contribute to the antioxidant activity of pomegranate juice (Miguel *et al.*, 2004b).

Pomegranate has accompanied mankind over hundreds of years and it is a symbol of life, health, longevity, femininity, fecundity, knowledge, immortality, and spirituality (Mahdihassan, 1984). In alternative medicines, pomegranate is considered as an antiparasitic agent (Naqvi *et al.*, 1991), a blood tonic, and to help in aphthae, diarrhea, and ulcers (Caceres *et al.*, 1987). Modern uses of pomegranate include treatment of acquired immune deficiency syndrome, hormone replacement therapy, cardiovascular protection, oral hygiene, and adjunct therapy to increase bioavailability of radioactive dyes during diagnostic imaging, as well as cosmetic beautification and enhancement (Curry, 2004).

The pomegranate fruit could be considered a functional food, because it has valuable compounds in different parts of the fruit that display functional and medicinal effects. These can act as antioxidant (Cam *et al.*, 2009), as antitumoral (Hamad and Al-Momene, 2009) or antihepatotoxic (Celik *et al.*, 2009) agents, and improves cardiovascular health (Davidson *et al.*, 2009). They have antimicrobial (Duman *et al.*, 2009), anti-inflammatory (Lee *et al.*, 2010), antiviral (Haidari *et al.*, 2009), antidiabetic (Xu *et al.*, 2009) properties, and they can improve oral (Di Silvestro *et al.*, 2009) and skin (Aslam *et al.*, 2006) health. They help to prevent Alzheimer's disease (Singh *et al.*, 2008) and erectile dysfunction in male patients (Forest *et al.*, 2007).

Table 4 : Nutritional composition of pomegranate fruit

Nutrients	Units	Value $100g^{-1}$
Proximates		
Water	g	77.93
Energy	kcal	83
Energy	kJ	346
Protein	g	1.67
Total lipid (fat)	g	1.17
Ash	g	0.53
Carbohydrate, by difference	g	18.70
Fiber, total dietary	g	4.0
Sugars, total	g	13.67
Minerals		
Calcium, Ca	mg	10
Iron, Fe	mg	0.30
Magnesium, Mg	mg	12
Phosphorus, P	mg	36
Potassium, K	mg	236
Sodium, Na	mg	3
Zinc, Zn	mg	0.35
Copper, Cu	mg	0.158
Manganese, Mn	mg	0.119
Selenium, Se	µg	0.5

Contd.

Vitamins		
Vitamin C, total ascorbic acid	mg	10.2
Thiamin	mg	0.067
Riboflavin	mg	0.053
Niacin	mg	0.293
Pantothenic acid	mg	0.377
Vitamin B-6	mg	0.075
Folate, total	µg	38
Folate, food	µg	38
Folate, DFE	µg_DFE	38
Choline, total	mg	7.6
Vitamin E (alpha-tocopherol)	mg	0.60
Vitamin K (phylloquinone)	µg	16.4
Lipids		
Fatty acids, total saturated	g	0.120
12:0	g	0.006
14:0	g	0.006
16:0	g	0.070
18:0	g	0.038
Fatty acids, total monounsaturated	g	0.093
16:1 undifferentiated	g	0.012
18:1 undifferentiated	g	0.077
20:1	g	0.004
Fatty acids, total polyunsaturated	g	0.079
18:2 undifferentiated	g	0.079
Fatty acids, total trans	g	0.009
Campesterol	mg	1
Beta-sitosterol	mg	4

Source : USDA Nutrient Data Base

Table 5 : Principal constituents of different parts of pomegranate fruit

Fruit component	Constituents	Reference
Juice	Anthocyanins, glucose, organic acids, ascorbic acid, ellagic acid, hydrolysable tannins, gallic acid, caffeic acid, catechin, quercetin, rutin, minerals	Poyrazoglu *et al.* (2002); Ignaro *et al.* (2006); Lansky and Newman (2007); Heber *et al.* (2007); Mousavinejad *et al.* (2009); Jaiswal *et al.* (2010)
Seed oil	Conjugated linolenic acid, linoleic acid, oleic acid, stearic acid, punicic acid, eleostearic acid, catalpic acid	Ozgul-Yucel (2005); Fadavi *et al.* (2006); El-Nemr *et al.* (2006); Sessano *et al.* (2009)
Peel	Luteolin, quercetin, kaempferol, gallagic, ellagic acid glycosides, ellagic acid punicalagin, punicalin, pedunculagin	Van Elswijk *et al.* (2004); Amakura *et al.* (2000); Seeram *et al.* (2005)

5. Maturity Indices and Quality Components

Pomegranates do not ripen off the tree and should be picked when fully ripe to ensure their best flavor (Kader *et al.*, 1984). The pomegranate fruit reaches full maturity within 4.5 - 6 months after full bloom, depending on climatic conditions (Ben-Arie *et al.*, 1984; Kader, 2006). Pomegranates can be harvested when they reach a certain size and skin color. Other maturity indices are titratable acidity and soluble solids contents. Each pomegranate type requires a certain acid : soluble solids content at harvest. Acidity of pomegranates varies between 0.13 and 4.98% at harvest. The acidity is < 1% in sweet cultivars, 1 to 2% in sweet-sour cultivars and > 2% in sour cultivars (Onur and Ka°ka, 1985). TSS of pomegranates varies from 8.3 to 20.5% at harvest. Thus, maturity indices depend on cultivar. Juice tannin content < 0.25% is preferred, and red juice color equal to or darker than Munsell color chart 5R-5/12 (Fig. 7) is desirable (Crisosto *et al.*, 2000). The fruits should be harvested before they become overripe and crack (split) open, especially under rainy conditions.

Fig. 7 : Red color of juice equal to or darker than Munsell Color Chart 5R-5/12

Pomegranate fruit quality depends on the following indices:

- Fruits should be free from preharvest defects such as fruit cracking, sun burn, surface blackening etc.
- Careful harvesting is required (Fig. 8) and fruits should not be injured during harvesting and transportation. Fruits must be free from surface abrasions, impact bruising, vibration injury, cuts and wounds (Fig. 9) etc.
- Skin color should be characteristic to cultivar. Dark red or pink red color is preferred in Indian market.
- Large fruit size is preferred.
- Aril color should be dark red and intense with soft seed or seedless.
- Flavor depends on sugar/acid ratio, which varies among cultivars. Soluble solids contents above 17% and total phenolics content below 0.25% are desirable for optimal levels of sweetness and astringency, respectively (Kader *et al.*, 1984; Crisosto *et al.*, 2000).

Fig. 8 : Harvesting operations in pomegranate

Fig. 9 : Surface abrasions and bruising during postharvest handling

6. Respiration and Ethylene Production Rates

Pomegranate fruits showed low respiration rates and non-climacteric respiration pattern (Elyatem and Kader, 1984). Pomegranates produced trace amounts of C_2H_4 and showed no response to exogenous C_2H_4 treatments. Both CO_2 and C_2H_4 production rates increased with temperatures (Fig. 10 and 11). The Q_{10} values for respiration were 3.4 between 0 and 10°C, 3.0 between 10 and 20°C, and 2.3 between 20 and 30°C (Elyatem and Kader, 1984). Respiration rate in pomegranate fruit declined with time after harvest. At temperatures between 0 and 10°C, the respiration rate remained below 8 ml CO_2 kg^{-1} h^{-1} during storage for up to 3 months. Pomegranates produced only trace amounts of ethylene (<2 µl kg^{-1} h^{-1}) while being held at 20°C for 2 weeks (Kader *et al.*, 1984). However, Sayyari *et al.* (2011) observed increase in respiration rate (from 23.80 to 32,2 mg CO_2 kg^{-1} h^{-1}) during 84 days storage. The ranges of respiration rates for 'Wonderful' pomegranates were 2-4, 4-8, and 8-18 ml CO_2 kg^{-1} h^{-1} at 5, 10 and 20°C, respectively, while ethylene production rates remained below 0.2 µl kg^{-1} h^{-1}.

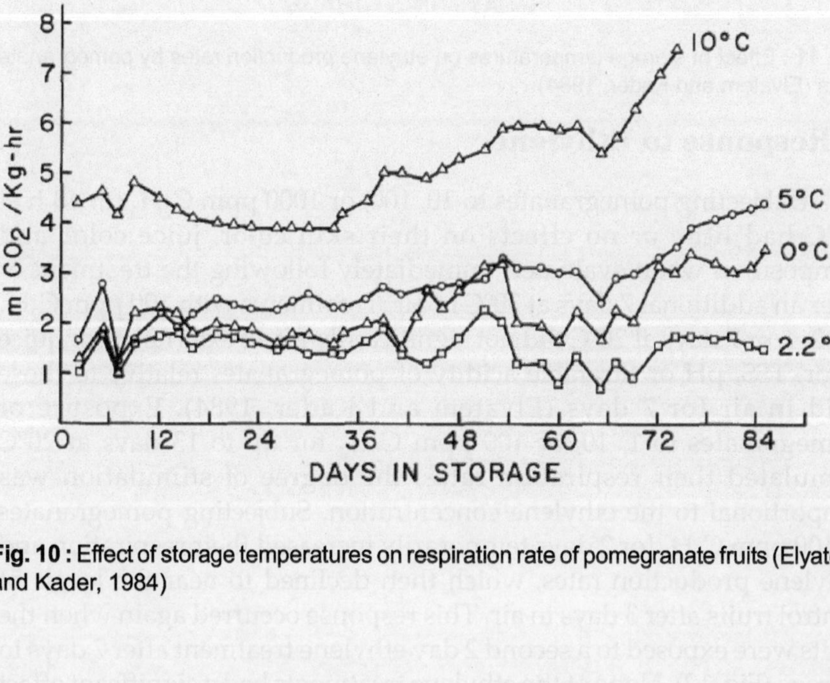

Fig. 10 : Effect of storage temperatures on respiration rate of pomegranate fruits (Elyatem and Kader, 1984)

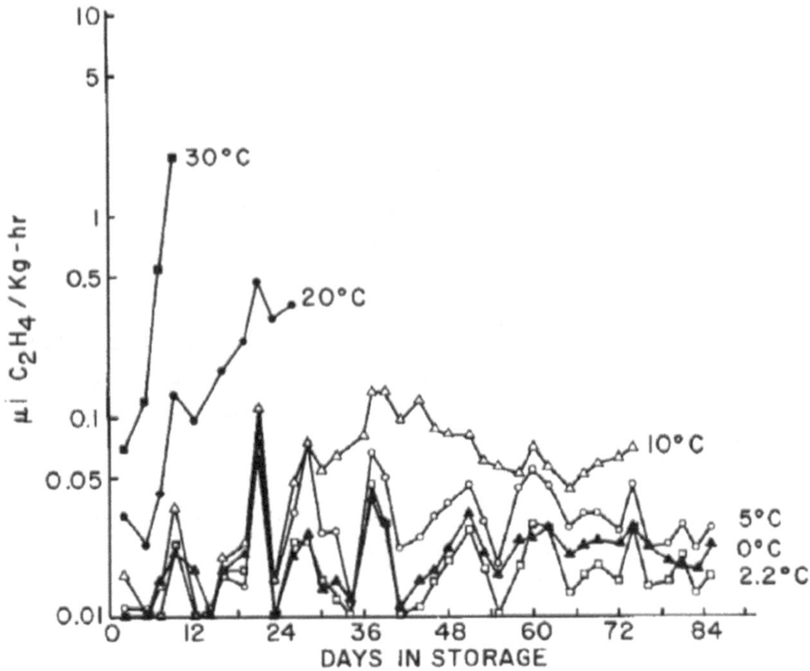

Fig. 11 : Effect of storage temperatures on ethylene production rates by pomegranate fruits (Elyatem and Kader, 1984)

7. Response to Ethylene

Subjecting pomegranates to 10, 100, or 1000 ppm C_2H_4 for 48 h at 20°C had little or no effects on their skin color, juice color and composition when evaluated immediately following the treatment or after an additional 7 days at 20°C in air. Treatments with 100 ppm C_2H_4 for 2, 4, or 7 days at 20°C did not significantly affect external color, juice color, TSS, pH or titratable acidity of pomegranates relative to those held in air for 7 days (Elyatem and Kader, 1984). Exposure of pomegranates to 1, 10, or 100 ppm C_2H_4 for up to 13 days at 20°C stimulated their respiration rates; the degree of stimulation was proportional to the ethylene concentration. Subjecting pomegranates to 100 ppm C_2H_4 for 2 days temporarily increased their respiration and ethylene production rates, which then declined to near the levels of control fruits after 3 days in air. This response occurred again when the fruits were exposed to a second 2 day ethylene treatment after 7 days in storage (Fig. 12). None of the ethylene treatments had a significant effect on skin color, juice color, TSS, pH, or acidity of the fruits (Kader *et al.*, 1984). Therefore pomegranates do not ripen once removed from the tree and should be picked when fully ripe to ensure the best eating quality for the consumer. Also, there is no value in treating harvested pomegranates with ethylene.

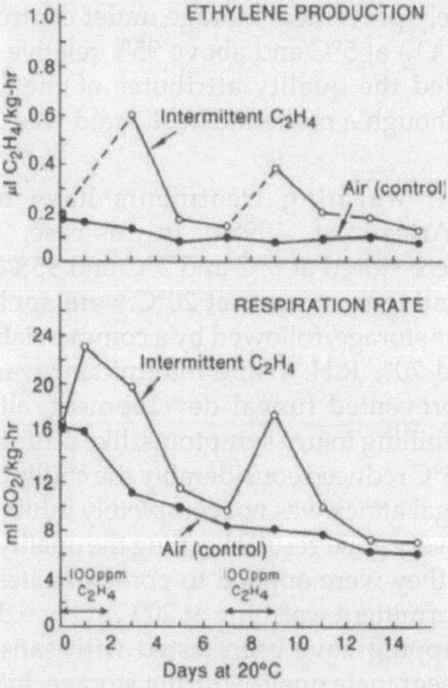

Fig. 12 : Subjecting pomegranates to 100 ppm ethylene for 2 days temporarily increased their respiration and ethylene production rates (Kader *et al.*, 1984)

8. Storage Temperatures

Temperature is the most important factor to control the respiratory activity, transpiration and the development of microbial pathogens. A fast pre-refrigeration using forced air is one of the simplest ways to extend pomegranate commercial life. This temperature has to be around 5°C to prevent the production of physiological disorders, during 2-3 months storage (Kader *et al.*, 1984; Artes, 1992). During their storage several physiological and enzymatic disorders, and fungal attacks, can affect seriously the quality attributes. Spanish 'sweet' pomegranates can suffer chilling injury if they are stored for more than two months at temperatures below 5°C (Artés, 1992). Some cultivars, *i.e.* 'Wonderful', can be stored without problems for 2 months at 5°C. The minimal safe temperature to store sweet pomegranates is 10°C (Kader *et al.*, 1984).

This temperature, however, does not prevent fungal development. Studies have demonstrated that the conventional storage of 'Mollar' pomegranate at 5°C and 90-96% R.H. up to eight weeks, leads to an acceptable decrease in fungal decay losses, but the risk of chilling injury was not completely prevented. Storage under controlled atmospheres (5% CO_2 and 5% O_2) at 5°C and above 95% relative humidity for two months, improved the quality attributes of the freshly harvested pomegranate, although a moderate husk scald was observed (Artes *et al.*, 1996).

Intermittent warming treatments have been applied to pomegranates (Artes *et al.*, 1998a). In this case, 'Mollar de Elche' pomegranates were stored at 0°C and 5°C and 95% R.H., for 80 days. Intermittent warming treatments at 20°C were applied in cycles, one day every six days storage, followed by a commercialization period of 7 days at 15°C and 70% R.H. While intermittent warming during the storage at 0°C prevented fungal development, although increased susceptibility to chilling injury symptoms, like pitting and husk surface scald, storage at 5°C reduced considerably the chilling injury symptoms although the fungal attack was not completely inhibited. The warming treatments led to very good results keeping the quality of pomegranates, especially when they were applied to pomegranates stored at 0°C. In pomegranate intermittent warming at 20°C every 6 days at 2 or 5°C, as well as film wrapping have been tested with satisfactory results in maintaining pomegranate quality during storage, in terms of retention of anthocyanin and titratable acidity, reduction of decay, and alleviation of chilling injury (Nanda *et al.*, 2001; Artes *et al.*, 2000). Effect of pre-storage heat treatment was investigated to maintain nutritive and functional properties during postharvest cold storage of pomegranate (Mirdehghan *et al.*, 2006). Pomegranate fruits were heat treated (dips at 45°C for 4 minutes) and stored at 2°C for 90 days. Arils from heat treated pomegranates exhibited higher total antioxidant activity than controls, which was correlated to the high levels of total phenolics and to lesser extent to ascorbic acid and anthocyanin contents. The levels of sugars (glucose and fructose) and organic acids (malic, citric and oxalic acids) remained also at higher concentrations in arils from treated fruits (Mirdehghan *et al.*, 2006)

Another technological treatment that is starting to be used with success in pomegranate is the preconditioning at moderate temperature (30-40°C) and high R.H. (90-95%) for a short period of time (1- 4 days), a technique also known as curing, that is applied before the conventional refrigerated storage. As reported, a pre-treatment at 35°C and 90-95%

R.H. applied for three days previously to a refrigerated storage during 80 days at 5 or 2°C and 90-95% R.H., reduced considerably the pitting and husk superficial scald (produced by the enzyme polyphenol oxidase) compared to control pomegranates without the pre-conditioning treatment. The effects observed were more marked when the conservation was carried out at 2°C than at 5°C, particularly during the additional period of one week at 15-20°C and 70-75% R.H., applied to simulate the retail sale period (Artes *et al.*, 1998b,c).

9. Relative Humidity

The relative humidity is the second factor in importance. The key concept is the deficit in vapour pressure or difference between relative humidity of the environment. Temperature and relative humidity are closely related and the objective is minimizing the weight loss without increasing the microbial development and decay. As pomegranates are commercialized without post-harvest treatments (washing, waxing, fungicides), the fruits must be gently brushed, and kept with a relative humidity around 90-95% if the fruits are healthy and clean (Artes, 1992).

10. Controlled and Modified Atmospheres

Artes *et al.* (2000a, b) investigated the effect of different packaging films, unperforated polypropylene (UPP) 25 μm thick, and perforated polypropylene 20 μm thick, on the quality attributes, physiological disorders, and decay of sweet 'Mollar de Elche' fruit, during cold storage and shelf life. They observed that at 2 and 5°C within the sealed UPP bags, traces of C_2H_2 were detected occasionally, while, within perforated bags the atmospheric composition was practically like air at the end of the 12 weeks of storage. At the end of the shelf life, all treatments maintained or increased in pH values, with the exception of perforated polypropylene treatment at 5°C which had slightly decreased values. The values of titratable acid (TA) at the end of 12 week storage at 2 or 5°C were lower than those at harvest except for perforated polypropylene at 5°C. Comparing the soluble solids content at harvest with those of the treatments at the end of refrigerated storage, no changes were detected, however, during the refrigerated storage at 2 and 5°C, the soluble solid content/titratable acid ratio increased significantly in MAP-stored fruits, due to the decrease in titratable acid based on their assessment, the best treatment for maintaining red colour of the arils at the end of cold storage was perforated polypropylene at 5°C and the lowest physiological disorders were found in modified atmosphere packaging treatments at both refrigerated-storage temperatures.

Another study on whole fruit 'Wonderful' pomegranate harvested from commercial orchards in the central coastal region of Israel was reported by Porat *et al.* (2009). They recommended the use of either Xtend® Easy-Tear or regular Xtend® modified atmosphere or modified humidity bags in 4–5 Kg export cartons. They observed that the Xtend® film had a higher moisture vapour transmission rates in comparison to conventional polyethylene and polypropylene films, which helps to eliminate excess moisture. Xtend® packaging reduced weight loss and scald incidence of the fruit and maintains internal quality and taste.

Kupper *et al.* (1995) found that 'Hicaz' pomegranate could be successfully stored for 6 months in controlled atmosphere containing 6 kPa CO_2 and 3 kPa O_2 at 6°C. Controlled atmosphere storage reduced weight loss, decay development, and husk scald in 'Mollar' stored in atmosphere containing 5 kPa O_2 and 0 or 5kPa CO_2 at 5°C for 2 months without any ethylene scrubber (Artes *et al.,* 1996). The anthocyanin content of arils in 'Wonderful' pomegranate increased during air storage at 10°C for 6 weeks, but the air enriched with 10 or 20 kPa CO_2 suppressed the anthocyanin biosynthesis during storage due to the inhibitory effect on phenylalanine ammonia lyase activity (Holcroft *et al.,* 1998). Hess-Pierce and Kader (2003) recommended 5 kPa O_2 + 15 kPa CO_2 as the optimal controlled atmosphere conditions for 'Wonderful' pomegranate at 7°C for 6 months (Fig. 12). Further studies on the controlled atmosphere storage of the same cultivar showed that 1 kPa O_2, or 1 kPa O_2 + 15 kPa CO_2 resulted in greater accumulation of fermentative volatiles than the CA treatment with 5 kPa O_2 + 15 kPa CO_2 (Defilippi *et al.,* 2006). Combination of controlled atmosphere storage with pre-storage antifungal treatment has been found very effective in controlling the *Botrytis* decay problem during long-term storage (Palou *et al.,* 2007). Synergistic effect of controlled atmosphere containing 5 kPa O_2 + 15 kPa CO_2 with antifungal treatment (potassium sorbate 3 minute dip in 3 per cent [w/v] solution at 21°C) reduced the decay development, maintained internal and external fruit quality during 15 weeks storage at 7.2°C (Palou *et al.,* 2007). It is evident from the published work that pomegranate is highly sensitive to low-oxygen atmospheres (<5 kPa), chilling injury and decay. Therefore, the beneficial effect of controlled atmosphere can be harnessed only if the optimal atmosphere conditions are worked out for each cultivar. The combination approach of pre-storage antifungal treatment with the optimal storage atmosphere (5 kPa O_2 + 15 kPa CO_2), and temperature (~ 7°C) suggested by Palou *et al.* (2007) seems quite promising and practical to encounter post harvest storage problems in pomegranate.

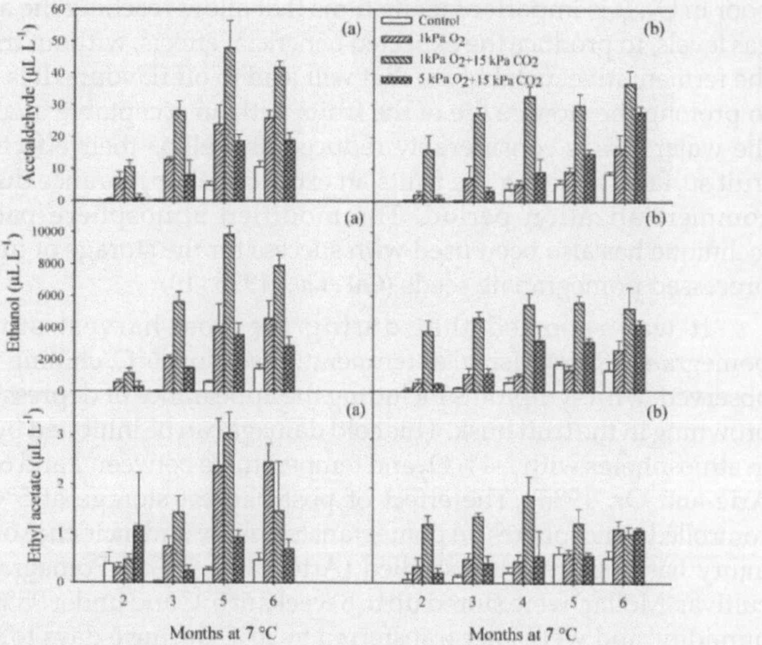

Fig. 13 : Effect of storage duration at 7°C and controlled atmosphere treatments on concentrations of fermentative metabolites in arils of (a) mid season and (b) late season harvested 'Wonderful' pomegranates

The individually shrink film (BDF-2001 and D-995) wrapped fruit of 'Ganesh' could be stored for 12, 9 and 4 weeks at 8°C, 15°C and 25°C, respectively, whereas non wrapped fruit could be kept for 7, 5 and 1 week under similar storage conditions (Nanda *et al.,* 2001). There was a tremendous reduction of weight loss in shrink-wrapped fruit in addition to greater firmness retention at the end of 12 weeks storage at 8°C. Individual shrink wrapping of pomegranate fruit maintained the harvest freshness for 1 month at ambient conditions and 3 months at 8°C. In conclusion, the beneficial effect of controlled atmosphere and modified atmosphere packaging in pomegranate include extended storage life, control of husk scald, chilling injury decay and weight loss. However, the only limitation is the inhibitions of the accumulation of anthocyanin in the aril during long-term storage of fruit in CO_2-enriched atmospheres.

The use of plastic packaging of whole pomegranates, or even pomegranate arils, in microperforated polypropylene bags, has also been studied. In these cases the respiration of pomegranates inside the bags, and the selective permeability of the polypropylene films for the different gases, in bags that are hermetically sealed, a modified atmosphere is generated around the fruit, atmosphere which is enriched in CO_2 and poor in O_2. It is important using films that allow reaching the adequate gas levels, to produce the expected beneficial effects, without triggering the fermentative metabolism that will lead to off flavours. It is possible to prolong the storage life of the fruits with an acceptable quality, and the water loss is considerably reduced as well as their effects on the fruit surface, showing the fruits an exceptional appearance during the commercialization period. The modified atmosphere packaging technique has also been used with success for the storage of minimally processed pomegranate seeds (Gil *et al.*, 1996a,b).

It was reported that during the post-harvest storage of pomegranates from Israel at temperatures below 6°C, chilling injury is observed, with symptoms including the appearance of depressions and browning in the fruit husk. This cold damage can be inhibited by storage in atmospheres with 2-4% O_2 and temperatures between 2 and 6°C (Ben-Arie and Or, 1986). The effect of post-harvest storage at 5°C under controlled atmospheres on pomegranate quality and incidence of chilling injury has been recently studied (Artes *et al.*, 1996). Pomegranates of cultivar 'Mollar' were stored up to 8 weeks at 5°C and under 95% relative humidity, and were later transferred to 20°C during 6 days to simulate the commercialisation period. Different controlled atmospheres were tested and the quality of the fruits was compared to that of the fruits stored in air. Storage under controlled atmospheres (10% O_2 and 5% CO_2; 5% O_2 and 5% CO_2; 5% O_2 and 0% CO_2) reduced weight loss, fungal decay and chilling injuries (husk surface scald). The storage in controlled atmospheres, however, reduced the vitamin C and sugars contents. Other studies on 'Wonderful' pomegranates stored in air enriched with 10 or 20% CO_2, showed that the colour of the seeds stored in air increased during refrigerated storage (10°C, 6 weeks), while the seed pigmentation was less intense in those of pomegranates stored in 10% CO_2, and even decreased in those fruits stored in 20% CO_2. This increase in seeds pigmentation was well correlated with the PAL activity of pomegranate seeds, a key enzyme of phenolic metabolism (Holcroft *et al.*, 1998). Nevertheless, these authors found that storage under atmospheres with a moderate CO_2 composition (10%) prolonged the storage life of pomegranates and kept the original quality, including an adequate seed colour.

Hess-Pierce and Kader (2003) reported that, in intact fruit, CO_2 enriched atmospheres up to 10 kPa resulted in higher concentrations of fermentative metabolites, including acetaldehyde and ethanol, especially after 4 and 5 months at 5°C, 7.5°C or 10°C. Accumulation of these volatiles was greater at 7.5°C and 10°C than at 5°C. However the highest concentrations were below the threshold values for off-flavors detection in fruits like pomegranates containing more than 10% total sugars (Ke *et al.*, 1991).

11. Weight Loss

According to Elyatem and Kader (1984), weight loss of 'Wonderful' pomegranates during cold storage is largely due to water lost through natural porosity of the skin and recommended storage in 95% or higher relative humidity. Shrivelling symptoms on fruit are noticeable only when weight loss exceeds 5% or more of the initial weight (Fig. 14). In general, pomegranates that are marketed without waxing should be kept at 90% to 95% RH to prevent water loss. However, use of plastic liners and waxing can reduce water loss, especially under conditions of lower RH.

Fig. 14 : Water loss in pomegranate fruits at different temperatures

12. Grading

Grading is desirable for fresh fruit marketing and export purpose. Table 6 provide the information on different grades of pomegranate fruit in India.

Table 6 : Different grades of pomegranate fruit and their description

Grade designation	Grade requirement	Grade tolerances
Extra class	Pomegranate in this class must be of superior quality. They must have the shape, development and colouring that are typical of the variety and/or commercial type. They must be free of defects, with the exception of very slight superficial defects, provided these do not affect the general appearance of the produce, the quality, the keeping quality and presentation in the package.	5% by number or weight of pomegranates not satisfying the requirements of the grade, but meeting those of class I grade or, exceptionally, coming within the tolerances of that grade.
Class I	Pomegranates in this class must be of good quality. They must be characteristics of the variety and/or commercial type. The following slight defects may be allowed, provided these do not affect the general appearance of the produce, the quality, the keeping quality and presentation in the package.- a slight defect in shape.- a slight defect in colouring;-slight skin defects (i.e. scratches, scars, scraps and blemishes) not exceeding 5% of the total surface area.	10% by number or weight of pomegranates not satisfying the requirements of the class, but meeting those of class II or, exceptionally, coming within the tolerances of that grade.
Class II	This class includes pomegranates which do not qualify for inclusion in higher classes, but satisfy the minimum requirements. Following defects may be there provided the pomegranates retain their essential characteristics as regard the quality, the keeping quality and presentation:- - defects in shape; - defects in colouring - skin defects (i.e., scratches, scars, scrapes and blemishes), not exceeding 10% of the total surface area	10% by number or weight of pomegranates not satisfying the requirements of the grade, but meeting the minimum requirements.

13. General Postharvest Handling Protocol for Pomegranates

Production
↓
Monitoring pesticide/fungicide application program
↓
Inspection at farm level or final field survey
↓
Harvesting by hand and cutting stem properly
↓
Transport to packhouse
↓
Receipt at packhouse
↓
Rejection of unwanted fruits
↓
Washing with a disinfectant and then with clean water
↓
Removal of excess water
↓
Treatment with fungicide and wax
↓
Drying
↓
Sizing
↓
Grading for quality
↓
Nose cleaning
↓
Packed into cardboard boxes
↓
Precooling
↓
Palletisation
↓
Storage in cold store
↓
Container loading
↓
Transportation to port

Fig. 15 : Postharvest handling operations in pomegranate

14. Physiological Disorders

14.1 Chilling injury

Chilling injury, decay, and weight loss are most important problems limiting storability of pomegranate. Elyatem and Kader (1984) observed that 'Wonderful' pomegranates develop chilling injury symptoms, namely brown discoloration or scald of the skin and surface pitting, if stored in air (conventional cold storage) at 5°C or lower temperatures (Fig. 16 and 17). Several studies showed that, compared to air, storage of 'Wonderful' pomegranates in different controlled atmosphere conditions significantly extended their postharvest life, not only by delaying fruit senescence but also by inhibiting the growth of microorganisms causing decay (Ben-Arie and Ore, 1986; Holcroft *et al.*, 1998). Hess-Pierce and Kader (2003) suggested storage at 7.5°C in 5 kPa O_2 + 15 kPa CO_2 as the optimum condition to maintain the original quality of 'Wonderful' pomegranates. Under these conditions, carefully stored fruit were satisfactorily kept for up to 20 weeks.

Fig. 16 : External symptoms of chilling injury

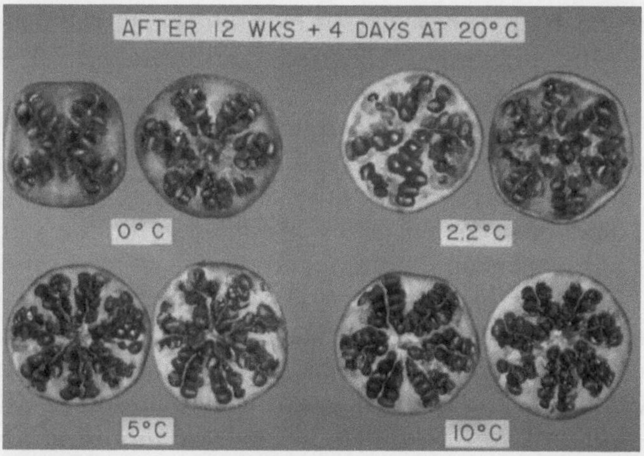

Fig. 17 : Effect of temperatures on chilling injury of pomegranates

To reduce CI incidence and to extend storability and marketing of pomegranates, good results were obtained with polyamines (Mirdehghan *et al.*, 2007a), heat (Mirdehghan *et al.*, 2007b), salicylic acid (Sayyari *et al.*, 2009), methyl jasmonate (Zolfagharinasab and Hadian, 2007), or methyl salicylate (Sayyari *et al.*, 2011) treatments prior to cold storage.

Pomegranate fruits were treated with salicylic acid at different concentrations (0.7, 1.4 or 2.0 mM), then stored at 2°C for 3 months to investigate CI. Salicylic acid treatments especially at 2mM concentration, were highly effective in reducing CI and electrolyte leakage in the husk of pomegranate, as well as ascorbic acid loss (Sayyari *et al.*, 2009). Methyl jasmonate and methyl salicylate have potential postharvest applications in reducing CI, maintaining quality and improving the health benefits of pomegranate fruit consumption by increasing the antioxidant capacity (Sayyari *et al.*, 2011a). The CI symptoms were significantly reduced by methyl jasmonate or methyl salicylate treatments. In addition, both treatments significantly increased total phenolics and anthocyanins with respect to controls. Hydrophilic and lipophilic total antioxidant activity decreased in control arils, but in both methyl jasmonate and methyl salicylate treated fruits hydrophilic total antioxidant increased while no significant changes occurred for lipophilic total antioxidant (Sayyari *et al.*, 2011a). Acetyl salicylic acid also reduced the CI in pomegranates. Acetyl salicylic acid could have potential postharvest application for reducing CI, maintain quality and improve the health benefits of pomegranate fruit consumption by increasing the antioxidant activity (Sayyari *et al.*, 2011b).

Reduction in CI of pomegranate fruit after heat treatment has been noticed by Mirdehghan *et al.* (2007b). Pomegranate fruit were heat treated by hot water dip at 45°C for 4 minute, and then stored at 2°C for 90 days. Fruit immersed in distilled water at 25°C for 4 minute served as controls. Pomegranates developed CI, manifested as increases in skin browning and electrolyte leakage, which were highly correlated. The severity of damage in control fruit was related to softening and loss of fatty acids with a concomitant reduction in the ratio of unsaturated/saturated fatty acids during storage. These CI symptoms were reduced in heat treated pomegranates. The heat treatment induced increases in free putrscine and spermidine during storage, which could have a role in the lower rate of fruit softening and in diminution of CI severity. These higher polyamine levels as well as maintenance of the unsaturated/saturated fatty acid ratio during storage could account for the maintenance of membrane integrity and fluidity. Thus, heat treatment could induce tolerance mechanism to low temperature through stimulation of polyamine biosynthesis (Mirdehghan *et al.*, 2007b). Exogenous application of putrscine or spermdine under pressure infiltration or immersion could induce acclimation of pomegranate to low temperature, and in turn protect the fruit from CI by increasing the levels of endogenous putrscien and spermdine, since the normal levels would not be high enough to induce this adaptation to cold storage. The polyamine treatment retarded the maturation process by reducing softening and the increase of the ratio TSS/acidity, as well as the loss of weight. Pressure infiltration of polyamines seems to be the best tool to increase the endogenous polyamines, although immersion also imparted beneficial effects which could be attributed to the waxy nature of pomegranate skin (Mirdehghan *et al.*, 2007a).

14.2 Husk scald

For long term storage, scald of the husk surface is another factor limiting storage life (Ben-Arie and Or, 1986). Scald symptoms appear as a superficial (skin) browning, similar to superficial scald of apples, and generally develops from the stem end of the fruit, spreading toward the blossom end as the severity increases (Fig. 18). Moreover, husk scald increases the susceptibility of the fruit to decay.

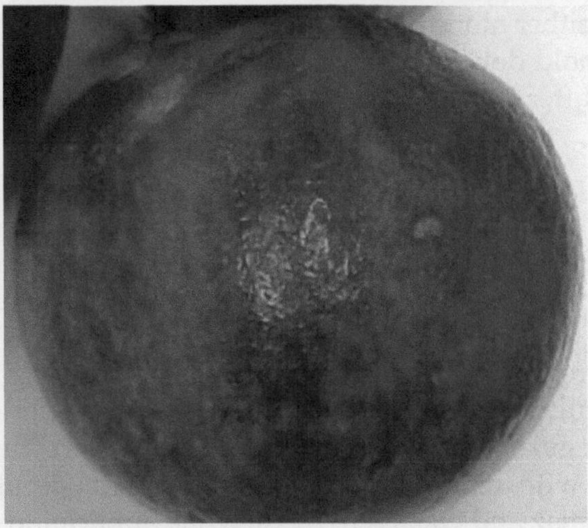

Fig. 18 : Superficial scald symptoms

Defilippi *et al.* (2006) evaluated the efficacy of pre-storage treatment with diphenylamine and/or 1-MCP and of atmospheric modification during storage on scald incidence and severity on 'Wonderful' pomegranates. Scald incidence and severity were greater on pomegranates harvested during late season than on those harvested during mid season, indicating that this disorder may be associated with senescence. All pomegranates from both harvests that were kept in air exhibited some scald after 4-6 months at 7°C. Neither diphenylamine, at 100 or 2200 $\mu L\ L^{-1}$, nor 1-MCP at 1 $\mu L\ L^{-1}$, alone or together reduced scald incidence and severity. In contrast, the three controlled atmosphere storage conditions tested (1 kPa O_2, 1 kPa O_2 + 15 kPa CO_2 and 5 kPa O_2 + 15 kPa CO_2) significantly reduced scald incidence and severity on pomegranate from both harvest dates for up to 6 months at 7°C (Defilippi *et al.*, 2006).

D'Aquino *et al.* (2010) assess the effectiveness of individual film packaging, applied as stand alone treatment or in combination with fludioxonil, on reducing the occurrence of husk scald, weight loss and decay. Fruit were dipped in an aqueous mixture containing 600 mg L^{-1} fludioxonil, wrapped or not wrapped with a polyolephinic heat-shrinkable film and stored at 8°C and 90% RH for 6 or 12 weeks plus one additional week of simulated shelf life at 20°C and 65-70% RH. Respiration decreased both in cold storage and at 20°C. Ethylene production was not detected during cold storage; its rate increased upon transfer to 20°C. Film wrapping almost completely inhibited weight loss and husk scald and preserved fruit freshness for the whole storage time.

Fludioxonil either alone or in combination with wrapping, effectively developed mold development, resulting in 50 to 67% less decay than control fruit after 12 weeks at 8°C plus one week shelf life (D'Aquino *et al.*, 2010).

14.3 Fruit cracking

Fruit cracking may affect pomegranate fruits and sometimes cause significant commercial losses, but fruit cracking actually may be regarded as the last stage of the normal pomegranate development process where the fruit is spreading its arils. Cracking susceptibility among the cultivars vary in different parts of the world and can be reduced by regular irrigation (Prasad *et al.*, 2003). Spraying with GA3 at 150 ppm or with benzyladenine at 40 ppm can significant reduce cracking (Yilmaz and Ozguven, 2006), while application of boron may also reduce fruit cracking (Singh *et al.*, 2003).

Fig. 19 : Fruit cracking

14.4 Sunburn

The cause of sun burn is the combined action of high solar radiation, low humidity and high temperatures (Fig. 20). Yazici and Kaynak (2006) reported that fruit surface temperatures that cause sunburn vary between 41°C and 47.5°C and kaolin treatment proved to be the best method to reduce sunburn injury in 'Hicaznar' cultivar. Holland *et al.* (2009) reported in Israel, late cultivars such as 'Wonderful' are much more susceptible to sunburn than early cultivars such as 'Akko' and 'Shani-Yonay'.

Fig. 20 : Sunburn symptoms in pomegranate

15. Postharvest Diseases

Gray mold, caused by *Botrytis cinerea* Pers., is the most economically important postharvest disease of pomegranate (Tedford *et al.*, 2005). Other fungi causing fruit rot worldwide include *Aspergillus niger*, *Penicillium spp.*, *Alternaria spp.*, *Nematospora spp.*, *Coniella granati*, or *Pestalotiopsis versicolor* (Wilson and Ogawa, 1979; Snowdon, 1990). As it occurs with other hosts (Droby and Lichter, 2004), *B. cinerea* can cause postharvest decay in pomegranate from surface-borne inoculums that infects the fruit through injuries or microwounds located on any part of the skin, but decay originating from blossom latent infection is frequently more important. Typically, the pathogen infects the flowers or the crown (calyx containing the stamens and pistils) of young fruits in the field, remains latent, and after harvest develops from the crown to the rest of the mature fruit causing an apparent brown discoloration of the skin. Palaou *et al.* (2007) propose this particular form of gray mold be called botrytis crown decay of pomegranate. *B.cinerea* is able to infect stored pomegranates by mycelia spread from infected fruit to adjacent healthy fruit, causing 'nests' of decay. Gray mold development is favored by usual pomegranate non-chilling storage conditions of 5-10°C and >90% RH, and losses due to this disease of up to 30% of harvested pomegranates when no postharvest fungicides were applied have been reported in California (Tedford *et al.*, 2005).

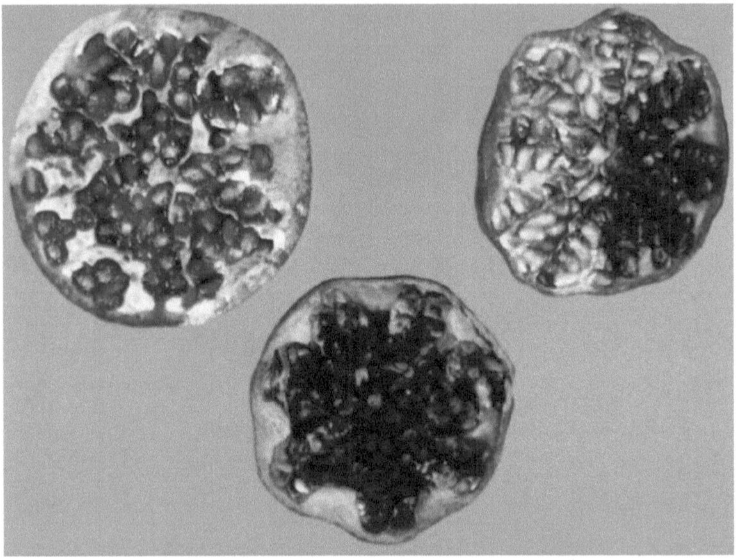

Fig. 21 : Alternaria rot in pomegranate

The application of fludioxonil has considerably reduced postharvest decay losses. However, general problems related to the use of fungicides such as the potential proliferation of resistant strains of the pathogens, concerns about public health, and environmental issues, make the search for alternative decay control methods. Common food additives (sodium bicarbonate, sodium carbonate, and potassium sorbate) were applied to the fungicide fludioxonil for the control of gray mold on 'Wonderful' pomegranates artificially inoculated with *B. cinerea* and stored at 7.2°C in either air or controlled atmosphere (5 kPa O_2 + 15 kPa CO_2) conditions (Palou *et al.*, 2007). Fludioxonil was superior to the other treatments. Potassium sorbate was the most effective additive. Synergistic effects between antifungal treatments and controlled atmosphere storage were observed. After 15 weeks of storage at 7.2°C, the combination of potassium sorbate treatment (3 minute dip in 3% solution at 21°C) and CA storage was as effective as combination of heated fludioxonil (30 s dip in 0.6 gL^{-1} of active ingredient at 49°C) and air storage (Palou *et al.*, 2007).

References

Adsule, R.N. and Patil, N.B. 1995. Pomegranate. In: Handbook of Fruit Science and Technology (Eds. Salunkhe, D.K. and Kadam, S.S.). Marcel Dekker, New York, 455.

Al-Maiman, S.A. and Ahmad, D. 2002. Changes in physical and chemical properties during pomegranate (*Punica granatum* L.) fruit maturation. *Food Chemistry*, 76 : 437-441.

Amakura, Y., Okada, M., Tsuji, S. and Tonogai, Y. 2000. High-performance liquid chromatographic determination with photodiode array detection of ellagic acid in fresh and processed fruits. *Journal of Chromatography*, 896 : 87–93.

Artes, F. 1992. *Factores de calidad y conservacion frigorifica de la Granada.* In: *Journadas Nacionales del Granado. University of Politecnica de Valencia.*

Artes, F., Gil, M.I. and Martinez, J.A. 1995. *Procedimiento para la conservacion de semillas de Granada en fresco.* Patent No. 9502362.

Artes, F., Marin, J.G. and Martinez, J.A. 1996. Controlled atmosphere storage of pomegranate. *European Food Research and Technology*, 203 : 33-37.

Artes, F., Tudela, J.A. and Gil, M.I. 1998a. Improving the keeping quality of pomegranate fruit by intermittent warming. *European Food Research and Technology*, 207 : 316-321.

Artes, F., Tudela, J.A. and Villaescusa, R. 2000. Thermal postharvest treatments for improving pomegranate quality and shelf life. *Postharvest Biology and Technology*, 18 : 245-251.

Artes, F., Tudela, J.A., Marin, J.G. and Gil, M.I. 1998c. *Nuevos tratamientos termicos para mejorar la calidad de la Granada conservada en fresco. Symposium Internacional Sobre el Granado*, Orihuela.

Artes, F., Villaescusa, R. and Tudela, J.A. 1998b. Improving pomegranate quality and shelf life by curing and intermittent warming during cold storage. *International Conference on Advances in Refrigeration Systems, Food Technologies and Cold Chain*, Sofia, Bulgaria.

Aslam, M.N., Lansky, E.P. and Varani, J. 2006. Pomegranate as a cosmeceutical source: pomegranate fractions promote proliferation and procollagen synthesis and inhibit metrix metalloproteinase-1 production in human skin cells. *Journal of Ethnopharmacology*, 103: 311-318.

Aviram, M., Dornfeld, L., Rosenblat, M., Volkova, N., Kaplan, M., Coleman, R., Hayek, T., Presser, D. and Fuhrman, B. 2000. Pomegranate juice consumption reduces oxidative stress, atherogenic modifications to LDL, and platelet aggregation: studies in humans and in atherosclerotic apolipoprotein E-deficient mice. *American Journal of Clinical Nutrition*, 71 : 1062–1076.

Barzegar, M., Fadavi, A., and Azizi, M.H. 2004. An investigation on the physico-chemical composition of various pomegranates (*Punica granatum* L.) grown in Yazd. *Iranian Journal of Food Science and Technology*, 2 : 9–14.

Ben-Arie, R. and Or, E. 1986. The development and control of husk scald on 'Wonderful' pomegranate fruit during storage. *Journal of American Society of Horticulture Science*, 111 : 395-399.

Ben-Arie, Segal, N. and Guelfat-Reich, S. 1984. The maturation and ripening of the 'Wonderful' pomegranate. *Journal of American Society of Horticulture Science*, 109 : 898-902.

Caceres, A., Giron, L.M., Alvarado, S.R. and Torres, M.F. 1987. Screening of antimicrobial activity of plants popularly used in Guatemala for the treatment of dermatomucosal diseases. *Journal of Ethnopharmacology,* 20 : 223–237.

Cam, M., Hisil, Y. and Durmaz, G. 2009. Classification of eight pomegranate juices based on antioxidant capacity measured by four methods. *Food Chemistry,* 112 : 721–726.

Celik, I., Temur, A. and Isik, I. 2009. Hepatoprotective role and antioxidant capacity of pomegranate (*Punica granatum*) flowers infusion against trichloroacetic acid-exposed rats. *Food and Chemical Toxicology,* 47 : 145–149.

Cemeroglu, B., Artik, N. and Erbas, S. 1992. Extraction and composition of pomegranate juice. *Fluessiges Obst,* 59 : 335-340.

Crisosto, C.H., Mitcham, E.J. and Kader, A.A. 2000. Pomegranates, Produce Facts. http://postharvest.ucdavis.edu/Produce/ProduceFacts/Fruit/Pomegranate.html.

Curry, S.C. 2004. Breast enhancement system. *US Patent 6673366.*

D'Aquino, S., Palma, A., Schirra, M. and Alberto Continella, A. 2010. Influence of film wrapping and fludioxonil application on quality of pomegranate fruit. *Postharvest Biology and Technology,* 55 : 121-128.

Davidson, M.H., Maki, K.C., Dicklin, M.R., Feinstein, S.B., Witchger, M.S., Bell, M., McGuire, D.K., Provos, J.C., Liker, H. and Aviram, M. 2009. Effects of consumption of pomegranate juice on carotid intima-media thickness in men and women at moderate risk for coronary heart disease. *American Journal of Cardiology,* 104 : 936–942.

Defilippi, B.G., Whitaker, B.D., Hess-Pierce, B.H. and Kader, A.A. 2006. Development and control of scald on Wonderful pomegranates during long-term storage. *Postharvest Biology and Technology,* 41 : 234-243.

Di Silvestro, R.A., Di Silvestro, D.J. and Di Silvestro, D.J. 2009. Pomegranate extract mouth rinsing effects on saliva measures relevant to gingivitis risk. *Phytotherapy Research,* 23 : 1123–1127.

Droby, S. and Lichter, A. 2004. Postharvest *Botrytis* infection: etiology, development and management. In: Botrytis: Biology, Pathology and Control (Eds. Elad, Y., Williamson, B., Tudzynski, P. and Delen, N.). *Kluwer Academic Publishers,* Dordrecht, The Netherlands. pp. 349-367.

Du, C.T., Wang, P.L. and Francis, F.J. 1975. Anthocyanins of pomegranate (*Punica granatum*). *Journal of Food Science,* 40 : 417-418.

Duman, A.D., Ozgen, M., Dayisoylu, K.S., Erbil, N. and Durgac, C. 2009. Antimicrobial activity of six pomegranate (*Punica granatum* L.) varieties and their relation to some of their pomological and phytonutrient characteristics. *Molecules,* 14 : 1808–1817.

El-Nemr, S.E., Ismail, I.A. and Ragab, M. 2006. Chemical composition of juice and seeds of pomegranate fruit. *Die Nahrung,* 34 : 601–606.

Elyatem, S.M. and Kader, A.A. 1984. Postharvest physiology and storage behaviour of pomegranate fruits. *Scientia Horticulturae,* 24 : 287-298.

Fadavi, A., Barzegar, M. and Azizi, H.M. 2006. Determination of fatty acids and total lipid content in oilseed of 25 pomegranates varieties grown in Iran. *Journal of Food Composition and Analysis,* 19 : 676–680.

Fadavi, A., Barzegar, M., Azizi, M.H. and Bayat, M. 2005. Physicochemical composition of ten pomegranate cultivars (*Punica granatum* L.) grown in Iran. *Food Science and Technology International,* 11 : 113–119.

Forest, C.P., Padma-Nathan, H. and Liker, H.R. 2007. Efficacy and safety of pomegranate juice on improvement of erectile dysfunction in male patients with mild to moderate erectile dysfunction: a randomized, placebo-controlled, double-blind crossover study. *International Journal of Impotence Research*, 19 : 564–567.

Gil, M.I., Tomas-Barberan, F.A., Hess-Pierce, B., Holcroft, D.M. and Kader, A.A. 2000. Antioxidant activity of pomegranate juice and its relationship with phenolic composition and processing, *Journal of Agriculture and Food Chemistry*, 48 : 4581-4589.

Gil, M.I., Martinez, J.A. and Artes, F. 1996b. Minimally processed pomegranate seeds. *LWT Food Science and Technology*, 29 : 708-713.

Gil, M.I., Sanchez, R., Marin, J.G. and Artes, F. 1996a. Quality changes in pomegranates during ripening and cold storage. *European Food Research and Technology*, 202 : 481-485.

Gil, M.I., Artes, F. and Tomas-Barberan, F.A. 1996c. Minimally processing and modified atmosphere packaging effects on pigmentation of pomegranate seeds. *Journal of Food Science*, 61 : 161-164.

Graham, S.A., Crisci, J.V. and Hoch, P.C. 2000. Cladistic analysis of the Lythraceae *sensu lato* based on morphological characters. *Botanical Journal of Linnean Society*, 113 : 1-8.

Haidari, M., Ali, M., Casscells, S.W. and Madjid, M. 2009. Pomegranate (*Punica granatum*) purified polyphenol extract inhibits influenza virus and has a synergistic effect with oseltamivir. *Phytomedicine*, 16 : 1127-1136.

Hamad, A.W. and Al-Momene, W. 2009. Separation and purification of crude ellagic acid from white flesh of pomegranate fruits as a potent anti-carcinogenic. *New Biotechnology*, 25 : 286.

Heber, D., Seeram, N.P., Wyatt, H., Henning, S.M., Zhang, Y., Ogden, L.G., Dreher, M. and Hill, J.O. 2007. Safety and antioxidant activity of a pomegranate ellagitannin-enriched polyphenol dietary supplement in overweight individuals with increased waist size. *Journal of Agricultural and Food Chemistry*, 55 : 1050–1054.

Hernandez, F., Melgarejo, P., Tomas-Barberan, F.A. and Artes, F. 1999. Evolution of juice anthocyanins during ripening of new selected pomegranate *(Punica granatum)* clones. *European Food Research and Technology*, 210 : 39-42.

Hess-Pierce, B. and Kader, A.A. 2003. Responses of 'Wonderful' pomegranates to controlled atmospheres. *Acta Horticulturae*, 600 : 751-757.

Hodgson, R.W. 1917. The Pomegranate, Bulletin No. 276. *University of California Press*, Berkeley.

Holcroft, D.M., Gil, M.I. and Kader, A.A. 1998. Effect of carbon dioxide on anthocyanins, phenylalanine ammonia lyase and glucosyltransferase in the arils of stored pomegranates. *Journal of American Society of Horticulture Science*, 123 : 136-140.

Holland, D., Hatib, K.and Bar-Yaakovi, I. 2009. Pomegranate: botany, horticulture, breeding. *Horticultural Reviews*, 35 : 127-191.

Ignarro, L.J., Byrns, R.E., Sumi, D., de Nigris, F. and Napoli, C. 2006. Pomegranate juice protects nitric oxide against oxidative destruction and enhances the biological actions of nitric oxide. *Nitric Oxide*, 15 : 93–102.

Jaiswal, V., Der Marderosian, A. and Porter, J.R. 2010. Anthocyanins and polyphenol oxidase from dried arils of pomegranate (*Punica granatum* L.). *Food Chemistry*, 118 : 11–16.

Kader, A.A. 2006. Postharvest biology and technology of pomegranates. In: Pomegranates: Ancient Roots to Modern Medicine (Eds. Seeram, N.P., Schulman, R.N. and Heber, D.). *CRC Press*, Boca Raton, Florida. Pp. 211-220.

Kader, A.A., Chordas, A. and Elyatem, S. 1984. Response of pomegranates to ethylene treatment and storage temperature. *California Agriculture*, 38 : 14-15.

Kulkarni, A.P. and Aradhya, S.M. 2005. Chemical changes and antioxidant activity in pomegranate arils during fruit development. *Food Chemistry*, 93 : 319-324.

Kumar, G.N.M. 1990. Pomegranate. In: Fruits of Tropical and Subtropical Origin *(Eds.* Nagy, S., Shaw, P.E. and Wardowski, W.F.). *Florida Science Source*, Lake Alfred, Florida. Pp. 328-347.

Kupper, W., Pekmezci, M. and Henze, J. 1995. Studies on CA storage of pomegranate (*Punica granatum* L. cv. Hicaz). *Acta Horticulturae*, 398 : 101-108.

Lansky, E.P. and Newman, R.A. 2007. *Punica granatum* (pomegranate) and its potential for prevention and treatment of inflammation and cancer. *Journal of Ethnopharmacology*, 109 : 177–206.

LaRue, J.H. 1980. Growing pomegranates in California, University of California, Division of Agricultural Sciences Leaflet 2459.

Lee, C.J., Chen, L.G., Liang, W.L. and Wanga, C.C. 2010. Anti-inflammatory effects of *Punica granatum* Lin *in vitro* and *in vivo*. *Food Chemistry*, 118 : 315–22.

Lee, S.W., Kim, K.S. and Kim, S.D. 1974. Studies on the compositional changes of pomegranate fruit during maturation. I. Changes in sugars, organic acids, amino acids, and the respiration rate. *Journal of Korean Society of Horticulture Science*, 15 : 57-63.

Mahdihassan, S. 1984. Outline of the beginnings of alchemy and its antecedents. *American Journal of Chinese Medicine*, 12 : 32–42.

Malhotra, N.K., Khujuria, H.N. and Jawanda, J.S. 1983. Studies on physic-chemical characters of pomegranate cultivars. II. Chemical characters. *Punjab Horticulture Journal*, 23 : 158.

Melgarejo, P., Salazar, D.M., Amoros, A. and Artes, F. 1995. Total lipids content and fatty acid composition of seed oils from six pomegranate cultivars. *Journal of the Science of Food and Agriculture*, 69 : 253-256.

Melgarejo, P., Salazar, D.M. and Artes, F. 2000. Organic acids and sugars composition of harvested pomegranate fruits. *European Food Research and Technology*, 211 : 185-190.

Miguel, G., Dandlen, S., Antunes, D., Neves, A. and Antunes, D. 2004a. The effect of two methods of pomegranate (*Punica granatum* L.) juice extraction on quality during storage at 4°C. *Journal of Biomedicine and Biotechnology*, 5 : 332-337.

Miguel, G., Fontes, C., Antunes, D., Neves, A. and Martins, D. 2004b. Anthocyanin concentrations of 'Assaria' pomegranate fruits during different cold storage conditions. *Journal of Biomedicine and Biotechnology*, 5 : 338-342.

Mirdehghan, S.H. and Rahemi, M. 2007. Seasonal changes of mineral nutrients and phenolics in pomegranate (*Punica granatum* L.) fruit. *Scientia Horticulturae*, 111 : 120–127.

Mirdehghan, S.H., Rahemi, M., Castillo, S., Martinez-Romero, D., Serrano, M. and Valero, D. 2007a. Pre-storage application of polyamines by pressure or immersion improves shelf life of pomegranate stored at chilling temperature by increasing endogenous polyamine levels. *Postharvest Biology and Technology*, 44 : 26-33.

Mirdehghan, S.H., Rahemi, M., Martinez-Romero, D., Guillen, F., Martinez-Romero, D. and Valero, D. 2007b. Reduction of pomegranate chilling injury during storage after heat treatment: Role of polyamines. *Postharvest Biology and Technology*, 44 : 19-25.

Mirdehghan, S.H., Rahemi, M., Martinez-Romero, D., Guillen, F., Martinez-Romero, D. and Valero, D. 2007c. The application of polyamines by pressure or immersion as a tool to maintain functional properties in stored pomegranate arils. *Journal of Agricultural and Food Chemistry*, 55 : 755-760.

Mirdehghan, S.H., Rahemi, M., Serrano, M., Guillen, F., Martinez-romero, D. and Valero, D. 2006. Pre-storage heat treatment to maintain nutritive and functional properties during postharvest cold storage of pomegranate. *Journal of Agricultural and Food Chemistry*, 54 : 8495-8500.

Moneam, N.M., El Sharaky, A.S., and Badreldin, M.M., 1988. Oestrogen content of pomegranate seeds, *Journal of Chromatography*, 438 : 438-442.

Mousavinejad, G., Emam-Djomeh, Z., Rezaei, K. and Khodaparast, M.H.H. 2009. Identification and quantification of phenolic compounds and their effects on antioxidant activity in pomegranate juices of eight Iranian cultivars. *Food Chemistry*, 115 : 1274–1278.

Nanda, S., Rao, D.V.S. and Krishnamurthy, S. 2001. Effects of shrink film wrapping and storage temperature on the shelf life and quality of pomegranate fruits cv. Ganesh. *Postharvest Biology and Technology*, 22 : 61-69.

Naqvi, S.A.H., Khan, M.S.Y. and Vohora, S.B. 1991. Anti-bacterial antifungal and anthelmintic investigations on Indian medicinal plants. *Fitoterapia*, 62 : 221–228.

Noda, Y., Kaneyuki, T., Mori, A. and Packer, L. 2002. Antioxidant activities of pomegranate fruit extract and its anthocyanidins: delphinidin, cyanidin, and pelargonidin, *Journal of Agricultural and Food Chemistry*, 50 : 166-171.

Onur, C. and Kaska, N. 1985. Akdeniz bölgesi narlarýnýn (*Punica granatum* L.) seleksiyonu (Selection of Pomegranate of Mediterranean region). *Turkish Journal of Agriculture and Forestry*, 9 : 25-33.

Ozgul-Yucel, S. 2005. Determination of conjugated linolenic acid content of selected oil seeds grown in Turkey. *Journal of American Oil Chemical Society*, 82 : 893–897.

Palaou, L., Crisosto, C.H. and Garner, D. 2007. Combination of postharvest antifungal chemical treatments and controlled atmosphere storage to control gray mold and improve storability of 'Wonderful' pomegranates. *Postharvest Biology and Technology*, 43 : 133-142.

Patil, A.V. and Karale, A.R. 2001. Pomegranate. (In) Fruits: Tropical and Subtropical (Eds. Bose, T.K. and Mitra, S.K.). Naya Prokash, Calcutta, India, pp. 614-620.

Porat, R., Weiss, B., Fuchs, Y., Sandman, A., Ward, G. and Kosto, I. 2008. Keeping quality of pomegranate fruit during prolong storage and transport by MAP: new developments and commercial applications. *Acta Horticulturae*, 804 : 115-120.

Poyrazoglu, E, Gokmen, V. and Arték, N. 2002. Organic acids and phenolic compounds in pomegranates (*Punica granatum* L.) grown in Turkey. *Journal of Food Composition and Analysis*, 15 : 567–575.

Prasad, R.N., Bankar, G.J. and Vashishtha, B.B. 2003. Effect of drip irrigation on growth, yield and quality of pomegranate in arid region. *Indian Journal of Horticulture*, 60 : 140-142.

Roy, S.K. and Waskar, D.P. 1997. Pomegranate. In: Postharvest Physiology and Storage of Tropical and Subtropical Fruits. CAB International, Wallingford, UK, 365.

Santagati, N.A., Duro, R. and Duro, F. 1984. Study on pigments present in pomegranate seeds, *Journal of Commodity Science*, 23 : 247, 1984.

Sassano, G., Sanderson, P., Franx, J., Groot, P., van Straalen, J. and Bassaganya-Riera, J. 2009. Analysis of pomegranate seed oil for the presence of jacaric acid. *Journal of Science of Food and Agriculture*, 89 : 1046–1052.

Sayyari, M., Babalar, M., Kalantari, S., Martinez-Romero, D., Guillen, F., Serrano, M. and Valero, D. 2011b. Vapour treatments with methyl salicylate or methyl jasmonate alleviated chilling injury and enhanced antioxidant potential during postharvest storage of pomegranates. *Food Chemistry*, 124 : 964-970.

Sayyari, M., Babalar, M., Kalantari, S., Serrano, M. and Valero, D. 2009. Effect of salicylic acid treatment on reducing chilling injury in stored pomegranates. *Postharvest Biology and Technology*, 53 : 152-154.

Sayyari, M., Castillo, S., Valero, D., Diaz-Mula, H.M. and Serrano, M. 2011b. Acetyl salicylic acid alleviates chilling injury and maintains nutritive and bioactive compounds and antioxidant activity during postharvest storage of pomegranates. *Postharvest Biology and Technology*, 60 : 136-142.

Seeram, N.P., Adams, L.S., Henning, S.M., Niu, Y., Zhang, Y., Nair, M.G. and Heber, D. 2005. *In vitro* antiproliferative, apoptotic and antioxidant activities of punicalagin, ellagic acid and a total pomegranate tannin extract are enhanced in combination with other polyphenols as found in pomegranate juice. *Journal of Nutritional Biochemistry*, 16 : 360–367.

Singh, D.B., Sharma, B.D. and Bhargava, R. 2003. Effect of boron and GA_3 to control fruit cracking in pomegranate (*Punica granatum*). *Current Agriculture*, 27 : 125-127.

Singh, M., Arseneault, M., Sanderson, T., Morthy, V. and Ramassamy, C. 2008. Challenges for research on polyphenols from foods in Alzheimer's disease: bioavailability, metabolism and cellular and molecular mechanism. *Journal of Agricultural and Food Chemistry*, 56 : 4855–4873.

Syed, D.N., Afaq, F. and Mukhtar, H. 2007. Pomegranate derived products for cancer chemoprevention. *Seminars in Cancer Biolology*, 17 : 377–385.

Tedford, E.C., Adaskaveg, J.E. and Ott, A.J. 2005. Impact of Scholar (a new postharvest fungicide) on the California pomegranate industry. Online. *Plant Health Progress* doi: 10.1094/PHP-2005-0216-01-PS.

Tezcan, F., Gultekin-Ozguven, M., Diken T, Ozcelik, B. and Erim, F.B. 2009. Antioxidant activity and total phenolic, organic acid and sugar content in commercial pomegranate juices. *Food Chemistry*, 115 : 873–877.

Turk, G., Sonmez, M., Aydin, M., Yuce, A., Gur, S., Yuksel, M., Aksu, E.H. and Aksoy, H. 2008. Effects of pomegranate juice consumption on sperm quality, spermatogenic cell density, antioxidant activity and testosterone level in male rats. *Clinical Nutrition*, 27 : 289–296.

Van Elswijk, D.A., Schobel, U.P., Lansky, E.P., Irth, H. and van der Greef, J. 2004. Rapid dereplication of estrogenic compounds in pomegranate (*Punica granatum*) using on-line biochemical detection coupled to mass spectrometry. *Phytochemistry*, 65 : 233–241.

Wilson, E.E. and Ogawa, J.M. 1979. Fungal, Bacterial and Nonparasitic Diseases of Fruit and Nut Crops in California. University of California, Division of Agricultural Sciences, Berkeley, USA. Publication 4090.

Xu, K.Z.Y., Zhu, C., Kim, M.S., Yamahara, J. and Li, Y. 2009. Pomegranate flower ameliorates fatty liver in an animal model of type 2 diabetes and obesity. *Journal of Ethnopharmacology*, 123 : 280–287.

Yazici, K. and Kaynak, L. 2006. Effect of kaolin and shading treatments on sun burn on fruits of Hicaznar cultivar of pomegranate (*Punica granatum* L. cv. Hicaznar). *ISHS International Symposium on Pomegranate and Minor Mediterranean Fruits*, Adana, Turkey.

Yilmaz, M. and Ozguven, A.I. 2006. The effect of some plant nutrients, gibberelic acid and pinolate treatments on the yield, fruit quality and cracking in pomegranate. *ISHS International Symposium on Pomegranate and Minor Mediterranean Fruits*, Adana, Turkey.

Zolfagharinasab, R. and Hadian, J. 2007. Influence of methyl jasmonate on inducing chilling tolerance in pomegranate fruits. *Pakistan Journal of Biological Science*, 10 : 612-616.

11

Probiotic Functional Foods in Recent Times

Renu Agrawal

Recent awareness of consumers towards innovative probiotic fermented food products containing friendly bacteria and healthier foods has coined the term functional foods. Food development in the world is taking place mainly in three strategic directions as functional foods, organic foods and national foods. Every year the global demand is increasing at the rate of 15-20 %.

Fermentation by newer technologies involves specific strain of lactic acid bacteria. This brings specific fermentation under controlled conditions resulting in a specific fermented product with enhanced organoleptic, nutritional and therapeutic qualities.

The present society has the knowledge how to be healthy. It is important to provide nutritionally balanced food with the goodness of healthy bacteria. Probiotic foods seem to be the best option. The functional food concept was first developed in Japan in 1990's when faced with side effects of medicines and escalating health care costs. The ministry of health and welfare initiated a regulatory system to approve certain foods with documented health benefits to improve health. These are now recognized as Foods for specialized health use (FOSHU). According to American dietic association (ADA) these foods are potential healthful food or food ingredients that may provide health benefits beyond the traditional nutrients. Probiotics (for life) are fermented foods cultured by beneficial microorganisms. Probiotic foods as yogurt and sauerkraut are very familiar. There are many more which have been enjoyed for centuries in different parts of the world. Fermented foods have values beyond their original states. Fermentation increases nutrients, gives a tasty zing, and increases the shelf life without the use of preservatives. Fermented foods support the immune system. The beneficial bacteria in fermented foods helps to keep the digestive tract healthy and protect them against virulent pathogens that cause food borne illnesses. The typical American diet makes the body vulnerable to these infections. The problem is compounded by our over-reliance on antibiotics and our increasing resistance to them. Unfortunately, antibiotics kill the beneficial microorganisms along with the pathogens. Probiotic foods can prevent or alleviate many health disorders ranging from allergy and asthma to yeast infection and heart diseases that result from food borne infections and antibiotic resistance. Probiotic foods provide special nutritional and therapeutic properties over the traditional foods. For commercialization extreme modifications are to be done. There are fermented foods based on vegetables, beans, and cereal grains. Fermented foods are nutritious, add variety and can fit easily into the daily diet. These are recognized as most healthy diets and carriers for microbes that reach the intestine when consumed and are health promoting (Russo *et al.*, 2005; Anonymous, 2010, Timmerman *et al.*, 2001).

Probiotic Functional Foods and Health

Probiotic functional foods contain significant levels of biologically active components that provide health benefits beyond basic nutrition. On fermentation, dairy products, meat and fish products, cereal products and many others contain lactic acid bacteria (LAB) that can enhance gastrointestinal function. These bacteria can improve lactose digestion, reduce cholesterol, prevent diaarhoea and can act on the immune system helping the body to resist and fight infection. It is also known to prevent/slow the growth of colon cancer, urogenital infections, alleviating constipation and treating food allergy. Down regulation of the allergic reaction by food has been studied (Kalliomaki *et al.*, 2004). They should also have good hydrophobicity which enables them to adhere to the intestinal epithelium and colonize. They are also involved in potential health beneficial roles like immunomodulation, pathogen exclusion, production of antimicrobial substances, anticarcinogenic and cholesterol lowering activities (Ziemer and Gibson, 1998).

For LAB to exert a probiotic effect they have to survive the stomach acid (pH 1.5) and high bile acid. They must arrive in the intestine in large quantities and adhere to the intestinal wall. They must also colonize in order to impart the beneficial effect. Finally, the bacteria must show some beneficial effects on human health. These probiotics and its beneficial effects can be utilized by supplementation in various foods (Parvez *et al.*, 2006). The gut flora can be altered to maintain the natural balance and the animal / human would return to better nutrition, growth and health status (Russo *et al.*, 2005). These bacteria cause fermentation and coagulation in milk by producing different optical forms of lactic acid (Henneberg, 1904) derived by the reduction of pyruvic acid. In vitro methods have been standardized to determine the transit tolerance (Charteris *et al.*, 1998).

These are recognized as gram positive, non-sporing, carbohydrate fermenting lactic acid producers, acid tolerant, non- aerobic, catalase negative and non- motile.

LAB'S were first referred as probiotics by Lily and Stillwell (1965). These organisms are naturally available in many food materials (Rodriguez *et al.*, 2000). The development of probiotic food incorporates LAB as starter cultures in which they perform fermentation by production of lactic acid. Apart from lactic acid they also produce low molecular weight compounds like acids, alcohols, hydrogen peroxide, diacetyls, esters and other metabolites. They inhibit the growth of spoilage organisms in food, contributing to the flavor and textural properties of the fermented foods. These properties have enabled these strains to be used for the production of wide range of fermented foods and have been given the GRAS (generally recognized as safe) status (Agrawal, 2005). The most commonly used probiotic lactic acid bacteria (LAB) in the production of fermented foods include *Lactobacillus* and *Bifidobacterium* sp.

A food can be regarded as functional if it affects beneficially one or more functions in the body beyond providing basic nutrition that improves health and well being or reduces the risk of disease. The market of functional foods is growing rapidly the world over. Functional foods are also known as nutraceuticals, designer foods, medicinal foods, therapeutic foods, super foods or even medifoods.

According to Fuller (1989) probiotic is a live microbial food supplement which beneficially affects the host by improving its intestinal microbial balance. According to Charteris *et al.* (1997) probiotic lactic acid bacteria (PLB) when ingested may have positive effects in the prevention and treatment of a specific pathogenic condition. Growing application of probiotic microorganism's that reside in the intestinal tract are frequently associated with health promoting attributes. The market for probiotic foods is growing globally very rapidly. Recently, a lot of research is being carried out to incorporate probiotic microorganisms in non- dairy food products as ice creams, chocolates and juices. Non dairy probiotic foods have become popular in European countries (Stanton *et al.,* 2001). Many companies are utilizing the probiotic cultures in food products (Table 1).An overview is well described earlier (Sanders, 1998; 2008).

Probiotic Lactic Acid Bacteria (PLB) in Meat Products

PLB have been used as starter cultures in the production of fermented sausages (Diebel *et al.,* 1961). Similar roles are contributed for flavours in cured meats which grew over 80% after storage at 20°C for 7 days (Kitchell and Ingram, 1963).

Dairy Foods

Milk is a rich source of nutritive compounds and can be enriched. Milk fat has saturated mono and polyunsaturated fatty acids. Conjugated linoleic acid prevents many diseases. Similarly milk proteins and bioactive peptides protect against many risk factors. Lactose derivatives are utilized to relieve constipation and in the modulation of intestinal flora. Milk minerals are used in the replacement of sodium in salt which prevents hypertension (Shahani and Chandan, 1979).

In developing functional dairy products various groups of experts are needed as scientific researchers, medical experts, nutritionists and microbiologists. Food technologists are needed for product development, process technologists and biotechnologists for processing the compounds, chemists to analyze the compounds and experts for marketing the products. Nutri marketing is needed to explain the results of R&D to health care professionals and to convince them of the benefits of functional foods.

Dairy products form the major part of probiotic foods. Milk is natural and forms a highly nutritive part of the balanced diet. The principal carbohydrate in milk is lactose. It is a disaccharide formed from galactose and glucose. It is 54 % of the total non- fat milk solids and provides 30 % energy of milk. It provides high value protein, vital minerals and vitamins. In earlier days the seasonal variation in milk production led the farms to preserve milk for the cold winter in different forms. Slowly the scaling up was done on bigger scale (Leporanta, 2001). Cultured butter milks and fermented milk products are on the top of dairy foods. Dairy-Based ingredients are used in milk-derived products. Dairy ingredients include fluid milk, milk concentrates, cheese and cheese products, frozen desserts, whey products and lactose, fermented milks, milk fat concentrates. Dairy-Based Ingredients is designed for maximum convenience to the consumers.

To prepare any probiotic milk products it is essential to know the properties of milk ,its components, composition, constituents, physical properties like color, flavor, density, specific gravity, surface tension, foaming, viscosity, specific heat, electrical conductivity, freezing point, boiling point and refractivity (FAO/WHO, 2001;2002). Procedures for basic milk processing, raw milk handling, separation, fat standardization, pasteurization, homogenization, packaging and storage must be known.

Fermented Milk

Mesophilic LAB like *Lactococcus lactis* and *Leuconostoc cremoris* strains are fermented at 30^{0}C for 16-20 h (Walstra *et al.*, 2005). Starter cultures other than mesophilic like *Lactobacilli* are also utilized in milk products. Kefir is a traditional fermented milk drink containing *Lactococcus, Leuconostoc, Lactobacillus, Acetobacte*r and yeasts. These also provide a special flavor and aroma. Yogurts are prepared by fermenting with *Streptococcus thermophilus* and *Lactobacillus delbruekii* sub sp. bulgaricus growing in synergy. The fermentation is carried out at 30-43^{0}C for 20 h (factors affecting the growth and survival of probiotic in milk (Swidan, 2009). Studies in probiotic fermented milk have been done to check the levels of lipoproteins in plasma (Richelsen *et al.*, 1996).

The selection of strain is specific to the fermentation time and the flavor of the product. Cheese where whey is separated after milk coagulation is also made by fermenting with different starter cultures. The production processes vary and the products contain live cultures. The production processes vary and the products contain live LAB. All fermented milk products contain live microorganisms unless they are pasteurized after fermentation. The total consumption of fermented milks and yogurt in the EU was 6.35 million tons (Bulletin of the International Federation, 2001). Addition of selected, well documented health effective strains is an easy and natural way of enhancing the functionality of the products.

Dairy probiotic foods can be divided into fermented milk, cheese, ice- cream, added value of functions where the milk composition is changed, functional dairy products with proven health benefits with prebiotic. Texture or the taste of a milk product fermented with a probiotic does not meet the consumer approval therefore it is used together with standard cultures. Probiotics are either added before the fermentations or a little part of the milk is fermented separately and then mixed. Sometimes the probiotic strain is added to the fermented product after fermentation. Commercially, the products yakult (*Lb. casei*; shirota), danone actimel (*Lb. casei* imunitass) butter milk, sweet milk, yogurt, fermented whey based drinks, fermented milks, cheeses, juices and mixture of milk and juice contain Lactobacillus species. Milk is a protective food matrix for probiotics and improves the survival of the strain in the intestine. Probiotics are also used in desserts like ice creams. Recently, a yogurt enriched with *Lb.gasseri* was shown to inhibit *Helicobacter pylori* (Fillipo *et al.*, 2001; Wang *et al.*, 2004).

Probiotic Fermented Dairy Beverages

Many drinking type probiotic dairy foods are prepared by adding fruit preparations or fruit essences plus coloring matter (Tamime, 2002). Kailasapathy *et al.*, (2008) reported that yogurts containing mango or strawberry reduced the pathogen levels as compared to plain yogurts. Fortification of products with ascorbic acid, amylase or inulin had demonstrated significant growth and higher viability of probiotic cultures (Donkor *et al.*, 2007).

There are many fermented milk beverages in India like chhash, lassi and butter milk. Probiotic drinks in the European market include A-fi 1, actimel. Aktifit, AB-piima, Bella Vita, Bifidus, Biofit, Biola, casilus, Cultura, Emmifit, Everybody, Fit and aktiv, Fundo, gaio, Gefilac, Gefilus, kaiku actif, Onaka, proviva, yakult, yoco activit.

Now in India there is a lot of awareness for better health by probiotic drinks. Yakult Danone India has launched "yakult", Nestle India has launched a new low fat Nesvita dahi. Mother dairy has introduced a probiotic curd with dietary fibre 'b-Active Plus' and "Nutrifit", Amul has launched probiotic lassi and ice cream.

Successful attempts have been made for carbonated probiotic fermented milk. *St. thermophilus* (MTCC 5460) and *Lb. rhamnosus* were used to prepare fermented milk from double toned milk. This was carbonated at a pressure of 15 kg/cm^2 (Shah, 2009).Effect of whey protein concentrate was studied on the survival of *Lb. acidophilus* in lactose hydrolysed yogurt when kept in refrigerator (Kailaspathy and Supraidi, 1996).

Cheeses

Cheese making process consists of lactic acid fermentation of milk a complex biological substrate containing enzymes and varying in its physical and chemical properties. During cheese maturation, break down of the curd due to proteolysis, lipolysis and other enzymic processes gives rise to changes in texture and flavor in the cheese (Kosikowski *et al.,* 2002).

Cheddar cheese when analysed by GCMS led to the identification of a large variety of compounds as ketones, aldehydes, alcohols, fatty acids and volatile sulphur compounds which contributed to its aroma and flavor. Flavor quality was consistent for a particular starter and any off flavours present were reproducible and characteristic of the starter strain used. The starter culture has a clearly defined role in cheddar flavor development.

Souring of milk is a potent mechanism for the preservation of the growth of pathogens without heating the product Jindal *et al.,*(2004) have done a lot of work on the use of chhanna and paneer whey. Whey is an excellent growth medium for lactic acid bacteria to ferment lactose in whey to form lactic acid. It replaces lost organic and inorganic salts to the extracellular fluid and is used to treat various ailments as arthritis, liver complaints and dyspepsia.

Table 1 : Probiotic cultures and sources

L. acidophilus NCFM® Danisco (Madison WI)	*B. infantis* 35264 Procter & Gamble (Mason OH)
L. fermentum VRI003 (PCC) Probiomics, Eveleigh, Australia	*L. rhamnosus* R0011, *L. acidophilus* R0052 Institute Rosell (Montreal, Canada)
L. acidophilus LA-1, *L. paracasei* Chr. Hansen (Milwaukee WI)	CRL 431, *B. lactis* Bb-12 *L. casei* Shirota, *B. breve* strain Yakult Yakult (Tokyo, Japan)
L. casei DN114001 DN173 010 ("*Bifidis regularis*™") Dannon (Tarrytown NY)	("L. casei Defensis ™", *B. animalis* Danone (Paris, France),
L. rhamnosus GR-1™ Urex Biotech (London, Ontario, Canada) and formerly *L. acidophilus* La-1)	*L. reuteri* RC-14™, Chr. Hansens (Milwaukee WI), *L. johnsonii* Lj-1 (same as NCC533 Nestlé (Lausanne, Switzerland)
L. plantarum 299V, *L. rhamnosus* 271 Probi AB (Lund, Sweden)	*L. reuteri* SD2112 Biogaia (Stockholm, Sweden)
L. rhamnosus GG ("LGG") Valio Dairy (Helsinki, Finland)	*L. rhamnosus* LB21, Essum AB (Umeå, Sweden)
Lactococcus lactis L1A *L. salivarius* UCC118 University College (Cork, Ireland)	*B. longum* BB536 Morinaga Milk Industry Co., Ltd.
(Zama-City, Japan) *L. rhamnosus* HN001 (DR20)	*B. lactis* HN019 (DR10), Danisco (Madison WI),

Bioactive Peptides in Fermented Milk

The occurrence of various bioactive peptides in fermented milks like yogurt, sour milk and dahi has been reported in many studies. ACE-inhibitory, immunomodulatory and opiod peptides have been found in yogurt and in milk fermented with a probiotic *L. casei* ssp. rhamnosus strain. Also, immune stimulating hexa- peptides have been detected in human casein (Parker *et al.,* 1984).Fermented functional foods with health benefits like Evolus and Calpis available in the market are based on bioactive peptides released by probiotic organisms (Shah,2007).There have been many clinical studies where consumption of probiotics have shown health improvement like eczema (Kalliomaki *et al.,* 2001; Marteau *et al.,* 2002) and reductions in antibiotic associated diarrhea (Plummer *et al.,* 2004) Fermented functional foods with health benefits like Evolus and Calpis available in the market are based on bioactive peptides released by probiotic organisms (Demin *et al.,* 1994; Flemming *et al.,* 1973). (Table 2, 3).

Table 2 : Commercial dairy products and ingredients with health or function claims based on bioactive peptides

L. casei shirota	6.5×10^9	*L. rhamnosus GG* $10^9 \times 10^{10}$
L. plantarum 299 v	5×10^8	*L. acidophilus NCFB 1748* 3×10^{11}
L. reuteri	1×10^8-10^{11}	*L. rhamnosus DSM 6594* 16×10^9

Source: Anonymous (2011)

Table 3 : Commercial dairy products and ingredients with health or function claims based on bioactive peptides

Brand names	Products	Functional bioactive peptide	Health/ Function claim
Calpis	Sour milk	Val-Pro-pro, Ile-pro-pro from casein and k-casein	Reduction blood pressure
Biozate	Hydrolysed whey protein isolate	β-lactoglobulin fragments	Reduction blood pressure
Product F200/ Lactium	Flavored milk, drink , confectionery, capsule	As 1- casein (91-100) (Try-Leu-Gly-Tyr-Leu-Glu-Gln-Leu-Leu-Arg)	Reduction of stress effects
Capolac	Ingredient	Casein phosphor peptide	Helps mineral absorption

Cereal Based Probiotic Fermented Dairy Beverages

Addition of cereals into milk enriches the mineral value and fibre. Probiotic fermentation enhances the nutritive value, palatability and functionality of cereals. It also reduces antinutritional factors. Millets like sorghum, ragi and jowar are utilized. Cereal solids are incorporated into milk before and after fermentation (Gupta *et al.*, 2007).

Cereals in Probiotic Functional Foods

Cereals offer another alternative for the production of probiotic functional foods. The multiple beneficial effects of cereals can be exploited in different ways leading to the design of novel cereal foods or cereal ingredients that can target specific populations. Cereals can be utilized as substrate for fermentation for the growth of PLB. However, it is important to know the composition and processing of cereal grains, the formulation of the substrate, the growth pattern and viability in the form of colony forming units of the starter culture, stability of the culture during the shelf life, organoleptic and sensory properties and the nutritional value of the final product (Wood, 1997).

Cereals contain water soluble fibre like β- glucan and arabinoxylan, oligosaccharides like gluco and fructo oligosaccharides and resistant starch which act as prebiotics.

The development of non- dairy probiotic product is a challenge to the food industry in its effort to utilize the abundant natural resources by producing high quality functional products. In this regard baby foods and confectionery formulations have been developed with supplementing probiotic culture (Saarela *et al.*, 2000).

Lactic acid fermentation of cereals has been established for a very long time especially being used in Asia and Africa for producing beverages, gruels and porridge. The preparation procedure is usually general by taking grains like maize, sorghum or millet grains which are soaked in clean water for 1-2 days. The grains become soft and are then milled into slurry. Hull, bran are removed after sieving and then fermented. The pH decreases and the acidity increases due to acid production by PLB.

In the west usually wheat and rye are used for sourdough preparation. The population of culture in fully fermented sourdoughs is more than 10^9 cfu/ml. The good growth of PLB in cereals suggests that the incorporation of a human derived prebiotic strain in a cereal substrate under controlled conditions would produce a fermented food with defined and consistent characteristics with health promoting properties. However, it is important to look into many of the technological aspects like composition and processing of the cereal grain and PLB. High cell growth rate and acidification may result in reduction of fermentation times and enhance the viability of the specific strain by preventing the growth of undesirable microorganisms present in the raw material (Marklinder and Lonner, 1992). Adaptability of the probiotic in the substrate is a very important criterion in the selection procedure of a suitable strain (Oberman and Libudzisz, 1998).

Table 4 : Recommended dose of *Lactobacilli*

L. casei shirota	6.5×10^9	L. rhamnosus GG 10	9×10^{10}
L. plantarum 299 v	5×10^8	L. acidophilus NCFB 1748	3×10^{11}
L. casei shirota	$1 \times 10^8 - 10^{11}$	L. rhamnosus DSM 6594	16×10^9

The culture strains have very particular nutritional requirements for carbohydrates, amino acids, peptides, fatty esters, salts, nucleic acid derivatives and vitamins (Severson, 1998). Many homofermentative and heterofermentative lactic acid bacteria were tested and oats were found to be a suitable substrate for the growth of Probiotic lactic acid bacteria. *Lb. acidophilus* was found to grow slowly. The highest viability was found with *Lb. plantarum* and *Lb. reuteri* (Mills *et al.,* 2011).The recommended dose of lactobacilli is given in table-4.

Using malt, barley or wheat as media for probiotic culture the malt medium was found best (Charalampopoulos *et al.,* 2002). It was also found to be strain specific. *Lb. plantarum* had the ability to tolerate low pH by maintaining a proton and charge gradient.

Survival of the probiotic strains during gastric transit is also influenced by the physico chemical properties of the food carrier used for delivery. The buffering capacity and the pH of the carrier medium are significant factors, as food formulations with pH ranging from 3.5 to 4.5 and high buffering capacity will increase the pH of the gastric tract and thus enhance the stability of the probiotic strain (Kailaspathy and Chin, 2000).

The grains of corn, sorghum, millet, barley, rye and oats contain a lot of crude fibre and lack gluten like proteins of wheat. The traditional foods from these grains lack flavor and aroma (Babiker, 2006). The fermentation with PLB improves the sensorial value, volatiles as higher alcohols, aldehydes, ethyl acetate and diacetyl are formed (Agrawal *et al.,* 2000). In general a single strain of LAB is not very acceptable and therefore in order to enhance the aromatic profile of the final product many strains are incorporated to bring out the desired flavor. It is important to study the metabolism of all the culture strains so that they grow in the cereal substrate and do not act antagonistically towards each other.

Cereals are generally suitable substrates for the growth of human derived probiotic strains. A systematic approach is needed in order to identify the intrinsic and processing factors that could enhance the growth and survival of the probiotic strain *in vitro* and *in vivo*. The improvement in the organoleptic properties can be done using supporting cultures that act synergistically on the probiotic strains. The functionality of colonic strains could be improved by the presence of specific non digestible components of the cereal matrix that could act as prebiotic. The possibility of separating specific fractions of non digestible soluble fibre from different types of cereals or cereal by products through processing technologies as pearling and sieving or enzymatic modifications will be helpful.

Fermentation of Sauerkraut

New cultivars of cabbage are constantly being developed to confirm the ever changing agricultural and ecological demands. Sauerkraut fermentation is a model system for the study of growth of mixed cultures and their biological control. Environmental factors like substrate composition, anaerobiosis, temperature, pH, salt concentration serve as a means to modify and direct the interactions of a microbial flora.

For the commercial production a controlled fermentation and information on the role and limitations of the undesirable or abnormal end product formation is necessary to establish conditions which are optimal for the growth of selective cultures (Stamer *et al.*, 1971).

Brined Cucumbers

The process is favoured by the introduction of starter cultures with the control of undesirable microbial activity. Brining treatments, environmental conditions and initial microbial populations are primary factors which decide the course of microbial activity. Mainly the fermentation is by homofermentative LAB.

Probiotics in Production of Soy sauce

Soysauce and sourdough both utilize the interaction between yeasts and Lactobacilli which contribute to the flavor of the final product (Yong and Wood, 1974).

In the sourdough bread fermentation lactic acid is produced by the bacterium which contributes to a distinctive flavor and texture of the finished bread. In the sour fermentation there is co-existence of bacteria and yeast. Enhanced growth of *Lactobacilli* on immobilization has been studied in soy milk (Ahmed *et al.,* 2010).

Probiotic Tempeh

Tempeh is a fermented soybean product covered with white mold, a compact cake produced by the fermentation of dehulled, hydrated, soaked and precooked soyabean cotyledons. The fermenting organism is *Rhizopus oligosporus*. The fungus improves the flavor and aroma of the product. Tempeh is a traditional fermented food in Indonesia. Stimulation of growth on *Rhizopus* with potassium ions is studied earlier (Pefialoza *et al.,* 1991). It is rich in calories, proteins, aminoacids, vitamins, minerals, antioxidants, antimicrobials and dietary fibres. Biotechnologically, it is nutritionally rich with nutty taste, odor of fresh mushroom and nougat like texture. It can be consumed from infants to aged people.

Today, it is very common in countries like Malaysia, Holland, Canada and West Indies. The substrates overgrown with desirable, edible microorganisms become resistant to invasion by spoilage, food poisoning and toxin producing microorganisms. It is beneficial for health as it decreases risk of heart diseases, strokes, osteoporosis, cancer ,digestive disorders and reduction of fat.

There are other cereals and legume substrates like chick pea, horse gram, lupin, common bean, ground nut, wheat and maize which are used for tempeh preparation. Soya milk has also been utilized for the tempeh production. Tempeh when fresh is perishable due to sporulation by fungi. Therefore, there is a practice to blanch the tempeh product, dry and powder it which can be stored for longer time. However, tempeh contains small amounts of oxalates which if accumulated can cause health problems.

Fermentation of soybean improves digestibility by reduction of antinutritional factors like tannin and phytate in addition to production of acids which inhibit the production of pathogenic bacteria which is important in the manufacture of food. They are low cost nutritious food which can be consumed by all socio economic groups.

For developing new foods large marketing campaigns are required and the consumer needs to adapt to the new product. Developing new and innovative products or reformulations of the ones which are existing enables the manufacturer to meet the expectations of the consumers who are very health conscious today.

Cereals are gaining a lot of importance for preparing probiotic foods as they have the ability to grow and deliver probiotic LAB to the human gut along with prebiotic compounds (Kalui *et al.*, 2006).

Clinical Studies of Probiotic Products

There have been many clinical studies where consumption of probiotics have shown health improvement like eczema (Kalliomaki *et al.*, 2001; Marteau *et al.*, 2002) and reductions in antibiotic associated diarrhoea (Plummer *et al.*, 2004).

Market of Probiotic Products

In Brazil, the sales of functional foods amounts to US $ 500 million per year representing about 1% of the total sales of the Brazilian food processing industry. The interests are mainly in dairy products, nutraceuticals and soy- based products. ABIA forecasts a substantial growth of 4.5- 5% for this segment of the market (Stanton *et al.*, 2001).

In order to exert the health benefits beyond inherent basic nutrition, it is necessary that the activity of the probiotic culture be maintained at sufficiently high levels throughout the shelf life of the product. There is a lot of interest to research microbiologists and food scientists who are employed in academics, industrially employed dieticians, microbiologists, nutritionists, and food scientists because this market is growing. Regulators in U.S. Food and Drug Administration, U.S. Department of Agriculture and people in state and city health departments are also looking ahead for such foods as there is increasing evidence as well as interest in the real and/or postulated health benefits and uses in disease treatment. Technologically, the probiotic cultures should have oxygen tolerance, acid and bile tolerance and heat tolerance. Many scientists are working in this direction to get good probiotic cultures. Commonly used cultures are *Lactobacillus* sp. and *Bifidobacterium* sp. (Stanton *et al.*, 2003 a, b).

The factors responsible for consumers inclination towards functional foods are, baby health conscious movement, chronic health diseases in aged people, changes in food regulations, market opportunity, increased health care costs, changing consumer demands and social attitudes, new opportunities to add value to the existing products with higher profits (Anonymous 2002, 2006, 2010). Presently, the main emphasis is on clinical applications (benefits and/or lack thereof) as well as future biomedical therapeutic uses in animal model studies focused on therapies and data for the probiotics to be used as complementary and alternative medicines. Product development with probiotics is on the forefront. Information on pre-and probiotics as important sources of micro-and macronutrients aids in the development of methods of bio-modification of dietary plant molecules for health promotion. Pre-and probiotics provides defense of food selections for individual consumption based on health needs and current status (Anonymous, 2011). It is assumed that global probiotics market will exceed US$28.8 billion by 2015. The market is set to witness impressive growth as consumers become more conscious about their health and switch to preventive healthcare due to rising healthcare cost (Table 5).

Table 5 : Probiotic products with targeted health benefits available in the US

Infant diarrhea	*L. rhamnosus* GG Culturelle (capsule)	Danimals (drinkable yogurt)
L. casei DN-114001	(aka "Immunitas™")	DanActive (fermented milk)
Inflammatory bowel	conditions	8-strain combination of 3
Bifidobacterium strains, 4	*Lactobacillus* strains and *S.*	*Thermophilus*
VSL#3 (powder)	Antibiotic associated	diarrhea; *C. difficile*
S. boulardii Florastor (powder)	*L. rhamnosus* GG Culturelle (capsule)	Danimals (drinkable yogurt)
L. casei DN114001 DanActive (fermented milk)	Gut transit time *B. animalis* DN-173 010 (aka)	"Bifidus regularis™"
Activia (yogurt)	Keeping healthy *L. reuteri* ATCC 55730 Stonyfield yogurt	*L. casei* DN-114001 DanActive (fermented milk)
L. casei Shirota Yakult	Allergy (atopic dermatitis in infants)	
L. rhamnosus GG Culturelle (capsule)	Danimals (drikable yogurt)	Lactose intolerance *L. bulgaricus* and/or *S.*
thermophilus (most strains)	All yogurts with live, active cultures	Colic in infants *L. reuteri* ATCC 55730 Reuteri drops
Immune support *B. lactis* HN019 (aka)	HOWARU™ or DR10	Naked Juice Probiotic Juice
Smoothie Strain sold as an	ingredient for dairy and supplement	products - contact Danisco
B. lactis Bb-12 Good Start Natural Cultures (infant formula) Nestle; Strain also sold as	an ingredient for dairy and supplement products - contact Chr.	Hansens (800-558-0802)

Contd.

L. casei DN114001 DanActive (fermented milk)		
L. rhamnosus GG Culturelle (capsule)	Danimals (drikable yogurt)	*L. reuteri* ATCC 55730 Stonyfield yogurt
Vaginal applications *L. rhamnosus* GR-1, *L. reuteri*	RC-14	Fem-Dophilus (capsules)
Irritable bowel syndrome	symptoms	*B. infantis* 35264 aka

India is also becoming an attractive destination for investments in this sector. The growing health-conscious in newer generation has paved the way for varieties of weight reducing, low-cholesterol, high fibre and nutrient-rich food products. Demand of these products is increasing manifold.

Companies Preparing Probiotic Products

1. Probiotic curd; Heritage Foods (India) Ltd.
2. 'b-Activ' probiotic curd (*L. acidophilus* and *B. lactis,* strain *BB12*); Mother Dairy
3. 'Nesvita' probiotic yoghurt; Nestle
4. Probiotic ice creams, 'Amul Prolife''Prolite' and 'Amul Sugarfree' Amul (Brand of Gujarat Cooperative Milk Marketing Federation Ltd.)
5. Yakult, Probiotic curd with*L. casei* strain *Shirota* Yakult; Danone India (YDI)Private Limited
6. Probiotic drugs Ranbaxy; Binifit
7. Probiotic drugs; Dr. Reddy's Laboratories
8. Probiotic drugs; Zydus Cadila
9. Probiotic drugs; Unichem
10. Probiotic drugs; JB Chem
11. Probiotic drugs; GlaxoSmithKline
12. Fructo-Oligo Saccharides; Probiotic drugs Glenmark Alkem Labs

Some of the dairy beverage products can be enriched with active ingredients such as omega-3 fatty acids, cholesterol-lowering ingredients and probiotics besides micronutrients. The factors responsible for consumer's inclination towards functional foods are baby health movement, chronic health diseases in aged, changes in food regulations, market opportunities, increased health care costs, changing consumer demands and new opportunities to add value to the existing products with higher profits.

References

Agrawal R; Rati ER; Vijayendra SVN; *et al.*,2000 Flavour profile of idli batter prepared from defined microbial starter cultures. *World Journal of Microbiology & Biotechnology* 16(7): 687-690.

Agrawal, R.2005.Probiotics: An emerging food supplement with health benefits. *Food Biotechnol.* Ahmad 19 (3): 227-246.

Ahmad, R., Abdullah, W.N.W., Liong, M.T. Enhanced Growth of *Lactobacilli* in Soymilk upon Immobilization on Agro wastes. 2010. JF S. 75(3).

Anonymous 2010. The benefits of probiotics for your pet. http:// www.Flintriver. Com/ product Info. Asppi=probiotics- overview. Htm. 8th jan. 2010.

Anonymous, 2002. Agriculture: a vision for the future- Probiotics "What's the hype" Bioag Enews letter. http:// www. Bio-ag.com/info/newsletters/enews/enews 2. Html.accessed 22 nd jan. 2008.

Anonymus 2006.Functional foods Japan. Product Report. Http;// www.functional foods. Japan.com/reports/functional foods 2006. Html.Jan.2008.

Anonymus 2011.Probiotics market needs more awareness http://www.business-standard.com/common/news_article.php?leftnm=5&autono 311953.

Bulletin of the International Federation 2001.

Charalam popoulos,D., Wang,R., Pandiella, S.S., Webb,C. 2002. Application of cereals and cereal components in functional foods : a review. *Int. J Fd. Microbiol.* 79: 131-141.

Charteris, W.P., Kelley, P.M., Morelli,L., and Collins,J.K. 1997.Selective detection, enumeration and identification of potentially probiotic *Lactobacillus* and *Bifidobacterium* species in mixed bacterial populations, Int.J. Fd. Microbiol. 35: 1-27.

Charteris, W.P., Kelly, PM., Morelli,L., and Collins,JK. 1998. Development and application of an in vitro methodology to determine the transit tolerance of potentially probiotic *Lactobacillus* and *Bifidobacterium* species in the upper human gastro intestinal tract. *J Appl Microbiol.* 84: 759-768.

Christine M. Kalui, Julius M. Mathara and Philip M. Kutima. African J Biotechnol. Probiotic potential of spontaneously fermented cerealbased foods - A review. Vol. 9(17), pp. 2490-2498, 26 April, 2010.

Demin,A.A., Malinin,V.V., Shataeva, L.K ., and Chernova, I.A. 1994. Chromatographic methods for isolation of immunostimulating peptides of milk whey. *Appl. Biochem. Microbiol.* 30: 255-259.

Diebel, R.H and Niven, C.F.Jr. 1961. Comparative study of Gaffkya homari, *Aerococcus viridians,* tetrad - forming cocci from meat curing brines and the genus *Pediococcus. J Bact.* 79: 175-180.

Donkor, O.N., Nilmini, SL I., Stolic,P., Vasiljevic,T and Shah,N.P. 2007. Survival and activity of selected probiotic organisms in set- type yogurt during cold storage. *Int. Dairy J.* 17: 657-665.

FAO/WHO 2001. Evaluation of health and nutritional properties of powder milk with live lactic acid bacteria. Report FAO/WHO expert consultation, 1-4 October, 2001; Cordoba, Argentina.

FAO/WHO 2002. Guidelines for the evaluation of probiotics in food. Food and Agriculture Organization of the United Nations and World Health Organization Working Group Report.

Filippo, C., Di Caro, S., Santarelli,L., Armuzzi, A., Gasbarrini, G and Gasbarrini,A. 2001. *Helicobacter pylori* treatment: a role for probiotics. *Dig Dis.* 19: 144-147.

Fleming,H.P., Walter,W.M Jr., and Etchells,JL. 1973. Antimicrobial properties of oleuropein and products of its hydrolysis from green olives. *Appl. Microbiol.* 26: 777- 782.

Fuller, R.1989. probiotics in man and animals. *J Appl bacterial.* 66: 365-378.

Gupta,V., Sharma,A., and Nagar,R. 2007. Preparation, acceptability and nutritive value of rabadi- A fermented mothbean food. *J Fd Sci Technol.* 44(6): 600-601.

Hassan A B., Ahmed I. A. M, Osman N M., Eltayeb, M. M., Osman, G. A. and Babiker E. E. 2006. Effect of processing treatments followed by fermentation on protein content and digestibility of pearl millet(*pennisetum typhoideum*) cultivars. *Pakistan Journal of Nutrition* 5 (1): 86-89, 2006.ISSN 1680-5194. © Asia Network for Scientific Information

Henneberg, W.C. 1904 Zur Kenntris der Milchsaurebakterien der brennerei- Maische, der Milch, des Bieres, der Presshefe, der Melasse, des Sauerkohls, der sauren Gurke und des Sauerteigs; sowie einige Bemerkungen uber die Milch saure bakterien des mens chlichen Magens. Zentbl. Bakt. Parasitkde, Abt II, 11, 154-170.

Jindal,A.R., Shore,M., Shukla,F.C., and Singh,B. 2004. Studies on the use of chhana and paneer whey in the preparation of puras (pan cakes). *Int. J Dairy Technol.* 57: 221-225.

Kailasapathy,K., Harmstorf,I and Phillips,M. 2008. Survival of Lactobacillus *acidophilus* and *Bifidobacterium animalis*. Ssp. Lactis in stirred fruit yogurts. LWT Fd *Sci Technol.* 41: 1317-1322.

Kailasapathy, K., and Supraidi, D. 1996. Effect of whey protein concentrate on the survival of *L.acidophilus* in lactose hydrolysed yogurt during refrigerated storage. *Milchwissenschaft.* 51: 565-568.

Kailaspathy,K., and Chin,J. 2000. Survival and therapeutic potential of probiotic organisms with reference to Lactobacillus and *Bifidobacterium* species. Immunol. *Cell Biol.* 78: 80-88.

Kalliomäki M, Salminen S, Arvilommi H, Kero P, Koskinen P and Isolauri E 2001 Probiotics in primary prevention of atopic disease: a randomized placebo controlled trial. *Lancet*, 357(92620) : 1076-1079.

Kalliomaki,M.A, and Isolauri,E.2004.Probiotics and down regulation of the allergic response.Immunol. *Allergy Clin. North Am.* 24: 739-752.

Kitchell, A.G and Ingram, M. 1963. Vacuum packed sliced Wiltshire becon. *Fd. Process Packag.* 32: 1-7.

Kosikowski,F.V., and Mistry,V.V. 2002. Cheese and fermented milk foods, Vol.I.Origins and principles. 3rd ed. F.V. Kosikowski, L.L.C., Westport, Connecticut, 1997. Chap. 26.

Leporanta, K. 2001. Developing fermented milks into functional foods. Innov. *Food Technol.* 10: 46-47.

Liliy,D.M and Stillwell,R.H. 1965. Probiotics- growth promoting factors produced by microorganisms. *Science.* 147: 747-748.

Ling,K., Charles. Marketing Operations of dairy cooperatives, USDA rural business cooperative service, *RBS Res. Rep.* 201.

Marklinder,J. and Lonner,C. 1992. Fermentation properties of intestinal strains of *Lactobacillus* of a sourdough and of a yogurt starter culture in an oat -based nutritive solution. *Fd. Microbiol.* 9: 197-205.

Marteau,P., Seksik,P ., and Jian,R. 2002. Probiotics and health: new facts and ideas. Curr OpinBiotechnol. 13 (5): 486-489.

Mills S, Catherine S, Gerald F F., and Ross R P. 2011. Enhancing the stress responses of probiotics for a lifestyle from gut to product and back again. *Microbial Cell Factories,* 10(Suppl,1):S19 doi:10.1186/1475-2859-10-S1-S19.

Oberman,H., and Libudzisz,Z. 1998. Fermented milks. In; Wood,B.J.B (Ed) microbiology of fermeneted foods.Vol. I Blackie Academic and Professional. London,UK. 308-349.

Parker, F., Migliore- Samour, D., Floch,F., Zerial,A., Werner,G.H., Jolles,J., Casaretto, M., Zahn,H., and Jolles,P. 1984. Immunostimulating hexa peptide from human casein: amino acid sequence, synthesis and biological properties. *Euro. J. Biochem.* 145: 677-682.

Parvez,S., Malik,K.A., Kang,S.Ah., and Kim, H.Y. 2006. Probiotics and their fermented food products are beneficial for health. *J Appl. Microbiol.* 1171-1185.

Pefialoza W., C.L. Davey, J.N. Hedger and D.B. Kell. 7 : 260-268. Stimulation by potassium ions of the growth of *Rhizopus oligosporus* during liquid- and solid-substrate fermentations.

Plummer,S., Weaver, M.A., Harris, J.C., Dee,P., Hunter,J. 2004. *Clostridium difficile* pilot study: effects of probiotic supplementation on the incidence of *C. difficile* diarrhea. *Int. Microbiol.* 7 : 59-62.

Richelsen, B., Kristensen,K., Pedersen,S.B. 1996. Long term 96months) effect of a new fermented milk product on the level of plasma lipo proteins: A placebo-controlled and double – blind study. *Eur. J Clin. Nutr.* 50: 811-815.

Rodriguez, E., Gonzalis,B., Gaya,P., Nunez,M and Medina, M. 2000. Diversity of bacteriocins production by lactic acid bacteria isolated from raw milk. *Intl. Dairy J.* 10:7-15.

Russo, J. 2005. Nutraceuticals and functional foods. *Encyclopedia of physical pharmacy,* 6:32-40.

Saarela,M., Mogensen,G., Fonden,R., Matto,J. and Mattila- Sandholm, T. 2000. Probiotic bacteria: Safety functional and technological properties. *J Biotechnol.* 84: 197-215.

Sanders, M.E.1998.Overview of functional foods; emphasis on probiotic bacteria. Int.dairy J. 8:341-347.

Sanders, M.E. 2008. Products with probiotics. (www.usprobiotics.org) Sciences University of Wales Institute, Cardiff United Kingdom. 2009.

Shah, N.P. 2009. Development of artificially carbonated fermented milk. MSc thesis, Anand Agricultural Univ., Anand, Gujarat,India.

Shah,N.P. 2007. Functional cultures and health benefits. *Int. dairy J.* 17: 1262-1277.

Shahani,K.M., and Chandan, R.C. 1979. Nutritional and healthful aspects of cultured and culture containing dairy foods. *J Dairy Sci.* 62:1685-1694.

Stamer,J R., Stoyla,BO. and Dunckel,B.A. 1971. Growth rates and fermentation patterns of lactic acid bacteria associated with the sauerkraut fermentation. *J Milk Food Technol.* 34:521-525.

Stanton, C., Gardiner, G., Meehan,H., Collins,K., Fitzgerald,G., Lynch,P.B. and Ross, R.P. 2001. Market potential for probiotics. *Am J Clin Nutr.* 73 (25): 476 s-483 s.

Stanton,C., Desmond,C., Coakley,M., Collins,J.K., Fitzgerald,G., Ross, R.P. 2003a. Challenges facing development of probiotic containing functional foods. 27-58. In : handbook of functional fermented foods. Chap.11. Ed.Farnworth,E.R. Boca Raton, F.L: CRC press.

Stanton,C., Desmond, C., Fitzgerald, G., Ross, R.P. 2003 b. probiotic health benefits-reality or myth ? *Aust J Dairy Technol.* 58: 107-113.Leporanta 2001.

Swidan N. Thesis submitted for the degree of doctor of philosophy Cardiff School of Health.

Tamime,A.Y. 2002. Fermented milks: a historical food with modern applications- a review. *European J Clin. Nutr.* 56(S4): S2-S15.

Timmermann, R. Huys,G., Pot,B., and Swings,J.2001. Identification and antibiotic resistance of isolates from probiotic products. Abstracts of 101st ASM general meeting. The American Society for Microbiology, Washington D C, USA, p.c-289.

Walstra, P., Wouters, J.T.M. and Geurts, T.J .2005.Dairy Sci. Technol. (II edition). Chap.22. Fermented milks. 551-575.

Wang, K. Y. , Li, SN , Liu CS , Perng DS , Su YC , Wu DC , Jan CM , Lai CH , Wang, T.N. and Wang, W.M. 2004. *Amer. J Clin.Nutri.* 80(3): 737-741.

Wood,P.J. 1997.Functional foods for health: opportunities for novel cereal processes and products. In: Campbell, MG. Mckee,L.S., Webb,C. (Eds.), Cereals: Novel Uses and Processes.Plenum,NY. 233-239.

Yong, F.M. and Wood,B.J.B. 1974. Microbiology and biochemistry of the soy sauce fermentation. *Adv.Appl. Microbiol.* 17:157-194.

Ziemer,C.J. and Gibson, G.R. 1998. An overview of probiotics, prebiotics and synbiotics in the functional food concept: perspectives and future strategies. Int. Dairy J. 8: 473-479.

12

Recent Trends in Food Biotechnology for Value Addition with Special Reference to Probiotics

Renu Agrawal

Recently people are moving towards newer trends in food diversification for value addition. The beneficial effects could be either maintenance, promotion or well being of health or reduction of the risk of a pathogenic process or a disease. Genetically modified (GM) crops / foods, algae, mushrooms and probiotics are among the important functional foods which apart from improving health also protect against infections and/or stimulate immune systems. The interactive research between physiology, microbiology, food technology and molecular biology may produce many healthful, diversified value based food products for human consumption.The concept of value addition in food is changing significantly. Earlier the emphasis was on survival through hunger satisfaction but today it is aiming towards food safety and food sciences at developing foods which promote good health and reduce the risk of major diseases. With these efforts, these products have come into the market and are identified as functional foods. These include foods containing phytochemicals, dietary fibre, structural lipids, bioactive peptides, polyunsaturated fatty acids.

Genetic Modification

Genetic modifications are set of technologies that alter the genetic makeup of living organisms, plants or bacteria. When genes are combined from different organisms in recombinant DNA technology, a genetically modified organism is formed. GM products include medicines and vaccines, foods and food ingredients, feeds and fibres.

Locating genes for important traits as insect resistance or desired nutrients is in fact a limiting step in the process. Recently, genome sequencing of different organisms are generating detailed maps along with data analyzing technologies.When food crops are genetically modified one or more genes are incorporated into the crops genome, using a vector containing several other genes like viral promoters, transcription terminators and antibiotic resistance. This may lead to the development of unknown toxic/ allergic components.

GM crops are commercially grown in over 40 countries. The main ones being herbicide and insecticide resistant soybeans, corns, cotton and canola. Also antivirus sweet potato has been field tested which has high iron and vitamins. Through GM there are a variety of plants which can survive severe weather conditions.

There are bananas with activity against hepatitis –B and fruits with higher post harvest shelf life.

There are a number of controversies for GM crops and foods which include human and environmental safety, labeling and consumer choice, intellectual property rights, food security, ethics and environmental protection.

In 1990's there was an outbreak of "Mad Cow" disease and "Foot and Mouth disease" that made concern for GM organisms and this awareness has demanded the content and origin of food on labels.

Due to these inherent problems the marketing of GM foods is regulated. Directives, as 2001/18/EC has been designed to require authorization for the placing on the market of GMO.

GM oilseed market increased for soy, rape, peanut, palm and sunflower. According to USDA in 2002 both GM and non-GM products were produced globally with over 320 m tones and a value of $ 60 billion.

Industry outlook

Monsanto, the world's largest GM seed firm is developing new technologies for the use of oil seeds with value addition and agronomic benefits. They are also developing soybeans with high omega-3 fatty acids. DuPont is planning to launch high oleic acid soybeans in 2011.

Plants do not produce these healthy fatty acids and therefore, genes from algae, fungi and mosses are being identified, characterized and then transferred for best combination.

According to Hashimoto *et al.* (1999) the GM potatoes express soybean glycinin gene. Momma *et al.* (1999) developed GM rice with 20% more protein. Novak *et al.* (2000) have developed several lines of GM cotton plants using a gene from *Bacillus thuragensis* providing increased protection against major lepidopteran pests. Pusztai *et al.* (1999) fed GM peas to rats for over 10 days. The results were similar as in the control peas and had no harmful effects on the animals.

Hashimoto (1999) fed GM potatoes to rats and found no difference in growth, feed intake, blood cell count and composition with the control. In another study by Fares and EL-Sayed (1998) found that the mice developed villus epithelial cell hypertrophy and multinucleation, disrupted microvilli, mitochondrial degeneration, increased numbers of lysosomes and autophagic vacuoles.

One of the major health concerns with GM food is its potential to increase allergies and anaphylaxis in humans eating unlabelled GM food stuff. There are good animal models for nutritional / toxicological testing but no such models exist for allergenity testing.

Today, there is a need for better testing methods before making GM foods available for human consumption. Although, the GM crops and GM foods give a lot of value addition to food but the future will be dependent if the delivery is wholesome and safe.

Algae; Spirulina (Blue Green)

Health benefits of Spirulina, blue green algae which is used as a dietary supplement is well known for value addition to food. It has many beneficial properties as an antioxidant and for immune stimulation. It also has anti viral activity against a variety of harmful viruses and shows promise as a cancer preventative agent and in the treatment for tumors. It has cardiovascular benefits including improvement of blood lipid profiles, prevention of atherosclerosis and control of hypertension. It stimulates the production of antibodies and cytokines (Capelli and Cysewski, 2009).Recently a lot of people are consuming algae in food preparations. Consumption of Spirulina is common in countries like Japan which is becoming common in other parts of the world.

Mushrooms

Mushrooms are well known for nutrition and high vitamin. Recently, the trend is towards consuming different varieties of mushrooms in various food formulations. Agaricus is a common and popular mushroom which adds value to food and is a part of many natural health supplements. This mushroom increases the immunity and is healthful as it produces many therapeutic compounds. It increases the immune cells that attack viruses and bacteria in blood.

Maitake Mushroom is another common mushroom with high vitamin D content. Reishi mushroom has very high antioxidant activity and also vitamin content. It has substances to stimulate overall health and well being of the spleen. It is also called as king of mushrooms.

All the mushrooms have the ability to reduce the risk of cancer significantly especially breast cancer. These add value to food. These days the consumption of mushrooms is becoming a delicacy with high nutrition.

Probiotics

Lactic acid bacteria are regarded as a major group of probiotic bacteria (Collins *et al.,* 1998; Tannock, 1998). There is a long history of health claims concerning lactic acid bacteria in food. The term means "For life" and is derived from Greek language.

It was in 1908 that Metchnikoff suggested that long, healthy life of Bulgarian peasants results from the consumption of fermented milk products. Vergin (1954) compared the harmful effects of antibiotics add beneficial effects of LAB.

Sperti (1971) explained probiotics as tissue extracts that stimulates microbial growth. Salminen (1996) defined it as a live microbial culture which beneficially influences the health and nutrition of the host. According to Ouwehand *et al.* (2002) probiotics are non- pathiogenic microorganisms that when ingested in certain numbers exerts a positive influence on the host physiology and health beyond inherent general nutrition.

Mechanism of Action

Stress in modern day life disrupted the homeostasis in the gut through change in dietary pattern and eating habits. Another factor is consumption of pharmaceutical compounds especially antibiotics which destroy bacteria creating an imbalance in the gut microflora leading to a number of diseases. Therefore, there is an increasing demand for food products which support health beyond providing basic nutrition.

Consumption of probiotics is known to help in balancing normal intestinal microflora. Some mechanisms have been speculated (Jacobsen *et al.,* 1999; Boirivant and Strober,2007; Allan, 2008). It is possible due to biochemical effects through the production of bacteriocin, a short chain fatty acids. An unfavourable environment for pathogens is created by reduction of pH, production of antimicrobial compounds including organic acids, H_2O_2 or diacetyl,competition for nutrients, immune clearance or there can be blockage of adherence of enteropathogens by occupying the niche of the intestinal mucosa.

Functional Characteristics

Antibiotic resistance

Regular use of antibiotics causes depletion of vitamins causing an imbalance in the gut microbial flora.This causes loss of calcium, magnesium and potassium ions from the body which creates an environment for proliferation of opportunistic microbes causing abnormal health to the host. Antibiotic resistance has been reported in a large number of lactic acid bacteria strains (Kozlova *et al.,* 1992; Mathur and Singh, 2005; D' Aimmo *et al.,* 2007).

Immuno modulatory activity

Administration of probiotic LAB increases the immune responses (Gill *et al.,* 2001; Gill and Rutherford, 2001). These include enhancement of phagocytic activity of peripheral blood leukocytes and natural killer cell activity (Fernandes and shahani, 1990; schiffrin *et al.,* 1997; Esther *et al.,* 2007).

Anti-tumerogenic activity

Administration of LAB has shown to suppress bacterial enzyme activity as β- glucoronidase,urease, fecal glycocholate, nitroreductase and azoreductase (Lidbeck *et al.,* 1991; Spanhaek *et al.,* 1998; Gorbach, 2000). Fermented products reduce the risk of carcinogenesis (Challa *et al.,* 1987; Femia *et al.,* 2002). Antitumor effect has been reported by oral intake of LAB (Hirayama and Rafter, 2000). LAB can also bind to mutagens and detoxify, reducing the risk of mutagenesis (Lankaputra and Shah,1998; Vorobjeva and Abilev, 2002). Mechanism may be either through binding of carcinogens, stimulation of protective enzymes or increasing the immune resoponse (Burns and Rowland, 2000).

A large number of reports support that PLB reduce the activity of nitroreductase and increase β- glucosidase enzyme which release flavonoids that have antimutagenic, antioxidative and immune stimulatory effects (stoner and Mukhtar, 1995; Piotrowska and Zakowska).

Antioxidative activity

Free radical scavenging property is very important in many food manufacturing industries (Virtanen *et al.,* 2007).

Enzymatic activity

Cell is associated with proteinase, peptidase transferase and intracellular peptidase for hydrolysis of proteins and aminoacids (Di Cagno *et al.,* 2004).

β- glucosidases, a major group of glycosyl hydrolase enzyme - catalyze the selective cleavages of β-1,4- glycosidic linkages that play an important role in several biological pathways (Yan *et al.,* 1998). Urease is known to protect microorganisms from harmful effect of acidic condition by increasing the environmental pH through conversion of urea into ammonia and CO_2 (Vande Guchte *et al.,* 2002).

Nutraceutical property

Lactic acid bacteria increase niacin, riboflavin, folic acid content in fermented foods. Vitamins B_{12} and B_6 have also increased (Shahani and Chandan, 1979; Sauer *et al.,* 1998) and protection against cancer (Sybesma *et al.,* 2003; Jagerstad *et al.,* 2004). LAB have shown to increase conjugated linoleic acid and inhibit carcinogenesis and tumorigenesis (Ip *et al.,* 1991; Pariza *et al.,* 2001) reduces body fat content and increases muscle mass (Chin *et al.,* 1992; Akahoshi *et al.,* 20020, improves hyperinsulinemia (Houseknecht *et al.,* 1998).

Medical importance

Gastroenteritis is caused by viral, bacterial or parasitic infections like diarrhea, ulcers, crohn's disease, traveler's diarrhea, pouchitis and other inflammatory bowel diseases (Felley *et al.,* 2001). Characteristic adherence of the culture to mucosal surface contributes to their efficacy of being probiotic.

Probiotic Foods for Value Addition

LAB imparts health benefits in GIT tract and also in respiratory and urogenital tracts (Cadieux *et al.*, 2008; Reid, 2008). These replace chemical additives with value addition and provide the consumer with new, attractive food products with health benefits. They also enhance the shelf life, improve texture and contribute to the pleasant sensory profile of the end product. Oral doses of 1-10 billion colony forming units (cfu/mL) per dose are usually effective.

Market Potential

Probiotic lactic acid bacteria products are very common in Japan and Europe (Stanton *et al.*, 2001). Interest in acceptance of functional foods is gaining importance in areas of developing new food processing, retailing and distribution technologies. Benefits of PLB foods are opening new opportunities to value added products for increasing profit.

Summary

Recently, there is a lot of awareness for nutritive food which can impart beneficial and therapeutic properties to the consumer and there is a shift from regular food. People are ready for different types of food including GM, Algae, mushroom or beneficial microorganisms like probiotics. Since there is substantial demand for natural treatments with respect to various diseases these diversified food products are gaining importance. Moreover, these improved products could be achieved by altering the gene sequences. To capitalize on this emerging knowledge better molecular and metabolic studies are essential. These products are an attractive solution for disease management through diet.

References

Akahoshi, A., Goto, Y., Murao, K., Miyazaki, T., Yamasaki, M., Nonaka, M., Yamada, K. and Sugano, M. 2002. Conjugated linoleic acid reduces body fats and cytokine levels of mice. *Biosci Biotechnol Biochem* 66(4): 916-920.

Allan, W.W. 2008. Mechanisms of action of probiotics. *Clin Infectious Diseases* 46: 101-111.

Boirivant, M. and Strober, W. 2007. The mechanism of action of probiotics. *Curr Opin Gastroenterol* 23(6): 679-692.

Burns, A.J. and Rowland, I.R. 2000. Anti-Carcinogenicity of Probiotics and Prebiotics *Curr Issues Intest Microbiol* 1(1): 13-24.

Cadieux, P., Wind, A., Sommer, P., Schaefer, L., Crowley, K., Britton, R.A. and Reid, G. 2008. Evaluation of reuterin production in urogenital probiotic *Lactobacillus reuteri* RC-14.[My paper] *Appl Environ Microbiol* 6: 12-19.

Capelli,B. and Cysewski,G.2009. Potential health benefits of spirulina microalgae; a review of the existing literature.9: 16-26.

Challa, A., Rao, D.R., Chawan, C.B., and Shackelford, L. 1997. *Bifidobacterium longum* and lactulose suppress azoxymethane induced aberrant crypt foci in rats. *Carcinogenesis* 18: 517-521.

Chin, S.F., Liu, W., Storkson, J.M., Ha, Y.L. and Pariza, M.W. 1992. Dietary sources of conjugated dienoic isomers of linoleic acid, a newly recognized class of anticarcinogens. *J Food Composition Anal* 5: 185–197.

Collins, J.K., Thornton, G. and Sullivan, G.O. 1998. Selection of probiotic strains for human application. *Int Dairy J* 8: 487–490.

D'Aimmo, M.R., Modesto, M. and Biavati, B. 2007 International Antibiotic resistance of lactic acid bacteria and *Bifidobacterium* spp. isolated from dairy and pharmaceutical products. *J Food Microbiol* 115(1): 35-42.

Diana Garnica. 2011. Medicinal benefits of mushrooms.http://EzineArticles.com/?expert=Diana_Garnica

Di Cagno, R., De Angelis, M., Auricchio, S., Greco, L., Clarke, C., De Vincenzi, M., Giovannini, C., D'Archivio, M., *et al.*, 2004. Sourdough bread made from wheat and non toxic flours and started with selected lactobacilli is tolerated in celiac sprue patients. *Appl Environ Microbiol* 70: 1088-1096.

Domingo,J.L. 2000 health risks of genetically modified foods: Many opinions but few data. Science 288 : 1748-1749.

Esther, N., Julia, W., Sonia, G.M., Ligia, D., Javier, R. and Ascension, M. 2007. Immunomodulatory effects of probiotics in different stages of life. *Brit J Nutr* 98: s90-95.

Fares, N.H. and El-Sayed, A.K. 1998. Fine structural changes in the ileum of mice fed on delta- endotoxin-treated potatoes and transgenic potatoes. *Natural Toxins.* 6 : 219-233.

Felley, C.P. Corthesy-Theulaz, I., Blanco Rivero, J.L., Sipponen, P., Kaufmann, M., Bauerfeind, P., Wiesel, P.H., Brassart, D., Pfeifer, A., Blum, A.L. and Michetti, P. (2001). Favourable effect of an acidified milk (LC-1) on *Helicobacter pylori* gastritis in man. *Eur J Gastroenterol Hepatol* 13: 25–29.

Femia, A.P., Luceri, C., Dolara, P., Giannini, A., Biggeri, A., Salvadori, M., *et al.*, 2002. Antitumorigenic activity of the prebiotic inulin enriched with oligofructose in combination with the probiotics *Lactobacillus rhamnosus* and *Bifidobacterium lactis* on azoxymethane induced colon carcinogenesis in rats. *Carcinogenesis* 11: 1953-1960.

Fernandes, C.F. and Shahani. K.M. 1990. Anticarcinogenic and immunological properties of dietary *Lactobacilli*. *J Food Prot* 53: 704-710.

Gill, H.S. and Rutherfurd, K.J. 2001. Immune enhancement conferred by oral delivery of *Lactobacillus rhamnosus* HN001 in different milk based substrates. *J Dairy Res* 68: 611-616.

Gill, H.S., Rutherfurd, K.J., Prasad, J. and Gopal, P.K. 2000. Enhancement of natural and acquired immunity by *Lactobacillus rhamnosus* (HN001), *Lactobacillus acidophilus* (HN017) and *Bifidobacterium lactis* (HN019). *Brit J Nutr* 83: 167–176.

Gorbach, S.L. 2000. Probiotics and gastrointestinal health. *Am J Gastroenterol* 95(suppl): S2-S4.

Hashimoto, W., Momma,K., Katube, T., Ohkawa, Y., Ishige,T., Kito, M., Utsumi, S. and Murata, K. 1999. Safety assessment of genetically engineered potatoes with designed soybean glycinin: compositional analyses of the potato tubers and digestibility of the newly expressed protein in transgenic potatoes. *J Sci.Food Agric.* 79: 1607-1612.

Hashimoto, W., Momma,K., Yoon, H.J., Ozawa,S., Ohkawa, Y., Ishige,T., Kito, M., Utsumi, S. and Murata, K. 1999. Safety assessment of transgenic potatoes with soybean glycinin by feeding studies in rats. *Biosci. Biotechnol.Biochem.* 63:1942-1946.

Hirayama, K. and Rafter, J. 2000. The role of probiotic bacteria in cancer prevention. *Microbes and Infection* 2(6): 681-686.

Houseknecht, K.L., Vanden-Heuvel, J.P., Moya-Camarena, S.Y., Portocarrero, C.P., Peck, L.W. and Nickel, K.P. 1998. Dietary conjugated linoleic acid normalizes impaired glucose tolerance in the Zucker diabetic fatty rat. *Biochem Biophy Res Comm* 244: 678–682.

Ip, C., Chin, S.F., Scimeca, J.A. and Pariza, M.W. 1991. Mammary cancer prevention by conjugated dienoic derivatives of linoleic acid. *Cancer Res* 51: 6118–6124.

Jacobsen, C.N., Rosenfeldt, N.V., Hayford, A.E., Mille, P.L, Michaelsen, K.F., Paerregaard, A., Sandstrom, B., Tvede, M. and Jakobsen, M. 1999. Screening of probiotic activities of forty seven strains of *Lactobacillus* spp. by *in vitro* techniques and evaluation of the colonization ability of five selected strains in humans. *Appl Environ Microbiol* 65: 4949–4956.

Jagerstad, M., Jastrebova, J. and Svensson, U. 2004. Folates in fermented vegetables: a pilot study. *Lebensmittel-Wissenschaft Und-Technologie* 37: 603–611.

Kulper, H.A., Noteborn, H.P.J.M. and Peijnenburg, A.A.C.M. 1999. Adequacy of methods for testing the safety of genetically modified foods. *The Lancet* 354 : 1315-1316.

Kozlova, E.V., Pivovarenko, T.V., Malinovskaia, I.V., Aminov, R.I., Kovalenko, N.K. and Voronin, A.M. 1992. Antibiotic resistance of *Lactobacillus* strains. *Antibiot Khimioter* 37(6): 12-15.

Lankaputhra, W.E.V. and Shah, N.P. 1998. Antimutagenic properties of probiotic bacteria and of organic acids. *Mut Res* 397: 169–182.

Lidbeck, A., Geltner Allinger, U., Orrhage, K.M., Ottova, L., Brismar, B., Gustafsson, J.A., Rafter, J.J., and Nord, C.E. 1991. Impact of *Lactobacillus acidophilus* supplements on the faecal microflora and soluble faecal bile acids in colon cancer patients. *Microbial Ecol Health Dis* 4: 81-88.

Mathur, S. and Singh, R. 2005. Antibiotic resistance in food lactic acid bacteria: a review. *Int J Food Microbiol* 105(3): 281-295.

Momma, K., Hashimoto, W., Ozawa,S., Kawai, S., Katsube, T., Takaiwa, F., Kito, M., Utsumi, S. and Murata, K. 1999. Quality and safety evaluation of genetically engineered rice with soybean glycinin: Analyses of the grain composition and digestibility of glycinin in transgenic rice. *Bioscience Biotechnology Biochemistry* 63 : 314-318.

Nakamura,R. and Matsuda,T.(1996). Rice allergenic protein and molecular- genetic approach for hypoallergenic rice. *Biosci. Biotechnol. Biochem.* 60: 1215-1221.

Novak, W.K. and haslberger, A.G. 2000. Substantial equivalence of antinutrients and inherent plant toxins in genetically modified novel foods. *Food and Chemical Toxicology.* 38: 473-483.

Ouwehand, A.C., Salminen, S., Tolkko, S., Roberts, P., Ovaska, J. and Salminen, E. 2002. Resected human colonic tissue: new model for characterizing adhesion of lactic acid bacteria. *Clin Diag Lab Immuno* 9: 184–186.

Pariza, M.W., Park, Y. and Cook, M.E. 2001. The biologically active isomers of conjugated linoleic acid. *Prog Lipid Res* 40: 283–298.

Piotrowska, M. and Zakowska, Z. 2005. The elimination of Ochratoxin A by Lactic Acid Bacteria strains. *Polish J Microbiol* 54: 279–286.

Pusztai, A., Grant,G., Bardocz, S., Alonso, R., Chrispeeis,M.J., Schroeder, H.E., Tabe, L.M. and Higgins, T.J.V 1999. Expression of the insecticidal bean alpha - amylase inhibitor transgene has minimal detrimental effect on the nutritional value of peas fed rats at 30% of the diet. *JNutri.* 129: 1597-1603.

Pusztai,A.(2000) The need for rigorous risk assessment. *Chemistry and Industry.* 8: 280.

Reid, G. 2008. Probiotic Lactobacilli for urogenital health in women. *J Clin Gastroenterol* 5: 23-39.

Salminen, S. 1996. Uniqueness of probiotic strains. Paper given at IDF Nutrition Week in Potsdam, Germany. *Newsletter of the International Dairy Federation* 145, Brussels: IDF, pp. 18-19.

Sauer, U., Cameron, D.C. and Bailey, J.E. 1998. Metabolic capacity of *Bacillus subtilis* for the production of purine nucleosides, riboflavin, and folic acid. *Biotechnol Bioeng* 59: 227–238.

Schiffrin, E.J., Brassart, D., Servin, A., Rochat, F. and Donnet-Hughes, A. 1997. Immune modulation of blood leukocytes in humans by lactic acid bacteria: criteria for strain selection. *Am J Clin Nutr* 66: 515S-520S.

Shahani, K.M. and Chandan, R.C. 1979. Nutritional and healthful aspects of cultured and culture containing dairy foods. *J Dairy Sci* 62: 1685–1694.

Spanhaak, S., Havenaar, R. and Schaafsma, G. 1998. The effect of consumption of milk fermented by *Lactobacillus casei* strain Shirota on the intestinal microflora and immune parameters in humans. *Eur J Clin Nutr* 52: 899-907.

Sperti, G.S. 1971. Probiotics. West Point, CT: Avi Publishing. Surawicz, C.M., Elmer, L.W., Speelman, P., McFarland, L.V., Chinn, J. and van Belle, G. (1989). Prevention of antibiotic associated diarrhea by *Saccharomyces boulardii:* a prospective study. *Gastroenterol* 96: 981–988.

Stanton, C., Gardiner, G., Meehan, H., Collins, K., Fitzgerald, G., Lynch, P.B., Ross, R.P. 2001. Market potential for probiotics. *Am J Clin Nutr* 73(2 Suppl): 476S–83S.

Stoner, G.D. and Mukhtar, H. 1995. Polyphenols as cancer chemopreventive agents. *J Cell Biochem* 22: 169–180.

Sybesma, W., Starrenburg, M., Tijsseling, L., Hoefnagel, M. H. and Hugenholtz, J. 2003. Effects of cultivation conditions on folate production by lactic acid bacteria. *Appl Environ Microbiol* 69: 4542–4548.

Tannock, G.W. 1998. Studies of the intestinal microflora: a prerequisite for the development of probiotics. *Int Dairy J* 8: 527–533.

Van de Guchte, M., Serror, P., Chervaux, C., Smokvina, T., Stanislav, D., Ehrlich. and Maguin, E. 2002. Stress responses in lactic acid bacteria. *Antonie van Leeuwenhoek* 82: 187–216.

Vergin, F. 1954. Anti-und Probiotika (Anti- and probiotics). *Hippokrates* 25(4): 116–119.

Virtanen, T., Pihlanto, A., Akkanen, S. and Korhonen, H. 2007. Development of antioxidant activity in milk whey during fermentation with lactic acid bacteria. *J Appl Microbiol* 102: 106–115.

Vorobjeva, L.I. and Abilev, S.K. 2002 Antimutagenic Properties of Bacteria: Review. *Appl Biochem Microbiol* 38(2): 112-122.

Yan, T., Lin, Y. and Lin, C. 1998. Purification and characterization of an extracellular β-glucosidase II with high hydrolysis and transglucosylation activities from *Aspergillus niger*. *J Agri Food Chem* 46: 431–437.

13

Aflatoxin Contamination in Chilli and its Management

S. Sudha and M. K. Naik

Spices are defined as a strongly flavored or aromatic substance of vegetable origin, obtained from tropical plants, commonly used as a condiment. India is known as 'land of spices'. Chilli (*Capsicum annuum* L.) also known as hot peppers and Capsicums also referred as bell peppers or sweet peppers belong to the genus *Capsicum* which is a member of the Solanaceae family. The native home of chilli is considered to New Mexico with secondary origin at Guatemala. They are known from pre-historic remains of Peru. Chilli was introduced into India by Portuguese in 16th century. Chilli is considered as one of the major commercial crops of the world. Both green and dried chillies are the important components of our routine diet. Chilli also contains vitamin A, C and E. It will give the required pungency, colour, taste and flavour to our dishes. The presence of capsaicin in chilli is responsible for its pungency and it has medicinal value. The extracted capsaicin is used in pain balms, cosmetics and medicines related to heart diseases. Capsanthin, a pigment in chilli is used for natural colouration to Jams, Jellies and squashes, since it is a natural pigment and no harmful or side effects on human health. Oleoresin is used in pain balms, vapours, etc. Hence, there is lot of demand for chilli oleoresin in the world market. The largest producer of chillies in the world is India contributing 52 per cent of world production, accounting for 11.67 lakh tonnes of production from an area of 7.5 lakh ha with 1550 kg/ha of productivity followed by Nigeria (17%), China (16%), Peru and Bangladesh (6%). India exported 1.8 million tonnes of chilli during 2009 (Reddy, 2010).

Aflatoxins are toxins produced by moulds which occur throughout the world. The term "aflatoxins" was coined in the early 1960s when the death of thousands of turkeys and other domestic animals was attributed to the presence of *Aspergillus flavus* toxins in groundnut meal imported from South America (Blount, 1961). Aflatoxin contamination of agricultural commodities has gained global significance as a result of their deleterious effects on human as well as animal health and its importance to international trade. Aflatoxins are the most potent carcinogens of mycotoxins and are the most commonly found toxin in chilli and paprika spice. Aflatoxin contamination has posed serious problems in commerce and international trade because of stringent quality standards imposed on aflatoxin by many importing countries. According to European Union's Rapid Alert System for Foods and Feed (RASFF), three consignments of chilli powder were rejected by the United Kingdom on July 5, 13 and 31 for the presence of aflatoxin above permissible limits. The presence of aflatoxin in the consignments reportedly was at 6.5 ppb (Potty, 2006).

The consumption of chilli and its products due to occurrence of aflatoxin contamination has become a major problem as it causes health hazards to the consumers and it affects the global market as well. Hence, there is a need to address the problem of aflatoxin contamination and pesticide residues in chilli crop. The hazardous analysis critical control point (HACCP) is a biologically powerful tool to ensure quality produce to foreign countries under the era of World Trade Organization (WTO) (Naik *et al.*, 2004).

Prevalence and Distribution of Aflatoxin Contamination

Aflatoxin is noted with great concern since it is a potent carcinogenic, mutagenic and immunosuppressive agent produced as secondary metabolites by the fungi *Aspergillus flavus* and *Aspergillus parasiticus* on a wide range of crops and food products. Many agricultural commodities including cereals, oilseeds, spices, dry fruits and feeds are contaminated by aflatoxin. Aflatoxin contamination in chilli is a serious problem and is governed by the grades of produce, kind of product, the diversity of pathogenic strains and condition of infection that deteriorates the quality of the produce thereby hitting the export trade in the international market. The literature on aflatoxin contamination particularly with respect to chilli and other related crops are presented in the chapter.

Groundnut samples from twenty selected markets in 10 regions of Ghana yielded high levels of the aflatoxigenic fungus *Aspergillus flavus*. The fungus was found associated with 31.7 and 12.8 per cent of groundnut samples respectively of all damaged and undamaged kernels assayed (Awuah and Kpodo, 1996). Mc Donald and Castle (1996) conducted a survey on aflatoxins in retail herbs and species in U.K. Out of 157 retail samples including curry powders, pepper, cayenne pepper, chilli, paprika, ginger, cinnamon and coriander, 95 per cent of the samples contained < 10 µg/ kg of total aflatoxins and only nine samples had higher levels. The highest concentration in a retail sample was 48 µg/ kg in chilli powder.

Soil and maize plant samples from four regions in Northern Vietnam were analysed. *Aspergillus flavus* was found in 120 soil samples (out of 140), 46 fresh maize samples (out of 420) and 141 dried maize samples (out of 420) (Nguyen *et al.*, 2000).

Reddy *et al.* (2000) reported the occurrence of aflatoxins in selected cereals and spices. Spices like ginger rhizomes, turmeric, black pepper and coriander were analyzed for the presence of aflatoxin. Ginger rhizomes and black pepper contained aflatoxin with a range of 24-26 µg/kg and 16-20 µg/kg respectively. However, turmeric and coriander did not contain any aflatoxin. Incidence and range of aflatoxin in poultry feed samples (groundnut cake, maize, millets, mixed feeds, rice bran, sorghum, soybean and sunflower) were determined, the aflatoxin contamination occurred at high levels of 3300 µg/kg in maize and groundnut samples. That is a good cause of concern.

Reddy *et al.* (2001) assessed the aflatoxin contamination in major chilli growing areas and chilli samples representing all the three grades collected from the wholesale markets of Andhra Pradesh. Chilli Grade 3 samples contained over 25 per cent of discoloured pods and were the most contaminated by *Aspergillus flavus*. Out of 182 chilli samples tested for the presence of aflatoxin, 59 per cent of the samples were contaminated with AFB1 and 18 per cent contained the toxin of non-permissible levels. The maximum percentage of chilli pods showing AFB1 levels were higher than 30 µg/ kg (Non permissible levels) was in Grade 3 and also the highest AFB1 concentration was 969 µg/ kg.

A preliminary survey was carried out to evaluate the level of total aflatoxin (AFT) and aflatoxin B_1 contamination in baby foods, groundnut products and maize products from the markets of Indonesia. Out of 82 groundnut products, 35 per cent were contaminated with aflatoxins at various levels ranging from 5-870 μg/kg. Two out of 11 samples 18 per cent of maize-based products) were contaminated with AFT ranging 5.8 to 12.4 μg/kg. On the other hand, out of 12 baby food samples, no sample was found contaminated with aflatoxin B_1 (AFB_1) (Razzazi *et al.*, 2004). Ajith Kumar and Naik (2005) conducted survey in Northern Karnataka districts of Gulbarga, Bellary and Raichur to know the incidence and severity of aflatoxin contamination in chilli. The highest incidence of (6.83%) was recorded in Bellary district followed by Raichur (5.29%) and Gulbarga district (4.19%).

The survey was conducted by Farid Waliyar *et al.* (2005) in major maize producing districts of Andhra Pradesh to asses the aflatoxin situation in maize. Out of 1151 samples of rainy season crop analyzed, six per cent contained > 20 mg/kg aflatoxin (above permissible limit) ; out of 310 maize samples analyzed from post-rainy season 8.6 per cent samples contained > 20 mg/ kg aflatoxin, out of 100 maize samples analyzed from storages, 20 per cent samples contained > 20 mg/kg aflatoxin. A survey was conducted to assess the extent of market maize contamination with aflatoxin in Kenya. Forty five per cent of maize products had aflatoxin levels greater than the Kenyan regulatory limit of 209 ppb. About 35 per cent of the maize products had aflatoxin content of >100 ppb and seven per cent had >1000 ppb and the maize obtained from local farms had aflatoxin levels > 20 ppb (Lauren Lewis *et al.*, 2005).

Farid Waliyar *et al.* (2007) examined the natural occurrence of Aflatoxins (AFB) produced by *Aspergillus flavus* in sorghum and pearl millet grains collected from farmers' field at harvest and storages during 2005 and 2006 in Andhra Pradesh and Maharashtra. In 2005, out of 440 sorghum samples analyzed, aflatoxin concentration ranged between 1-561 μg/kg in 71.6 per cent samples. Whereas, in 505 pearl millet samples aflatoxin concentration range was 1-1047 μg/kg in 73 per cent samples. In 2006, aflatoxin concentration in 179 sorghum samples analyzed were 1-362 μg/kg in 36.3 per cent samples. In 66 pearl millet samples, it was 1-24 μg/kg in 22.7 per cent samples. In 30 stored sorghum samples aflatoxin concentration was 1-317 μg/kg in 43.3 per cent samples and of 77 stored pearl millet samples aflatoxin concentration was 1-32 μg/kg in 12.3 per cent samples. Navya *et al.* (2007) studied the incidence of *A. flavus* and aflatoxin on seed samples of groundnut among 40 different samples screened, the levels of *A. flavus* ranged from 0 to 72 per cent.

Assessment of Population of *Aspergillus flavus*

Aflatoxigenic fungi are common components of soil mycobiota and soil serves as a reservoir for *Aspergillus flavus* fungi that produce carcinogenic aflatoxins in agricultural commodities. Aflatoxigenic fungi reside in soil as conidia, sclerotia and hyphae, which act as primary inoculum for directly infecting crops and also invade developing seeds of crops.

The effect of corn and peanut cultivation on the soil population of *Aspergillus flavus* in soil was examined. Drought stress in corn plants greatly increased the soil population of *A. flavus* from 2000 cfu g^{-1} soil to 6400 cfu g^{-1} of soil (Bruce *et al.*, 1995).

Abbass *et al.* (2004) assessed the spatial variability of soil population of *Aspergillus flavus* in Mississippi Delta field under different crops. The highest propagule density of 794 cfu g^{-1} of soil of *Aspergillus flavus* was recorded in corn, 251 cfu g^{-1} of soil in cotton and 457 cfu g^{-1} soil in wheat crop. Bruce (2006) studied the relationship between soil density of *Aspergillus flavus* and the incidence of peanut colonization. The highest population density of *Aspergillus flavus* was recorded in cultivable land (111-1733 cfu g^{-1}) compared to forested (2-19 cfu g^{-1}) and fallow (6-61 cfu g^{-1}) soils and up to 92 per cent seed colonization was observed in peanut seeds.

Reddy *et al.* (2007) examined the effects of cotton-corn rotation and glyphosate use on the soil population of *Aspergillus flavus*. The soil population of *Aspergillus flavus* ranged from 1.4 to 5.8 X 10^3 (cfu) g^{-1} and the population was significantly greater in glyphosate resistant cultivars. Zablotowicz *et al.* (2007) assessed the density of *Aspergillus flavus* propagules and other soil microflora (*Fusarium* spp.) associated with Mississippi Delta soils. Propagule density of *Aspergillus flavus* ranged from 1.97 to 4.31X 10^3 (cfu) g^{-1} while, the total *Fusaria* ranged from 2.99 to 5.37 X 10^3 (cfu) g^{-1} soil and the frequency of aflatoxin production in isolates ranged from 13 to 81 per cent depending on soil.

Naik and Sudha, 2009, assessed the distribution of *Aspergillus flavus* population in the soil. Among three districts, Bellary recorded maximum population density (835.90 cfug^{-1} soil) followed by Raichur (677.00 cfug^{-1} soil) and the lowest population density of 501.20 cfug^{-1} soil was recorded in Gulbarga district.

Host Range of *Aspergillus flavus*

Mycoflora in storages are known to harbor toxigenic fungal species, aspergilli are the major contaminants produced under high humid conditions. The fungus *Aspergillus flavus* readily colonizes several important crops and is of significant importance because the fungus has potential to produce toxin. The fungus *Aspergillus flavus* has a wide host range and the high risk commodities for aflatoxin contamination are rice, maize, groundnut, sesame, fig and onion.

Grinstein *et al.* (1992) examined onion bulbs grown in Israel at markets to determine the presence of pathogens in a particular growing season; *Aspergillus* spp. was the only genera found on summer onion which were marketed shortly after harvest. Decomposition of onion bulb was investigated over three years, where the predominant fungus observed was *Aspergillus* spp. (Garcia *et al.*, 1997)

The incidence of *Aspergillus flavus* in the aerosphere of maize fields in three seasons was studied. The maximum incidence of *Aspergillus flavus* was recorded during the monsoon season with 63 per cent followed by summer having 45 per cent and winters with 24 per cent (Bilgrami, 1990). *Aspergillus flavus* group of fungi associated with maize, wheat and rice were isolated from different granaries. All the samples analysed showed high frequency of *Aspergillus flavus*, *A. niger*, *Alternaria* spp. and *Penicillium* spp. (Ramesh Chandra, 1997).

Verma *et al.* (1997) analyzed the samples of oilseeds for the incidence of aflatoxigenic fungi and aflatoxin contamination. The incidence of *Aspergillus flavus* and aflatoxin contamination was the highest in peanut seeds (94.16 ppm) followed by cotton (0.29 ppm), sesame (0.20 ppm) and castor (0.17 ppm). Per cent incidence of *Aspergillus flavus* on different varieties of oilseeds was in the order of peanut (11%), cotton (8%), sesame (9%) and castor (10%) respectively. Giridhar and Reddy (1998) reported the mold infestation on onion bulbs and noted that 35 fungal species representing 16 genera were associated with onion bulbs in which *Aspergillus* spp, *Fusarium* spp and *Penicillium* spp. were the most common fungi and about 50 per cent strains of *Aspergillus flavus* were found to be toxigenic.

Bedi and Rattan (2003) reported that the cultivars of *Aspergillus flavus* obtained from maize kernels showed wide variation in their cultural characters. The quantitative aflatoxin production by *Aspergillus flavus* isolate was correlated with the cultural characters. The sclerotia production and the aflatoxin contamination showed positive correlation. The isolate Af-4-2 produced the highest number of sclerotia $22cm^{-2}$ with two highest aflatoxin production of 1566 ppb.

Bagwan and Meshram (2004) estimated the mycoflora association and aflatoxin B_1 contamination in dried fig fruit samples using potato dextrose agar plate method using indirect competitive ELISA. *Aspergillus flavus* and *Aspergillus niger* were the common storage fungi associated with stored fig fruits. The aflatoxin contamination ranged from 0.6 to 39 mg/kg. Reddy *et al.* (2004) detected the *Aspergillus* contamination in rice samples, Among four *Aspergillus* spp. isolated, contamination by *Aspergillus flavus* (41%) dominated in all the collected samples followed by *Aspergillus niger* (22%), followed by *A. ochraceus* (18%). Singh *et al.* (2004) evaluated seeds of groundnut using the standard blotter (deep freeze) and agar plate methods to determine the associated fungal flora. Ten fungal species from seven genera were found associated with seeds, among them *Aspergillus flavus* and *Aspergillus niger* were found dominant and pathogenic.

Shazia Rasheed *et al.* (2004) examined the seed borne mycoflora of 12 groundnut seed samples collected from different localities of Pakistan using agar plate, blotter and deep freezing methods. Of the 14 genera and 28 species of fungi isolated, 18 fungal species were *Aspergillus flavus.* However *A. niger* was found predominant. All the seed samples were infected by *A. flavus*. Anuja Gupta and Vijayanand (2006) evaluated the associated fungal flora incidence of rice grains in storages. *Rhizopus stolonifer* (33.2%), *Drechslera oryzae* (10.2%) and *Aspergillus flavus* (6.8%) were the main fungi recorded. The infection of Aspergilli in rice grains was recorded in surface-sterilized seed, kernel, hull and kernel powder. The parrot green coloured colonies of *Aspergillus flavus* dominated in occurrence in all cultivars of rice (Mangala *et al.*, 2006).

Reddy *et al.* (2006) investigated the fungal contamination of rice seeds under storage conditions. *Aspergillus* was detected in greater number in unsterilized paddy seeds, kernel and kernel powder than in samples obtained after surface sterilization of paddy seeds, hull, separated kernel and kernel powder. *Aspergillus* spp. reached the starch in rice kernel. However, *Aspergillus flavus* and *Aspergillus niger* were dominating.

Dilek Hepren (2007) reported presence of aflatoxin contamination in figs after harvesting and during drying stage. Five to thirteen per cent of the samples were contaminated with aflatoxin and the highest level of aflatoxin was 700 ppb.

Study on the fungi and some mycotoxins (Aflatoxin B1 - AFB1, ochratoxin A - OTA and Zearalenone - ZEN) contaminating rice (*Oryza sativa*) in Niger State of Nigeria was carried out. One thousand and sixty two fungi were isolated and identified from one hundred and ninety six moldy rice samples collected from the state. The most prevalent fungal species on rice were, *Penicillium* spp., *A. flavus*, *A. parasiticus*, *A.*

niger, *Mucor* spp., *Rhizopus* spp. and *Alternaria* spp. The commonest fungal contaminant AFB1, was detected in 97 of the samples analyzed at concentrations between 20-1642 ug/kg (Hussaini *et al.*, 2007).

Aspergillus flavus colonized all different hosts used for the study with varied per cent of infection. The highest per cent infection (70.66) was recorded on fig fruit followed by onion (60.33), groundnut (52.00), maize (41.33), sesame (27.33) and the least infection was noticed on paddy grains (17.66), indicating a wide host range among field and horticultural crops (Sudha *et al.*, 2009).

Toxigenic Variability of *A. flavus*

The aflatoxin is chemically highly substituted coumarins and contained a fused dihydrofurofuron configuration, naturally occurring compounds. They are known to occur in two series aflatoxin B1 and its derivatives and aflatoxin G1 and its derivatives. Of all the aflatoxins, aflatoxin B1 has the highest potency as toxin and as a carcinogen (Wogan, 1975).

Several surveys were conducted to identify the toxin producing strains. Rao *et al.* (1965) reported incidence of toxigenic strains of *A. flavus* affecting groundnut in certain eastern coastal districts of India. All strains of *A. flavus* did not produce aflatoxin. The type culture *A. flavus* NRRL 482 was a non-producer even though it produces several compounds, which fluoresce blue and green which might represent aflatoxin precursors (Detroy *et al.*, 1971).

Diener and Davis (1966) reported that 23 out of 26 isolates of *A. flavus* produced aflatoxin B1 and the rest formed both Bl and G2. Eight *A. flavus* strains isolated in Denmark produced small amount of aflatoxin (Krogh *et al.*, 1966). Joffe (1969) found that 90 per cent of the 1626 isolates of *A. flavus* from peanuts and soil in Israel produced aflatoxin B1. But only about 8.4 per cent of the isolates produced all the four aflatoxins.

In experiments conducted with *A. flavus* strains isolated from groundnut fields at different periods of the year, it was possible for us to show significant differences in the intensity of the toxigenesis of these micromycetes. These differences appear to be dependent on the evolution of the mycoflora during growth of the plant (Lafont and Lafont, 1977)

A. flavus typically produces aflatoxins B1 and B2 and cyclopiazonic acid and is extremely variable in mycotoxin production, with strains ranging in toxigenicity from non producers to potent producers of aflatoxin (Horn and Dorner, 1999).

Ajith Kumar and Naik (2006) analyzed aflatoxin B_1 (AFB_1) content by an indirect competitive ELISA for 16 isolates of *Aspergillus flavus* obtained from chilli. The highest concentration of aflatoxin (1652.2μg/kg) was produced by AFL-16 isolate followed by AFL-8 (1081.9μg/kg). The lowest range of 6.6μg/kg was produced by the isolate AFL-10, indicating a wide variability in aflatoxin producing isolates and potency in production of aflatoxin. They also indicated the occurrence of virulent and non virulent isolates in the same geographical regions.

Sangit-Kumar *et al.* (2007) isolated twenty five of *Aspergillus flavus* isolates from maize. They were tested for aflatoxin detection and four isolates were found highly toxigenic while eight were moderately toxigenic and ten mildly toxigenic. Among all isolates tested, only three isolates, i.e. AF14, AF20 and AF25 were found non-toxigenic.

Twenty two isolates of *A. flavus* were isolated from forty different ground nut samples and analyses for their ability to aflatoxin following ammonia vapour and TLC analysis test, except AF15 and AF33 all the other isolates were found to be either AFB1 or AFB2. Polymerase chain reaction was done to differentiate aflatoxingenic and non-aflatoxigenic *A. flavus* isolates using three primer pair apa-2, vr-1 and adomp-1. Out of six *A. flavus* isolates four (AF1, AF5, AF2 and AF21) were found aflatoxigenic and two isolates (AF15 and AF17) were non-aflatoxigenic (Navya *et al.*, 2007).

Sixteen isolates of *Aspergillus flavus* varied in their ability to produce toxin. There was a wide range of aflatoxin production by different isolates. The range was from 99.96 to 1639.10 μg/kg. but AF10 isolate obtained from Laxminagar camp-1 of Bellary district showed the maximum potency of 1639.10 μg/kg followed by AF 13 obtained from Bhaghyanagar camp-2 of Bellary district with 1438.56 μg/kg and AF 15 obtained from Dhadesugur of Bellary district with 1237.70 μg/kg and AF 3 obtained from Madarki camp-1 of Gulbarga district with 1195.80 μg/kg and the least production of 99.96 μg/kg was noticed in AF 14 obtained from Somasamudra of Bellary district. The remaining isolates were in the range of 500-900 μg/kg (Naik and Sudha, 2009).

Detection of Aflatoxin in Chilli and other crops and their products

Seventy nine pre-packaged samples of 12 different types of spice powders (5 cardamom, 5 cayenne pepper, 8 chilli, 5 cloves, 7 cumin, 5 curry powder, 5 ginger, 5 mustard, 10 nutmerg, 12 paprika, 5 saffron and 7 white pepper) were estimated for the aflatoxin B_1 content by HPLC method. Aflatoxin B1 (AFB_1) was detected in 34 samples, all of the *Cayenne pepper* samples were contaminated with AFB_1 levels ranging from 2-32 µg AFB_1/ kg. Three nutmeg samples contained levels ranging from 6-20 µg/kg. Paprika contained levels of aflatoxin B1 ranging from 1-20µg/kg. Chillies, cumin, curry powder, saffron and white pepper samples had levels ranging from 1-5 µg/kg. (Martin *et al.*, 2001).

Reddy *et al.* (2001) analyzed aflatoxin B_1 (AFB_1) content by an indirect competitive ELISA, for the samples of the three grades of the chilli pod (Grade 1 to 3) and chilli powder. The highest AFB_1 concentration of 969 mg/kg was found in one sample representing grade 3. As much as 9 per cent of the chilli powder contained non-permissible aflatoxin levels. Yellamanda Reddy *et al.* (2003) analyzed the groundnut samples/ kernel and different varieties of groundnut for aflatoxin B_1 production by ELISA method. Very high level of aflatoxin was observed in 9.7 percent of samples ranging from 35 to 8172 mg/kg. In TMV-2 the widely grown groundnut variety recorded, > 30 mg/kg of aflatoxin contamination and other varieties *viz.*, JL-24, TAG-24, TG-26 and GG-2 had aflatoxin at very low level < 5 mg/kg.

The degree of aflatoxin contamination in different Indonesian food stuffs were determined by ELISA and HPLC methods. Eighty two groundnut products, 12 baby food products and 11 maize products were analyzed for total aflatoxin (AFT) and aflatoxin B_1 (AFB1) using ELISA. Thirty five per cent of the groundnut products were contaminated with aflatoxins ranging from 5 to 87 µg/kg. Eighteen per cent of the maize based products were contaminated with aflatoxin ranging from 5.8 and 12.4 µg/kg from 12 analyzed baby food samples, none of the sample was found to be contaminated with aflatoxin (Razzazi, 2004).

The mycoflora invading chillies (*Capsicum annuum* L.) kept in cold storage was studied and also analysed for the aflatoxin B_1 production by HPLC method. Species of *Aspergillus* were found to be dominant on stored chillies and the samples were contaminated with aflatoxin B_1 to the extent of 5.5 µg/ kg (Kiran *et al.*, 2005).

Fazekas *et al.* (2005) evaluated 91 spice samples (70 ground red pepper, 6 black pepper, 5 white pepper, 5 spice mix and 5 chilli samples) for the presence of aflatoxins B_1, B_2, G_1 and G_2 and Ochratoxin A by HPLC method (High performance liquid chromatography). Out of 70 ground red pepper samples, 7 of them contained AFB1 concentration exceeding the maximum level of 5 µg/kg (6.1 - 15.7 µg/kg) and one chilli sample exceeded 5 µg/ kg (8.1 µg/kg). Thirty of the 70 ground red pepper samples contained Ochratoxin (OTA), 8 of them in a concentration exceeding the maximum level of 10 µg/kg (10.6 - 66.2 µg/kg), one chilli sample contaminated with Ochratoxin A (2.1 µg/kg).

The kernel samples of discoloured rice cultivars were detected for the presence of aflatoxin B_1 by indirect ELISA. Aflatoxin B_1 contamination was high in rice cultivars of *Cotton, Dorasannalu* and BJ_1 160 –175 µg/kg) moderate in *Swathi* and *Vijaya Mahsuri* (33-45 µg/kg) and was absent in TKM 9 (Mangala *et al.*, 2006). The production of aflatoxin B_1 in rice cultivars was estimated by indirect competitive ELISA. Aflatoxin B_1 production by *Aspergillus flavus* on paddy and milled rice substrates of cultivars were estimated by indirect competitive ELISA. The aflatoxin production on paddy substrate was maximum in RH 12 (608mg/kg) and was at par (242-270mg/kg) in *Samba Mashuri* and *Ajaya*; it was low in rest of the cultivars. *A. flavus* inoculated on milled grains showed from a minimum of 4018 mg/kg to a maximum of 4655 mg/kg cv. *Ajaya* (Mangala *et al.*, 2007).

The chilli fruit samples from farmers field of Bellary district had higher contamination of 24.64 µg/kg followed by Raichur (7.58 µg/kg) and Gulbarga (2.84 µg/kg) districts. When aflatoxin was detected among various chilli products, only chilli powder contained 23.20 µg/kg of chilli products which was above permissible limit. The other indigenous chilli products likes chilli masala powder, sambar powder, puliogare powder and vangibath did not contain any aflatoxin. (Naik and Sudha., 2009).

Management of Aflatoxin Contamination in Chilli

Varietal screening under laboratory condition

Chemical, though necessary to control many diseases, is undesirable and often uneconomical as a long term solution to crop health. Therefore, efforts in identification of resistant genotypes would make other measures more effective. Indiscriminate use of chemicals in chilli has resulted in problem of pesticide residues on chilli and their products. The chilli crops containing pesticide residues above the permissible limit are banned for export. Nearly 20 per cent of chilli produced in India is surplus and can be well as export purpose. Hence, use of resistant varieties is essential.

Mehan (1987) at ICRISAT used a modified method to screen 850 germplasm accessions for their reaction to seed invasion and colonization by *A flavus*. Resistance of the three genotypes *viz.*, PI-337394F, PI-337409 and UF-71513 was confirmed and six new sources of resistance (Ah 7223, J-11, U-4-47-7, Var. 27, Faizpur and Monir 240-30) were identified. Resistance in three of them (PI-337394F, PI-337409 and J-11) was also confirmed by Kisyombe *et al.* (1985).

About 40 groundnut genotypes were screened for field resistance to seed invasion by the aflatoxigenic fungus, *A. flavus*. The significant varietal differences were observed at harvest in respect of levels of naturally occurring seed infestation by *A. flavus* in both field and laboratory screening, while four varieties PI-337409, PI-337394F, 73-30 and 73-33 had moderate levels of resistance (Waliyar and Bockelee, 1987).

Out of 25 groundnut lines, including germplasm, advanced breeding lines, and cultivars tested, cultivars 55-437, J11, and PI 337394 F were the least infected. Among the ICRISAT advanced breeding lines involving parents resistant to *A. flavus*, ICGV 87084, ICGV 87094, and ICGV 87110 were resistant. The results showed that some breeding lines possessed good level of resistance to *A. flavus*, reflecting presence of genes for resistance (Waliyar, 1994). The efficiency of selection for resistance to aflatoxin development by *Aspergillus flavus* in maize grain was studied in 3 CIMMYT low land tropical maize populations. Populations 22 (DMR) and 28 (DMR) were the most damaged and population 31 DMR suffered the lowest infection (Leon *et al.*, 1995).

Verma *et al.* (1996) studied the per cent inhibition of fungi on different varieties of peanut seeds and the results showed that *A. flavus* population was more in Dhokya White followed by G-51, Dhokya red and HB-11. Varma *et al.* (2001) evaluated 14 varieties of groundnut against aflatoxin contamination in glass house condition. Most of the varieties exhibited colonization comparable to the most susceptible variety TMV-2. Three varieties *viz.*, S-206, KRG-1 and GPBD-4 recorded relatively low level of colonization rating indicating their tolerance to *A. flavus.*

Rahmianna *et al.* (2004) evaluated 13 groundnut genotypes for tolerance to drought and aflatoxin contamination. Genotypes J11 and ICGV 93280 showed resistant reaction, ICGV 95494 showed tolerance to *in vitro* seed colonization by *Aspergillus flavus* and ICGV89104 showed tolerance to pre-harvest infection by *Aspergillus flavus*. Ajith Kumar *et al.* (2005) evaluated thirty genotypes of chilli against aflatoxin contamination fungus *Aspergillus flavus* under laboratory conditions using pinprick method of inoculation. Genotypes LCA-206, HTL, Goli, and 9623 showed immune reaction, whereas JCA-283, AAUDC-24, KA-2 and IND-10 showed resistant reaction and remaining ones were susceptible.

Sudha *et al.* (2009) screened 99 chilli genotypes under artificial conditions in the laboratory using pin prick method of inoculation against aflatoxin contamination caused by *Aspergillus flavus*. Only one genotype BK-21-SPS-06 showed resistant reaction. Genotypes IC-26, BK 32, BKXKubinal, SH-KC-23, LCA-206 and VNK-3C showed moderately resistant reaction. Twenty one genotypes showed moderately susceptible, 31 showed susceptible and 40 genotypes showed highly susceptible reaction.

Botanicals in Management of Aflatoxin Fungus *A. flavus*

In recent years, there is a change in trend towards organic produces with concern for quality food and care for protection of environment. Therefore, there is need and scope for exploiting sources of alternative chemotherapeutant which are reservoirs in many plant species. Infact, plant extracts have opened a new avenue for the control of plant diseases besides being safe and non phytotoxic. The plant extracts are known effective against various plant pathogens.

Bansal and Sobti (1990) reported that neem extract was the most effective plant extract in reducing *Aspergillus flavus* incidence on groundnut seeds by application of two per cent aqueous solution of neem. The incidence of *A. flavus* on groundnut seeds was 2.66 per cent as compared to control with 9 per cent.

Meena and Mariappan (1993) studied efficacy of different plant products on mycoflora of sorghum. Among different treatments, seeds treated with neem inhibited the mycelial growth (20.5 mm) of *A. flavus* effectively, so also seeds treated with *Pongamia pinnata* (61.50 mm) as compared to control (87.5 mm). The leaf extract of *A. indica* was used to treat the sunflower seeds. It was more effective in inhibiting seed mycoflora like *A. flavus* and *A. niger*, improving seed germination and seedling vigour (Thippeswamy and Lokesh, 1997). Similar observations were also made by Thirumala Rao and Sitaramaiah (2000).

Srivatsava *et al.* (1997) studied antifungal activity of neem (*Azadirachta indica* L.) and karanj (*Pongamia pinnata* L.) seeds and neem leaves against *A. flavus* and *A. niger*. Among different treatments, neem seeds treated with neem seed oil (NSO) was the best in inhibiting the mycoflora of neem seeds. The bioefficacy of neem seed kernal powder (NSKP), *karanj* seed kernel powder (KSKP) and neem leaf powder (NLP) were more effective when pelleted on wet seeds as compared to dry seeds. The NSO and NLP inhibited the growth of *A. flavus*.

The effects of neem fresh leaves, dried leaves, pathogens, oil and seeds was determined on the aflatoxin producing fungus, *Aspergillus flavus*. Neem oil at 0.5% dilution gave the highest inhibition of aflatoxin B_1 (18.54), the per cent inhibition by fresh neem leaves, neem oil at 1% and neem seeds was 11.46, 21.47 and 20.69 per cent respectively (Sinha *et al.*, 1999).

Reddy *et. al.* (2004) evaluated different botanicals/ plant extracts for the inhibition of fungal growth of *Aspergillus flavus*. Among different plant extracts *Allium sativum* (bulb), *Azadirachta indica* (leaf), *Eucalyptus terticornis* (leaf), *Pongamia pinnata* (kernel) extracts and garlic bulb extract (5%) completely inhibited the fungal growth of *Aspergillus flavus*. Pongamia kernel extract gave control of 57 per cent of *Aspergillus flavus* at 20 per cent concentration. Eucalyptus (40%) and neem extract (28%) were less effective in inhibiting the growth of *A. flavus*. Ajith Kumar *et al.*, (2005) reported 100 per cent inhibition of *A. flavus* by use of neem seed kernel extract, nimbicidin and pongamia oil at five per cent concentration under *in vitro*.

Crude ethanolic extracts of kaffir lime leaf, bitter cucumber fruit and tobacco leaf were examined for their ability to control the growth of *A. flavus*. The results showed that the ethanolic extracts of all herbs had an inhibitory effect on fungal growth. Kaffir lime at 10 % showed inhibition of 89.62 to 41.6 mm and tobacco at 8-10 % showed inhibition of 87.0 to 22.6 mm. However, the bitter cucumber at 10 % gave inhibitory effect on the fungus (89.6 to 53.8 mm) (Dusanee Thanaboripat, 2006). In order to develop an eco-friendly method of prevention of *A. flavus* infection and aflatoxin contamination, aqueous extracts obtained from leaves of 30 medicinal plants belonging to different families were evaluated for their ability to inhibit the growth *of A. flavus in vitro*. Among them, the leaf extract of zimmu (*Allium sativum* L. x *Allium cepa* L.) was the only one that showed antifungal activity against *A. flavus* and recorded 73% inhibition of *A. flavus* growth (Sandosskumar *et al.*, 2007).

Sudha and Naik., (2009) reported Neem Seed Kernel Extract (NSKE) and nimbicidin (5%) were effective in reducing the *Aspergillus flavus* infection in chilli fruits. The per cent colonization of *A. flavus* under laboratory conditions was 2.20 in Neem Seed Kernel Extract (NSKE) treated chilli fruits and 7.57 in nimbicidin treated chilli fruits as compared to control (38.33%). Under field conditions least incidence was recorded in NSKE treated plot (1.65%) as against control (7.41%).

Biological Control of Aflatoxin Fungi *A. flavus*

It is now realized that lasting success cannot be achieved by poisoning the environment and the mankind is increasingly turning to natural control by restoring the biological balance favourable to the crop.

In crops sensitive to aflatoxin, bio-control will have a long lasting solution from the point of view of food safety and health hygiene.

The co-inhabiting mycoflora associated with *A. flavus* on individual maize grains are evaluated for their influence on aflatoxin synthesis. All 13 types of associations of different fungal species inhibited (*A. flavus)* aflatoxin B1 and G1 production at different levels (34.3-100%). Inhibition of aflatoxin B1 was consistently less than that of G1. Inhibition of radial growth of *A. flavus* by *Fusarium moniliforme* [*Gibberella fujikuroi*] (59.8%), *Trichoderma viride* (72.5%) and *Rhizopus nigricans* [*R. stolonifer*] (42%) could be directly correlated to the percentage inhibition of aflatoxin production. High levels of inhibition of aflatoxin elaboration were noted in competition of *A. flavus* with other toxigenic molds (Choudhary, 1992).

Experiments were designed to determine whether certain non-toxigenic fungi commonly isolated from maize kernels can affect aflatoxin B1 development when inoculated with *A. flavus* onto individual unsterilized and autoclaved maize kernels. *Trichoderma viride* and *A. niger* were strong antagonists inhibiting growth of *A. flavus* by 87 and 66% respectively, whereas *A. versicolor*, *Fusarium moniliforme* [*Gibberella fujikuroi*], *Paecilomyces variotii* and *Emericella quadrilineata* inhibited growth of *A. flavus* by <51%. Less aflatoxin B1 was detected when *A. flavus* was paired with *A. niger* or *T. viride* than with the other test fungi. When *A. niger* or *T. viride* was introduced onto the kernels 72 h before inoculation with *A. flavus,* no aflatoxin B1 was detected in unsterilized kernels and the levels of aflatoxin B1 were greatly reduced from 700 ppb to 160 and 140 ppb. in autoclaved kernels, respectively (Aziz and Shahin, 1997).

A study of the fungal growth in dual cultures revealed that out of the nine isolates, two *Trichoderma* spp. (*T. harzianum* and *T. viride*) inhibited growth of seed associated fungi, *A. flavus* (Claudia Calistru *et al.,* 1997). Twenty six *Trichoderma* isolates were tested for their ability for Production of volatile and non-volatile chemicals inhibitory to *A. flavus* using Dennis and Webster (1971a, 1971b) method. The production of non-volatile chemicals inhibited growth of *A. flavus* and colony diameter was just 27 mm as compared to 85 mm in the control (Desai *et al.,* 2000).

Manwar *et al.* (2000) identified the antagonistic nature of two strains of *Pseudomonas fluorescens*, wherein it could effectively inhibit the mycelial growth of *A. flavus* in plates under *in vitro*. Praveen Kumar *et al.* (2000) tested a talc powder based formulation of an antagonistic strain of *P. fluorescens* for its efficacy against *A. flavus* and *A. niger,* which inhibited growth on seeds. The treated seeds exhibited increased germination, seedling vigour and emergence compared to control.

Bhuvaneshwari and Subba Rao (2001) reported that mango fruits inoculated with *T. viride* remained free from the infections of *A. niger* and *A. flavus* indicating their suppression by antagonistic nature of *T. viride*. Thakur *et al.* (2003) studied the effect of two biocontrol agents *Trichoderma* and *Pseudomonas* as seed treatment against *Aspergillus flavus* in groundnut crop. Two isolates of *Trichoderma viride,* one isolate of *Trichoderma harzianum* and one isolate of *Pseudomonas fluorescence* (Pf) provided greater protection to seed infection by A f 11-4. The highest reduction (73%) was recorded with Pf153 followed by Tv-20 (65%) and Th-23 (63%) and the least with *Trichoderma spp.* (40%).

The antagonistic potential of 212 *Trichoderma* isolates, obtained from soil samples collected from 4 districts in Andhra Pradesh, India, and 2 districts in Karnataka, India, against a highly aggressive and toxigenic strain of *A. flavus* (Af 11-4) was investigated. One-hundred and forty-five isolates were antagonistic to Af 11-4. Among these, only 39 showed clear inhibition zone against Af 11-4. These isolates were identified and further evaluated for antagonism, involving production of volatile and non-volatile antibiotics, and hyphal interaction with Af 11-4 (Srilakshmi *et al.,* 2001).

Of the *Trichoderma* isolates evaluated against *Aspergillus flavus,* culture filtrate of *Trichoderma virens* showed complete inhibition (100%) of growth of *Aspergillus flavus* at 15% concentration. *Trichoderma harzianum* showed a moderate inhibition (50%) of *Aspergillus flavus* followed by *T. reseii* (38%), *T. viridae* (25%) and *T. koningii* (25%) (Reddy *et al.,* 2004).

Colonization of *Aspergillus flavus* (2.00%) was recorded in *Pseudomonas fluorescens* (1X10^8 cfu/ml) treated chilli fruits as compared to *Trichoderma harzianum* (5.17%). However in control, the per cent infection was to the extent of 38.33 under laboratory conditions and under field conditions the per cent reduction of incidence of *A. flavus* was 2.42 and 2.60 in *Pseudomonas fluorescens* and *Trichoderma harzianum* treated plots which were on par with each other as against control (7.41%) (Sudha and Naik., 2009).

Fungicides in Management of Aflatoxin fungus, *A. flavus*

The use of fungicide is a quick and effective method of managing diseases when the resistance sources are not available. These fungicides have to be judiciously used at the minimum level according to the need and at appropriate time ensuring avoidance of residues on fruits/grains.

Six fungitoxicants were tested for their efficacy in controlling *A. flavus* as seed dressers in their respective doses. Among the fungicides, thiram and dithane M-45 were effective in reducing the incidence of *A. flavus* on groundnut, whereas Bavistin was found to increased the per cent seed germination (Bansal and Sobti, 1988).

Thiram and dithane M-45 (mancozeb) were used at a concentration of 0.3% and 0.2% respectively for management of *Aspergillus flavus* on groundnut seeds. The incidence of *Aspergillus flavus* on groundnut seeds on using thiram (0.3%) and dithane M-45 (0.2%) was 1.03% and 3%, respectively, as compared to control (9%) (Bansal and Sobti, 1990). The efficacy of thiram, dithane M-45 and captan against seed borne fungi of soybean with 18-28 per cent of seed germination under *in vitro* and 22-23 per cent under *in vivo* was reported (Ganacharya, 1979 and Tripathi and Singh, 1991). Meena and Mariappan (1993) evaluated the efficacy of captan against *Aspergillus flavus*. Captan was highly effective in inhibiting mycelial growth (20.09 mm).

Solanke *et al.* (1997) studied the effect of fungicides in response to soybean seeds MACS-13 and PK-472 to *Aspergillus flavus* infection. Thiram dry seed treatment gave effective results. Thiram besides improving the germination percentage (74%and71%) of soybean seeds MACS-13 and PK-472 respectively, controlled pre and post emergence mortality caused by *A. flavus*. Savitri *et al.* (1998) in their studies obtained an effective control of seed-borne fungi (*Aspergillus* spp.) of groundnut with mancozeb @ 3.0 g/kg of seeds, carbendazim @ 2.0 g/kg of seeds and their mixture.

Pankaj Sharma and Champawat (2000) treated the external and internal mycoflora of jojoba seeds with different seed dressing fungicides. Among seed dressing fungicides, bavistin, captan, indofil M-45, thiram and vitavax were effective against *A. flavus*. The pre and post-emergence mortality due to *A. flavus* was effectively checked by application of vitavax and bavistin. Of the fungicides tested, carbendazim was effective at 100 ppm in inhibiting *Aspergilli*. Tricycalzole completely reduced the growth of *Aspergillus flavus* (Reddy *et al.*, 2004). Ajith Kumar *et al.*, (2005) evaluated the fungicides *in vitro* against the infection of *A. flavus* in chilli. Among the non-systemic fungicides the maximum inhibition of radial mycelial growth (91.1%) was observed with 0.3 per cent mancozeb followed by captan (85.2%).

The efficacy of non-systemic fungicides viz., mancozeb and captan (0.3%) as tested against *Aspergillus flavus* infection in chilli under laboratory conditions. Captan treated fruits recorded the least incidence of *A. flavus* infection (1.57%) and was on par with mancozeb treated fruits (1.90%). However in control, the per cent infection was to the extent of 38.33. Under field conditions the least incidence of *A. flavus* infection (2.16%) was recorded in captan treated plot as against control (7.41%) Sudha and Naik., 2009).

Application of neem cake (250 kg/ha) and Trichoderma enriched Farm Yard Manure (25 kg/ha) resulted in significant reduction of *Aspergillus flavus* population, reducing the chances of aflatoxin contamination in chilli (Sudha and Naik., 2010).

Aflatoxins have world wide occurrence, Awareness about aflatoxin contamination is to be created since it not only affects the food crops, food products, human and animal health but also affect the global market. The detection of aflatoxin in chilli is a matter of greater concern which needs to be addressed carefully by the producers, consumers and traders in a holistic approach utilizing good agricultural practices (GAP) as preventive measures. Additionally, creation of awareness among the consumers about the presence of aflatoxin in chillies may induce producers as well as traders to market aflatoxin free chilli fruits. Therefore, better understanding is necessary on aflatoxin research and exchange of information.

References

Abbaas, H. K., Zablotowicz, R. M. and Locke, M. A., 2004, Spatial variability of *Aspergillus flavus* soil populations under different crops and corn grain colonization and aflatoxins. *Can. J. Bot.* 82(12): 1768-1775.

Ahuja, A. K. and Pande, B. N., 2004, Dissipation pattern of mancozeb and propineb residues in chilli fruits. *Indian J. Pl. Prot.* 32(2): 93-95.

Ajith Kumar, K. and Naik, M. K., 2005, Prevalence and distribution of aflatoxin contamination of chilli (*Capsicum annuum* L.) field and market. *Karnataka J. Agric. Sci.*, 18(2): 520-523.

Ajith Kumar, K. and Naik, M. K., 2006, Detection of aflatoxin producing *Aspergillus flavus* isolates from chilli and their management by post harvest treatments. *J. Food Sci. Technol.* 43(2): 200-204.

Ajith Kumar, K. Naik, M. K., Allolli, T. B. and Hosamani, R. M., 2005, Evaluation of genotypes, fungicides and plant extracts for the management of aflatoxin contamination in chilli caused by *Aspergillus flavus* Link. Fers. *Indian. J. Pl. Prot.* 33 (1): 115-118.

Anjaiah, V., Thakur, R P. and Koedam, N., 2006, Evaluation of bacteria and *Trichoderma* for biological control of pre-harvest seed infection by *Aspergillus flavus* in groundnut. *Biocon. Sci. Technol.* 16(3/4): 431-436.

Anonymous, 2008, Drop in domestic and global output boosts red chilli. An online article. *http://www.capitalmarket.com.*

Anuja Gupta and Vijayanand, K. G., 2006, Mycoflora associated with rice grains stored in rice mills and warehouses. *Seed Res.* 34(2): 232-234.

Awasti, 2002, Pesticide residues in vegetables. In: *International conference on vegetables for sustainable food and nutritional security in the new millennium.* Bangalore, November 11-14, 2002, pp.274-275.

Awuah, R. T. and Kpodo, K. A., 1996, High incidence of *Aspergillus flavus* and aflatoxins in stored groundnut in Ghana and the use of a microbial assay to assess the inhibitory effects of plant extracts on aflatoxin synthesis. *Mycopathologia,* 134 (2): 109-114.

Aziz, N. H. and Shahin, A. A. M., 1997, Influence of other fungi on aflatoxin production by *Aspergillus flavus* in maize kernels. *J. Food Safety,* 17(2): 113-123.

Bagwan, N. B. and Meshram, B. M., 2004, Mycoflora and aflatoxin contamination in dried fig fruits. *J. Mycol. Pl. Pathol.* 34 (2): 671-673.

Bansal, R. K. and Sobti, A. K., 1988, Control of *Aspergillus flavus* associated with groundnut seeds. *Indian Phytopath.* 41: 643-645.

Bansal, R. K. and Sobti, A. K., 1990, An economic remedy for the control of two species of *Aspergillus* on groundnut. *Indian Phytopath.* 43: 451-452.

Baquar Imam Zaidi, S., Mohd. Imran Khan and Saxena, S. K., 1991, Effect of fungicides on mycoflora of chick pea seeds. *Indian Phytopath.,* 44: 394-395.

Bedi, J. S. and Rattan, G. S., 2003, Variability in aflatoxigenic potential and seed colonization of *Aspergillus flavus* in maize. *J. Mycol. Pl. Pathol.* 33(2): 200-203.

Bhuvaneshwari, V. and Subba Rao, M., 2001, Evaluation of *Trichoderma viride* antagonistic to post-harvest pathogens on mango. *Indian Phytopath.* 54(4): 493-494.

Bilgrami, K. S. and Choudhary, A. K., 1990, Incidence of *Aspergillus flavus* in the aerosphere of maize fields at Bhagalpur. *Indian Phytopath.* 43(1): 38-42.

Blount, W. P., 1961, Turky X disease. *Turkeys,* 9: 52.

Bruce W. Horn., 2006, Relationship between soil densities of *Aspergillus* species and colonization of wounded peanut seeds. *Can. J. Microbiol.* 52: 951-960.

Bruce W. Horn., Greene, L. R. and Dorner, J. W., 1995, Effect of corn and peanut cultivation on soil populations of *Aspergillus flavus* and *Aspergillus parasiticus* in South Western Georgia. *Appl. Environ. Microbiol.* 61(7): 2472-2475.

Choudhary, A. K., 1992, Influence of microbial co-inhabitants on aflatoxin synthesis of *Aspergillus flavus* on maize kernels. *Lett. Appl. Microbiol.* 14(4): 143-147.

Claudia Calistru, Michelle McLean and Patricia Berjak, 1977, *In vitro* studies on the potential for biological control of *Aspergillus flavus* and *Fusarium moniliformae* by *Trichoderma* species. *Mycopathologia,* 139: 115-121.

Cotty, P. J., 1997, Aflatoxin producing potential of communities of *Aspergillus* section flavi from cotton producing areas of the United States. *Mycol. Res.,* 101: 968-704.

Desai, S. Thakur, R. P. and Anjaiah, V., 2000, Characterization of isolates of *Trichoderma* for potential biocontrol against *Aspergillus flavus* infection in groundnut. *International Arachis Newslett.* 20: 57-59.

Detroy, R. W., Lillehej. E. B. and Cieglar, A., 1971, 'Aflatoxin and related compounds", in *Microbial Toxins* Vol. VI, Fungal Toxins, Eds., Cieglar, A., Kadis, S. and Aji, S. J., Academic Press, New York, pp. 3-178.

Devi, K. T., Mayo, M. A., Reddy, K. L. N., Delfose, P., Reddy, G., Reddy, S. V. and Reddy, D. V. R., 1999, Production and characterization of monoclonal antibodies for aflatoxin B1. *Lett. Appl. Microbiol.,* 29: 284-289.

Diener, U. L. and Davis, N. D., 1966, Aflatoxin production by isolates of *Aspergillus flavus. Phytopathology.* 56: 1391-1393.

Dilek Heperen, Funda Karbancioglu Guler and Nesrin Mecik, 2007, Mycotoxin problems in Turkey-case study on dried figs. In: 2nd *Asian Congress on Microbial Diversity for Asian Prosperity* held by Indian Society of Mycology and Plant Pathology and Osmania University Hyderbad, December 19-22, 2007 pp. 288.

Dusanee Thanaboripat, Sittichai Chareonsettasilp, Kanokwan Pandee and Kittipun Udomwongsup, 2006, Inhibitory effect of kaffir lime, bitter cucumber and tobacco extracts on the growth of *Aspergillus flavus*. *KMITL Sci. Tech. J.* 6(1):18-24.

Farid Waliyar, Reddy, S. V. and Lava Kumar, P., 2005, Aflatoxin contamination in maize in Andhra Pradesh. *J. Mycol. Pl. Pathol.* 35(3): 498.

Farid Waliyar, Reddy, S. V., Lava Kumar, P., Belum, V. S., Reddy, K. N., Rai, Alur Ashok, S. and Ravinder Reddy, 2007, Surveillance for natural contamination of mycotoxins in sorghum and pearl millet grains in Andhra Pradesh and Maharastra states, India. In: 2nd *Asian Congr. on Microb. Divers.for Asian Prosperity* held by Indian Society of Mycology and Plant Pathology and Osmania University Hyderbad, December 19-22, 2007 pp. 294.

Farid Waliyar, Hassan, H. and Bosc J. P., 1994, Sources of resistance to *Aspergillus flavus* and aflatoxin contamination in groundnut genotypes in West Africa *Plant Dis.* 78(7): 704-708

Fazekas, B., Tar A. and Kovacs, M., 2005, Aflatoxin and ochratoxin A content of spices in Hungary. *Food Addit. Contam.* 22(9): 856-863.

Ganacharya, N. M., 1979, Effect of fungicidal seed treatment on emergence, nodulation and grain yield of soybean in Marathwada. *J. Maharashtra Agric. Univ.* 4(1): 112-113.

Giridhar, P. and Reddy, S. M., 1998, Incidence of fungi in relation to mycotoxins on garlic and onion bulbs in Andhra Pradesh. *J. Mycol. Pl. Pathol.* 28(3): 347-350.

Gracia, J. L., Lopez, M. T., Perera, E. J., Berkman, H. and Gonzalez, L. A., 1997, Investigations on chemical control during storage of onion bulbs. *Agrotecnia Cuba,* 27(1): 8-10.

Grinstein, A., Elad, Y., Temkin Gorodeiski, N., Rivan, Y. And Frankel, H., 1992, Reduced volume application of fungicides for the control of onion rots. *Phytoparasitica,* 20 (4): 293-300.

Hawksworth, D. L., Sutton, B. S. and Ainsworth, G. C., 1983, *Ainsowrth and Bisby's Dictionary of the Fungi*, VIII Edition, Commonwealth Mycological Institute, London, pp. 445.

Hesseltine, C. W., 1986, Global significance of mycotoxins. In: *Mycotoxins and Phytotaxins.* Eds.Steyn, P. S. and Vleggaar, R., Amstredam, Netherland, Elservier. P. 1-18.

Hobbs, H. A., Reddy, D. V. R., Rajeshwari, R. and Reddy, A. S., 1987, Use of direct antigen coating ELISA procedures for detection of plant viruses. *Plant Dis.,* 71: 747-749.

Horn, B. W. and Dorner, J. W., 1999, Regional Differences in Production of Aflatoxin B_1 and Cyclopiazonic Acid by Soil Isolates of *Aspergillus flavus* along a Transect within the United States. *Appl. Environt. Microbiol.* 65 (4): 1444-1449.

Horn, B. W., 2005, Ecology and population biology of aflatoxigenic fungi. (In) *Aflatoxin and Food Safety*. (Ed.) H. K. Abbas. CRC Press, Boca Raton, Fla. Pp. 95-116.

Horn, B. W., Greene, R. L. and Dorner, J. W., 1995, Effect of corn and peanut cultivation fron soil populations of *Asperfgillus flavus* and *A. parasiticus* in South Western Georgia. *Appl. Environ. Microbiol.,* 61: 2472-2475.

Hussaini, A. M., Timothy, A. G., Olufunmilayo, H. A., Ezekiel, A. S. and Godwin, H. O., 2007, Fungi and some mycotoxins contaminating rice (*Oryza sativa*) in Niger State, Nigeria. *African J. Biotechnol.*, 6(2): 99-108.

Jelinek, C. F., Pohland, A. E. and Wood, G., 1989, World wide occurrence of mycotoxin with food and feeds-update. *J. Assoc. Off. Anal. Chem.*, 72: 223-230.

Joffe, A. Z., 1969, Aflatoxin produced by 1626 isolates of *Aspergillus flavus* from groundnut kernels and soils in Israel. *Nature*. 221: 492-494.

Kharbade, S. B., Kale, V. D. and Dethe, M. D., 2003, Residues of imidacloprid on chillies. *J. Pestol.* 27(1): 18-19.

Kiran, D. R., Narayana, K. J. P. and Vijayalakshmi, M., 2005, Aflatoxin B_1 production in chillies (*Capsicum annuum* L.) kept in cold stores. *African J. Biotechnol.* 4(8): 791-795.

Kisyombe, C. T., Beute, M. K. and Payne, G. A., 1985, Field evaluation of peanut genotypes for resistance to infection by *Aspergillus parasiticus*. *Peanut Sci.* 12: 12-17.

Kole, R. K., Hemanta Banerjee and Anjan Bhattacharyya, 2002, Monitoring of pesticide residues in farm gate vegetable samples in West Bengal. *Pesticide Res. J.*, 14 (1): 77-82.

Krogh, P., Hasselagar, E., Hald, B. and Kloller, A., 1966, Incidence of aflatoxin in Denmmark, *Nature*. 204: 129-130.

Lafont, P. and Loafont, J., 1977, Toxigenesis of *Aspergillus flavus* isolated from groundnut fields. *Mycopathologia*. 62(3): 183-185.

Lauren Lewis, Mary Onsongo, Henry Njapau, Helen Schurz-Rogers, George Luber, Stephanie Kieszak, Jack Nyamongo, Lorraine Backer, Abdikher Mohamud Dahiye, Ambrose Misore, Kevin De Cock, Carol Rubin and the Kenya Aflatoxicosis Investigation Group, 2005, Aflatoxin contamination of commercial maize products during an outbreak of acute aflatoxicosis in Eastern and Central Kenya. *Enviro. Health Perspect.* 113(12):1763-1767.

Lee, L. S., Lee, L. V. and Russell, T. E., 1986, Aflatoxin in Arizona cotton seeds: field inoculation of bolls by *Aspergillus flavus* spores in wind driven soil. *J. Ann. Oil. Chem. Soc.*, 63: 530-532.

Leon, C de Kitbamroong, C., Buangsuwan, D. and Tanboon, P., 1995, Selection for resistance to aflatoxin formation in maize through seed inoculation. *Food Addit. Contam.* 12(3): 491-495.

Lillehoj, E. B., Kwolek, W. F., Zuber, M. S., Bockholdt, A. J., Cavert, O. H., Findley, W. R., Guthrie, W. D., Horner, E. S., Josephson, I. M., King, S., Manwiller, A., Sauer, D. B., Thompson, D., Turner, M. and Wildstrom, N. W., 1980, Aflatoxin in corn before harvest: Interactions of hybrids and locations. *Crop Sci.* 20: 731-734.

Lillehoj, E. B., Mc Milliam, W.W., Githrie, W. D. and Barry, D., 1980, Aflatoxin producing fungi in free harvest corn inoculum source in insects and soils. *J. Environ. Qual.*, 9: 691-694.

Macdonald, S. and Castle, K., 1996, A UK retail survey of aflatoxins in herbs and spices and their fate during cooking. *Food Addit. Contam.* 13(1): 121-128.

Madan, V. K., Beena Kumari, Singh, R. V., Rakesh Kumar and Kathpal, T. S., 1996, Monitoring of pesticide residues in farm gate vegetable samples in Haryana. *Pesticide Res. J.*, 8(1): 56-60.

Mahalingappa, P. B., Reddy, K. D., Reddy, K. N. and Subbaratnam, G. V., 2006, Dissipation of ethion and chlorpyriphos residues in/ on chilli (*Capsicum annuum* L.). *Pesticide Res. J.* 18(1): 72-73.

Mangala, V. N., Reddy, K. R. N., Reddy, C. S. and Muralidharan, K., 2007, Impact of *Aspergillus flavus* on rice seedling growth and aflatoxin B_1 production. *Indian J. Pl. Prot.* 35(1): 76-80.

Mangala, V. N., Reddy, K. R. N., Singotamu, L., Chary P. M. S., Reddy, C. S. and Muralidharan, K., 2006, Aspergilli colonize and produce aflatoxin B_1 in discoloured rice grains. *J. Mycol. Pl. Pathol.* 36(3): 418-426.

Manwar, A. V., Vaiganekar, P. D., Bhongale, L. S. and Chincholkar, S. B., 2000, *In vitro* suppression of plant pathogens by siderophores of Fluorescent pseudomonas. *Indian J. Microbiol.* 40: 109-112.

Martin, M. L., Martins, H. M. and Beqnarde, F., 2001, Aflatoxins in spices marketed in Portugal. *Food Addit. Contam.* 18(4): 315-319.

Meena, S. S. and Mariyappan, V., 1993, Effect of plant products on seed borne mycoflora of sorghum. *Madras Agric. J.*, 80 (7): 383-387.

Mehan, V. K., 1987, Screening groundnut for resistance to seed invasion by *Aspergillus flavus* and aflatoxin production. *Proc. of International Workshop* on *Aflatoxin contamination of groundnut*, 6-9 Oct., 1987, ICRISAT Centre, India, Patancheru, Andhra Pradesh, India 323-324.

Naik, M. K. and Rawal, R. D., 2002, Disease resistance in horticultural crops. In: *Resource Management in Plant Protection,* Ed. Sarath Babu, B., Varaprasad, K, S., Anitha, K., Prasad Rao, Chakrabarty, S. K. and Chandrurkar, P. S. Plant Protection Association of India, Hyderabad., pp: 64-68.

Naik, M. K., Ajith Kumar, K., Waliyar, F. and Reddy, S. V., 2003, Aflatoxin contamination in chilli-Biotechnological approaches in detection and management. In: Winter school on bioterchnological approaches in diognosis amd management of plant diseases of exporting horticultural crops. Samiyappan, R (ed.), Tamil Nadu Agriculture University, Coimbatore, India.

Naik, M. K., Raju, K., Ajith Kumar, K and Achari, R., 2004, SPS and HACCP measures for handling post harvest infections in export oriented horticulture produce under WTO era. In: *Symp. on Biolo, Biotechnol. Epidemiol. And Manag. Of Plant Diseases,* Gulbarga University, Gulbarga, December 9-10.

Naik, M. K., 2007, Aflatoxin and post harvest disease problems in exportable horticulture produce. In: *Microbial Diversity for Asian Prosperity* 2nd *Asian Congress of Mycology and Plant Pathology,* Osmania University, Hyderabad. December 19-22. pp296.

Naik, M. K. and Sudha, S., 2009, Current issues and future strategies on aflatoxin contamination and management in chilli. In: *National Workshop on Current Trends and Future Prospects in Production and Export of Spice with Special Reference to Chilli* held on 27th -28th February at Dharwad, Karnataka, pp-118-119.

Navya, H. M., Chandra Nayaka, S., Hariprasad, P., Udaishankar, A. C., Niranjana, S. R. and Prakash, H. S., 2007, Screening for the incidence of aflatoxigenic and non-aflatoxigenic isolates of *A. flavus* in ground nut and its effect on seed quality. *In: National Seminar on Molecular Plant Pathology and Biotechnology for Sustainable Crop Production.* Held at University of Mysore, organized by Indian Phytopathological Society and University of Mysore, November 28-29, 2007 pp. 25-26.

Nguyen, Q. A., Dau, N. H. and Nguyen, T. T. D., 2000, Results of a study on the presence of *Aspergillus flavus* in maize fields during harvest. *Khoa Hoc Ky Thuat Thu Y Veterinary Sciences and Techniques.* 7(2): 63-67

Pankaj Sharma and Champawat, R. S., 2000, Seed mycoflora of jojoba (*Simmondsia chinensis*), their pathogenic potential and control. *J. Mycol. Pl. Pathol.* 30(3): 398-401.

Panse, V. G. and Sukathme, P. V., 1985, *Statistical Methods for Agricultural Workers.* ICAR Publications, New Delhi, pp. 145-155.

Pontecorvo, G., 1956, The parasexual cycle in fungi. *Ann. Rev. Microbiol.* 128: 162-171.

Potty, V. H., 2006, Chilli exports- The aflatoxin blues. *Indian Food Industry*, 25: (5): 27.

Praveen Kumar, L., Niranjana, S. R., Prakash, H. S. and Shekar Shetty, H., 2000, Effect of *Pseudomonas fluorescens* formulation against *Pyricularia grisea* in rice. *Crop Improv.*, 27(2): 159-166.

Rahmianna, A. A., Taufiq, A. and Yusnawan, E., 2004, Evalution of ICRISAT Groundnut genotypes for end of season drought tolerence and aflatoxin contamination in Indonesia. *International Arachis News lett.* 24: 14 -17.

Rajbashar, D., Kumar, R. S. S., Regupathy, A. and Sridhar, K., 2002, Harvest time residues of Propiconazole (Tilt 25EC) in chillies. In: *Resource Management in Plant Protection during Twenty first century* by Plant Protection Association of India, 14-15 November, Vol 2, p55-57.

Ramesh Chandra and Sarbhoy, K., 1997, Production of aflatoxins and zearalenone by the toxigenic fungal isolates obtained from stored food grains of commercial crops. *Indian Phytopath.* 50(4): 458-468.

Rao, K. S., Mahadedvan, T. V. and Tulpule, P. G., 1965, Incidence of toxigenic strains of *Aspergillus flavus* affecting groundnut crop in certain coastal districts of India. *Indian J. Medical. Res.* 53: 1196-1202.

Razzazi Fazeli, E., Noviandi, C. T., Porasuphatana, S., Agus, A. and Bohm, J., 2004, A survey of aflatoxin B_1 and total aflatoxin contamination in baby food, peanut and corn products sold at retail in Indonesia analysed by ELISA and HPLC. *Mycotoxin Res.* 20(2): 51-58.

Reddy, B. N., 1987, Antifungal properties of some plant extracts. *Indian Phytopath.* 40: 551-553.

Reddy, C. S., Reddy, K. R. N., Raja Kumar, N., Laha, G. S. and Muralidharan, K., 2004, Exploration of aflatoxin contamination and its management in rice. *J. Mycol. Pl. Pathol.* 34(3): 816-820.

Reddy, D. V. R., Nambiar, P. T. C., Rajeshwari, R., Mehan, V. K., Anjaiah, V. and Mc Donald, D., 1988, Potential of Enzyme-Linked Immunosorbent Assay for detecting viruses, fungi, bacteria, MOL's mycotoxins and hormones. In: *Biotechnology in Tropical Crop Improvement.* Proceed. *Intern. Biotechnol. Workshop*, 12-15 January, ICRISAT, Patancheru, pp. 43-49.

Reddy, D. V. R., Thirumala Devi, K., Reddy, S. V., Waliyar, F., Mayo, M. A., Rama Devi, K., Ortiz, R. and Lenne, J. M., 2000, Estimation of aflatoxin levels in selected food and feeds in India. *Proceed. of the Interna. Workshop on Food Safety Management in Developing Countries*, CIRAD-FAO.CIRAD CD-ROM, Montpellier, France, 11-13, December, 2000. pp: 1-4.

Reddy, K. N., Abbass, H. K., Zablotowicz, R. M., Abel, C. A., Koger, C. H., 2007, Mycotoxin occurrence and *Aspergillus flavus* soil propagules in corn and cotton glyphosate-resistant cropping systems. *J. Food Addit. Contam.* 24(12): 1367-1373.

Reddy, K. R. N., Reddy, C. S., Mangala, U. N. and Muralidharan K., 2006, Sites of infection of *Aspergillus spp.* in seeds of rice cultivars. *J. Mycol. Pl. Pathol.* 36(2): 271-277.

Reddy, S. V., Kiran Mayi, D., Uma Reddy, M., Thirumala Devi, K. and Reddy, D. V. R., 2001, Aflatoxins B_1 in different grades of chilli (*Capsicum annuum* L.) in India as determined by indirect competitive ELISA. *Food Addit. Contam.* 18(6): 553-558.

Reddy, D. K., 2010, Chilli, Indian scenario. World Spice Congress 4th February, 2010, New Delhi.

Sandoss Kumar, R., Karthikeyan, M., Mathiyazhagan, S., Mohankumar, M., Chandrasekar, G. and Velazhahan, R., 2007, Inhibition of *Aspergillus flavus* growth and detoxification of aflatoxin B1 by the medicinal plant Zimmu (*Allium sativum* L. x *Allium cepa* L.). *World J. Microbiol. Biotechnol.* 23(7): 1007-1014

Sangit Kumar, Meena Shekhar, Khan, A. A. Pankaj Sharma, 2007, A rapid technique for detection of toxigenic and non-toxigenic strains of *Aspergillus flavus* from maize grains. *Indian Phytopath.* 60(1): 31-34.

Savitri, H., Sugunakar Reddy, M. and Muralimohan Reddy, B., 1998, Effect of seed treatment with fungicides and insecticides on seed-borne fungi, storage insect pests, seed viability and seedling vigour of groundnut. *Seed Res.* 26(1): 62-72.

Shazia Rasheed, Shahnaz Dawar, A. Ghaffar and Shahid Shaukat, S., 2004, Seed borne mycoflora of groundnut. *Pak. J. Bot.* 36(1): 199-202.

Singh, K. P., Pandey, S. Y. and Singh, S., 1982, Aldicarb residues in red chillies. *Indian J. Agril. Sci.* 52(1): 776-778.

Singh, S. D., Rawal, P. and Bhargava, N. K., 2004, Pathogenic potential and control of seed mycoflora of groundnut (*Arachis hypogaea*). *J. Mycol. Pl. Pathol.* 34(2): 687-690.

Sinha, B. K., Rajan, K. S. and Pandey, T. N., 1999, Effect of some plant extract on aflatoxin production neem (*Azadirachta indica*). *Phytomedica.* 20 (1/2) : 49-52.

Snedecor, V. G. and Cochran, W. G., 1967, *Statistical Methods.* 6th Edn., Oxford and IBH Publishing Company, Kolkata.

Solanke, R. B., Kore, S. S. and Sudewad, S. M., 1997, Detection of soybean seed borne pathogens and effect of fungicides. *J. Maharashtra Agric. Univ.* 22(2): 168-170.

Srilakshmi, P., Thakur, R. P., Prasad, K. S. and Rao, V. P., 2001 Identification of *Trichoderma* species and their antagonistic potential against *Aspergillus flavus* in groundnut. *Internat. Arachis Newslett.* 21: 40-43

Srivastava, K. K, Gupta, P. K., Tripathi, Y. C. and Sarvate, R., 1997, Antifungal activity of plant products on spermoplane fungi of *Azdardirachta indica*, neem seeds. *Indian Forester.* 123: 157-161.

Stoloff, K., 1977, Aflatoxin an over view. (In) J. V. Rodricks, CW Hesseline, M. A. and Mehlman (Eds.). Mycotoxin in human and animal health. Pathox publisher inc, Park forest South, Illinois, pp 7-28.

Sudha, S. and Naik, M. K., 2009, An integrated approach for management of aflatoxin contamination (*Aspergillus flavus*) in chilli (*Capsicum annuum* L.). (In) *International Conference on Horticulture*-2009, Bangalore, 9-11 November, pp 1116-1122.

Sudha, S., Naik, M. K. and R. M. Hosmani., 2010, Screening of chilli genotypes and host range studies against aflatoxin contamination caused by *Aspergillus flavus* (Link.) Fers., *Vegetable Science*, Vol. 36(2): 155-158.

Sudha, S. and Naik. M. K., 2010, Effect of soil amendment on population of *Aspergillus flavus*, the aflatoxin causing fungus in chilli rhizosphere soil, *J. Mycopathol. Res.* 48(1): 129-131.

Thakur, R. P., Rao, V. P. and Subramanyam, K., 2003, Influence of biocontrol agents on population densities of *Aspergillus flavus* and kernel infection in groundnut. *Indian Phytopath.* 56(4): 408-412.

Thippeswamy, T. and Lokesh, S., 1997, Effects of leaf extracts on seed mycoflora, germination and seedling vigour of sunflower variety ASPH-11. *International J. Trop. Pl. Dis.* 15: 53-58.

Thirumala Devi, K. T., Mayo, M. A., Gopal Reddy, Reddy, S. V., Delfose, P. and Reddy, D. V. R., 2000, Production of polyclonal antibodies against Ochratoxin A and its detection by ELISA. *J. Agric. Food Chem,* 48: 5079-5082.

Thirumala Rao, S. K. and Sitaramaiah, K., 2000, Stimulation of *Trichoderma* spp. and inhibition of *Aspergillus niger* in water extracts of neem cake amended soil *in vitro. Indian J. Mycol. Pl. Pathol.* 30(2): 236-238.

Tripathi, D. P. and Singh, B. R., 1991, Mycoflora of soybean and their control. *Madras Agric. J.* 78(1-4): 130-132.

Vanamala Anjaiah, Thakur, R. P., Rao, V. P., Sharma, K. K., Cornelis, P. and Koedam, N., 2001, A biological control approach making use of rhizobacteria and soil fungi for soilborne post harvest infection of *Aspergillus flavus* in groundnut. *Bulletin OILB/SROP.* 24(3): 151-155.

Varma, T. S. N., Eggtha, S., Naidu, G. K. and Gowda, M .V. G., 2001, Screening groundnut varieties for *in vitro* colonization with *Aspergillus flavus. International Arachis News Lett.* 21: 13-14.

Vasanthi, S. and Bhat, R. V., 1998, Mycotoxins in food occurrence, health and economic significance and food control measures. *Indian J. Medical Res.* 108: 212-224.

Verma, R. J., Kolhe, A. S. and Dube, H. C., 1996, Aflatoxin contamination in different varieties of peanuts. *Indian Phytopath.* 49: 62-66.

Verma, R. J., Mehta, D. N., Ravl, P. J. and Dube, H. C., 1997, Aflatoxin contamination in oilseeds, oil cakes and oil samples. *Indian Phytopath.* 50(1): 77-82.

Walier, F. and Bockelle Morvan, A., 1987, Resistance of groundnut varieties to *Aspergillus flavus* in Senegal. In: *Aflatoxin contamination of groundnut. Proceed. of the International Workshop,* 6-9 October, pp.305-310.

Waliyar, F., Hassan, H. and Bosc, J. P., 1994, Sources of resistance to *Aspergillus flavus* and aflatoxin contamination in groundnut genotypes in West Africa. *Plant Dis.* 78(7): 704-708.

Wani, T. A. and Paul, M. S., 2004, Effect on neem products and Dithane Z-78 on rhizosphere mycoflora of chilli (*Capsicum annuum*). *Environ. Ecol.* 22(1): 197-199.

Wogan, G. N., 1975, Mycotoxins. *A Review of Pharmacology.* 15: 437-451.

Yellamanda Reddy, T., Sulochanamma, B. N., Subramanyam and Anand Balaguravaiah, D., 2003, Influence of weather, dry spells and management practices on aflatoxin contamination in groundnut. *Indian Phytopath.* 56(3): 262-265.

Zablotowicz, R. M., Abbass, H. K. and Locke, M. A., 2007, Population ecology of *Asperpgillus flavus* associated with Mississippi Delta soils. *J. Food Addit. Contam.* 24: 1102-1108.

Zawiyah, S., Man, Y. B. C., Nazimah, S .A. H., Chin, C. K., Tusukamoto, I., Hamanyza, A. H. and Norhaizan, I., 2007, Determination of organochlorine and pyrethroid pesticides in fruit and vegetables using SAX/PSA clean-up column. *Food Chem.* 102 (1): 98-103.

14

Hybrid Vigour in Ash Gourd

L.Pugalendhi, V.Rajashree, S.Praneetha and S.Natarajan

Ash gourd (syn:wax gourd, winter melon, white gourd, white pumpkin, ash pumpkin, Chinese preserving melon and hairy melon) is a vegetable belonging to family Cucurbitaceae and genus *Benincasa* with only one cultivated species, *Benincasa hispida* (Thunb.) Cogn. (2n=2x=24). Ash gourd is an important vegetable especially in the Indo-China region, which is also the major producer of ash gourd (Rubatzky and Yamaguchi, 1997), but it is also a potential crop of tropical Africa and America (Sirohi *et al.*, 2005).

Grown extensively in tropical regions of the Old World, *Benincasa* is less common to the New World (Peter, 1991). This annual vine, native to the Asian tropics (including India), is grown primarily in the Old World tropics. To a lesser extent, it is also grown in the New World for its edible fruits (Walters and Deckers-Walters, 1989). The flesh is white to pale green, with a rather bland flavour (Marr *et al.*, 2007). In China and Southeast Asia, the mature fruits are used in preparation of soup, while fruits are cooked in curries particularly in South India (Sureja *et al.*, 2006).Valued mainly for its long storage life, it is used as a vegetable during lean period and has a good scope for value addition. The fruits are consumed as baked, fried, boiled, pickled or candied/preserved (Robinson and Decker-Walters, 1999).

It is a prized cucurbit vegetable because of its high nutritional and medicinal properties, industrial importance for processing, long storability, better transport qualities and higher yield per unit area. It is used mainly for treatment of diabetes and diuresis diseases in Korea and hypertension and inflammation in traditional Chinese medicine (Kalloo and Bergh,1993). Ash gourd juice has potential to improve the weak nervous system (Arora, 2003). The ability of anti- oxidation and angiotension converting enzyme (ACE) activity inhibition may provide protective effects against cardiovascular diseases and cancer (Huang *et al.*, 2004). The juice of ripe fruit is used for curing of insanity and epilepsy (Ramesh *et al.*,1989), urinary infection and biliousness (Nayar and More, 1998).It possesses anti-ulcer activity (Grover *et al.*, 2001), and is used as a tonic for heart (Nayar and More, 1998). In China, India and Cuba, mature fruits are "candied", commonly known as "petha". In China, a canned beverage is made from the fruit and the "hairy melon" is also sliced and eaten raw or cooked (Walters and Deckers-Walters, 1989).

Since *Benincasa* is a monotypic genus, its gene pool is not too diverse, yet the genotypes under cultivation are not well characterized, which makes breeding of good ash gourd cultivars difficult (Peter, *et al.*, 1991). The existence of wild populations is unconfirmed and as such the centre of *Benincasa hispida* is uncertain. Indo-China is the centre of diversity for ash gourd (Rubtzky and Yamaguchi,1997). Rind and seeds of a gourd discovered at the Kana site in Papua New Guinea are identified as remains of *Benincasa hispida*; therefore, it may possibly be domesticated at the Kana site, Indo-China region, America and Africa.

In spite of its economic importance and availability of a considerable genetic diversity in plant and fruit characters, the genetic potentialities of ash gourd are practically unexplored and very little attempt has so far been made for its genetic improvement (Verma *et al., 2007*).

Mandal *et al.* (2002) found relatively higher contribution of dominant gene action over the additive gene action and they suggested the heterosis breeding in ash gourd for improvement of yield traits.In India, a wide range of variability is available for different component characters in ash gourd, but very sporadic efforts have been made for its genetic improvement. A rapid improvement in ash gourd could be brought about by exploiting hybrid vigour because heterosis is as high as 37.6% over the best parent was recorded by Sureja *et al.* (2006). A wide range of genetic diversity among parents is an essential feature of any hybridization programme. Hence, plant breeders are interested to estimate the extent of genetic diversity among different genotypes which will help them to achieve the set goal through appropriate breeding strategy.

Crossing among parental lines is a very efficient method, for creating variability. However, selection of divergent parents is important because greater is the genetic divergence among the parents for the characters, better are the chances of obtaining a pool of genotypes of large diversity (Verma *et al.*, 2007, Pandey *et al.*, 2008). At the same time, of course, genotypes possessing desirable traits are important. Selection of highly productive cultivars of good quality within the available material is the most important for increasing the total production of ash gourd. Short vine length (bushy type), high flesh thickness of the fruit and fruit yield are the desirable characters of good quality of ash gourd. The least number of days to female flowering, especially if lesser than the number of days to male flowering, is another desirable character because it can be an evidence of earliness.

Being a monoecious and cross-pollinated crop, ash gourd provides ample scope for exploitation through hybrid vigour. A significant increase in productivity can be brought about by assessing the available genetic variability and the heterosis, *i.e.*, hybrid performance.

Indo-China region, being a centre of diversity, is endowed with great variability in terms of morphological characters especially, growth habit, maturity including shape, size and flesh thickness of fruits (Rubatzky and Yamaguchi ,1997). Although there is substantial variation in ash gourd vegetative traits, especially fruit characters, it is difficult to distinguish genotypes on their external morphology alone. Assessment of genetic diversity based on phenotype has limitations, since most of the morphological characters are greatly influenced by environmental factors and the development stage of the plant. In contrast, molecular markers based on DNA sequence polymorphism are independent of environmental conditions and show a higher level of polymorphism.

This necessitates the assessment of genetic diversity in ash gourd using the modern molecular approaches. This would allow a more efficient utilization of plant characters in developing suitable varieties for yield and stability. In addition, an unambiguous, reliable, fast and cost-effective assessment of genetic diversity is important for determining the uniqueness and distinctiveness of the phenotypic and genetic constitution of genotypes to protect breeder's intellectual property rights (Franco *et al.* 2001).

Hybrid Vigour

The term heterosis is the 'hybrid vigour' and it is defined as the manifestation of greater vigour in leaf area, growth, dry matter accumulation and higher yield. In heterosis, the F_1 population, obtained by crossing two genetically diverse parents, shows superiority over both parents in quantitative and qualitative characters *viz.*, vitality, reproductive capacity, stress resistance, adaptability etc.

Heterosis conventionally refers to the increased or decreased vigour of F_1 over its mid-parent value. Fonesca and Patterson (1968) coined the new term heterobeltiosis to denote heterosis over better parent. The heterosis is governed by allelic, non- allelic, plasmic and / or maternal interactions, which are otherwise functionally named either sheltering, inhibiting, stimulating, complementing and / or dosage adjusting. Heterosis can be understood in three different aspects namely directional, functional and transmission through sexual phase.

Heterosis is a complex biological phenomenon manifested in the superiority of hybrids over parental forms due to rate of development of one or more complex traits (Konarev, 1976). Comparing with the usual recombination effects, heterosis is exhibited mainly in F_1 progeny and passed on to the subsequent generations. Heterosis is mainly manifested in quantitative characters and is expressed in vigour of vegetative organs and also in increased yield and quality characters like resistance to diseases and tolerance to climatic fluctuations.

Exploitation of heterosis for various characters has become a potential tool in the improvement programme and RAPD markers have been used extensively in cucurbits to identify cultivars and hybrids (Meng *et al.*, 1996). Nine inbred lines of ash gourd were crossed by Sureja *et al.* (2006) to produce all possible (72) F_1 hybrids, including reciprocals, which were evaluated along with the nine parents for 13 growth and yield-related traits and 14 quality traits, in a replicated field trial. A high level of heterosis was observed among the hybrids for most of the traits examined, including iron and zinc content. Significant reciprocal differences in heterosis were observed.

The inbred lines were analysed using 130 RAPD primers, of which 26 were informative and amplified 163 DNA marker bands. A total of 47 polymorphic bands were obtained with a mean of 1.8 per primer which, in combination, discriminated all the inbreds from each other. Pair-wise genetic distance measurements ranged from 0.056 to 0.179, suggesting a narrow genetic base for the inbreds.

Verma *et al.* (2007) selected ten elite inbred lines of ash gourd (*Benincasa hispida*) based on their performance and crossed in a diallel mating system without reciprocal crosses to generate 45 F1 hybrids. The 10 parental lines used were DAG-2(P_1), DAG-4(P2), DAG-1(P_3), DAG-9(P_4), DAG-6(P_5), DAG-8(P_6), DAG-13(P_7), DAG-5(P_8), DAG-

12(P_9) and DAG-11(P_{10}). The hybrids along with the parents were grown to study heterosis in 10 yield related characters.

The range of heterosis for total number of male flowers varied from -36.64 (P6 × P7) to 66.67% (P4 × P5) for mid parent heterosis, -33.65 (P6 × P10) to 22.50% (P4 × P9) for better parent heterosis and zero (P5 × P10) to 131.58% (P1 × P10) for top parent heterosis. Out of 45 crosses, 21 crosses showed significant negative mid parent heterosis, ten crosses showed significant negative better parent heterosis, whereas none of them showed negative top parent heterosis. Out of five top crosses expressing significant negative mid parent heterosis (P6 × P7, P6 × P10, P3 × P6, P7 × P10 and P1 × P9) and better parent heterosis (P6 × P10, P6 × P2, P3 × P6, P1 × P6 and P1 × P7), most of them had P6 as one of the parents. The F1 hybrid P6 × P7 registered the highest negative better parent heterosis (-36.64%) and P6 x P10 showed the highest negative better parent heterosis (-33.65%) for this trait.

The total number of male flowers and total number of female flowers also had wide range of heterosis. The range varied from -53.33 (P2 × P8) to 100.00% (P9 × P10 and P4 × P5), -56.25% (P1 × P7) to 100% (P1 × P5) and -61.17 (P1 × P7) to 16.67% (P9 × P10) for mid parent, better parent and top parent heterosis respectively. Significant positive mid parent heterosis was observed in 21 crosses, better parent heterosis in 18 crosses and top parent heterosis in four crosses. The cross, P1 × P5 displayed the highest better parent heterosis (100%) and P9 × P10 exhibited the highest top parent heterosis (16.67%) for this trait. The range of heterosis varied from -36.84 (P1 × P2) to 87.50% (P8 × P10), -50.00 (P1 × P2) to 87.50 (P8 × P10) and -50.00 (P1 × P2) to 25.00% (P8 × P10) over the mid parent, better parent and top parent, respectively for number of fruits per vine.

Among the 45 F1 hybrids, four hybrids exhibited significant positive heterosis over the top parent (P2). Among the top five crosses showing high positive heterosis, over mid parent (P8 × P10, P6 × P8, P3 × P8, P9 × P10) and better parent (P2 × P5, P8 × P10, P6 × P8, P3 × P8 and P7 × P9), three had P8 as one of the parents. The cross P8 × P10 displayed the highest mid parent and better parent heterosis (87.5%) and top parent heterosis (25.00%).

In number of fruits per vine, wide range of heterosis was also noticed for individual fruit weight and it ranged from -48.94 (P2 × P9) to 120.51% (P5 × P10) for mid parent heterosis, -58.51 (P3 × P7) to 115.00% (P5 × P10) for better parent heterosis and 64.69 (P1 × P6) to 19.63% (P3 × P8) for top parent heterosis. Significant positive heterosis over mid parent, better parent and top parent were observed in 20 crosses, eight crosses and four crosses, respectively. The heterosis for yield per vine ranged from -72.46 (P1 × P2) to 166.67 % (P5 × P10) over mid parent, -77.25 (P1 × P2) to 164.29% (P5 × P10) over better parent and -83.43 (P1 × P6) to 23.36% (P3 × P8) over top parent. It showed the highest positive heterosis (166.49%) among all other traits.

The hybrid P5 × P10 (DAG-6 × DAG-11) exhibited the highest mid parent (166.67%) and better parent heterosis (166.49%), followed by P9 × P10 (141.88 and 128.23%) and P4 × P5 (107.62 and 94.64%), respectively for yield per vine. The highest positive significant heterosis over the top parent (DAG- 4) was detected in the hybrid P3 × P8 (23.35%), followed by the cross P2 × P10 (13.79%) for this trait. Among the five traits, the highest (more than 100%) level of heterosis was detected for yield per vine, followed by individual fruit weight and moderate range of heterosis (50 to 100%) was observed for number of fruits per vine. The cross P8 × P10 (DAG-5 × DAG-11) was the best heterotic combination (in terms of better parent, mid parent and top parent heterosis) for number of fruits per vine and P5 × P10 for individual fruit weight and fruit yield per vine.

Sureja *et al.* (2006) observed high better parent and mid parent heterosis for fruit yield per vine, fruit number per vine, fruit weight and TSS content. Mohanty and Mishra (2000) observed high heterosis over mid parent and better parent for fruits per vine (156.3 and 145.6%), individual fruit weight (92.3 and 82.0%) and yield per plant (191.4 and 188.2%), respectively, in pumpkin.

The range of per cent heterosis of 45 F1 hybrids, number of heterotic crosses and three best F1 hybrids in terms of performance and heterosis for flowering traits are presented in Tables 1 and 2. Results of the study suggested that both nature and magnitude of heterosis differ from trait to trait depending on cross combinations. Considerable heterosis was observed in the positive and negative directions. Heterosis in negative direction is desired for characters such as vine length, days to opening of first male and female flowers, and days to fruit maturity. Among the 10 yield related traits, the range of heterosis was intermediate (50 to 100%) for number of fruits per vine, while it was narrow (<50%) for days to opening of first male and female flowers, and days to fruit maturity. High level of positive heterosis (>100%) was exhibited for yield per vine and fruit weight over the mid parent and better parent and a considerable amount of positive mid and better parent heterosis (between 50 and 100%) were also recorded for number of fruits per vine (Table 2). Days to opening of first male and female flower and days to fruit maturity demonstrated small negative heterosis over the mid parent and better parent. This is consistent with the observations of Sureja (2003).

The results showed that DAG-2 × DAG-8 was the best heterotic combination in terms of better parent, mid parent and top parent heterosis for first flowering node of male and female flowers; DAG-13 × DAG-5 and DAG-6 × DAG-5 for days to opening of first female flowers; DAG-12 × DAG-11 for sex ratio; DAG-2 × DAG-9, DAG-2 × DAG-4 and DAG-2 × DAG-1 for days to fruit maturity and DAG-1 × DAG-11 for number fruits per vine. The best F1 hybrids having the highest mid-parent and better parent heterosis were observed in DAG-6 × DAG-11 and DAG-9 × DAG-6 for fruit weight and DAG-6 × DAG-11 and DAG-12 × DAG -11 for yield per vine. DAG-1 × DAG-5 and DAG-4 × DAG-11 produced 34.33 and 31.67 kg fruit yield per vine displaying positive top parent heterosis of 23.5 and 14.0%, respectively.

Table 1: Range of heterosis, heterotic crosses and best hybrids of ash gourd for different flowering traits

Particulars	Days to opening of first male flower	Days to opening of first female flower	Node of first female flower	Number of male flowers/plant	Number of female flowers/plant
Range of heterosis (%) over					
Better parents	-10.33 to 15.54	-12.31 to 7.75	-20.69 to 26.00	-33.65 to 82.50	-56.25 to 100
Mid parents	-11.57 to 8.44	-13.80 to 4.43	-22.47 to 14.29	-36.64 to 66.00	-53.33 to 100
Top parents	-7.08 to 12.62	-11.63 to 7.76	-5.47 to 27.41	-0.00 to 131.58	-61.17 to 16.67
No. of heterotic crosses over					
Better parents	24	25	20	14	18
Mid parents	35	35	24	22	21
Top parents	19	29	4	0	4
Three best F_1 hybrids with heterosis (%) over					
Better parents					
1st	-10.33 (D-1 x D-5)	-12.31 (D-13 x D-5)	-20.69 (D-2 x D-8)	-33.65 (D-8 x D-11)	100.00 (D-2 x D-6)
2nd	-0.06 (D-2 x D-4)	-9.23 (D-6 x D-5)	-17.05 (D-4 x D-6)	-25.81 (D-8 x D-13)	90.91 (D-12 x D-11)
3rd	-5.83 (D-6 x D-11)	-7.75 (D-4 x D-12)	-16.47 (D-4 x D-13)	-20.00 (D-1 x D-8)	87.50 (D-9 x D-6)
Mid parents					
1st	-11.57 (D-1 x D-5)	-13.80 (D-13 x D-5)	-22.47 (D-2 x D-8)	-36.64 (D-8 x D-13)	100.00 (D-12 x D-11)
2nd	-7.77 (D-4 x D-12)	-10.10 (D-6 x D-5)	-17.52 (D-4 x D-13)	33.65 (D-8 x D-11)	100.00 (D-9 x D-6)
3rd	-6.67 (D-2 x D-4)	-9.75 (D-4 x D-6)	-17.51 (D-4 x D-6)	-29.22 (D-1 x D-8	86.67 (D-2 x D-6)
Top parents					
1st	-7.08 (D-4 x D-120	-11.63 (D-2 x D-5)	-5.47 (D-2 x D-8)	-	16.67 (D-12 x D-11)
2nd	-6.06 (D-2 x D-4)	-8.52 (D-6 x D-5)	-5.47 (D-8 x D-5)	-	11.17 (D-5 x D-11)
3rd	-4.55 (D-2 x D-9)	-8.52 (D-4 x D-11)	-4.11 (D-5 x D-11)	-	11.17 (D-4 x D-6)
S.Ed	1.07	0.90	1.36	2.84	0.67

[1]Standard error (S.E.) was calculated based on the performance of 45 F_1 hybrids as per the heterosis (%) value.

DAG - Delhi ash gourd.

Table 2: Range of heterosis, heterotic crosses and best hybrids of ash gourd for sex ratio and yield traits

Particulars	Sex ratio	Number of fruits per vine	Days to fruit maturity	Fruit weight (g)	Yield per vine (kg)
Range of heterosis (%) over					
Better parents	-51.69 to 165.90	-50.00 to 87.51	-10.16 to 15.54	-58.51 to 115.00	- -
Mid parents	-54.95 to 105.90	-36.84 to 87.50	-10.28 to 12.98	-48.94 to 120.51	-72.46 - 166.29
Top parents	-8.71 to 165.78	-50.00 to 25.00	-5.08 to 15.53	-64.69 to 19.63	-82.43 – 23.36
No. of heterotic crosses over					
Better parents	20	15	11	8	11
Mid parents	24	21	22	19	14
Top parents	1	4	3	4	2
Three best F_1 hybrids with heterosis (%) over Better parents					
1st	-59.59 (D-12 x D-11)	11.67 (D-4 x D-6)	-10.16 (D-2 x D-9)	115.00 (D-6 x D-11)	164.69 (D-6 x D-11)
2nd	-37.75 (D-6 x D-11)	87.50 (D-5 x D-11)	-8.93 (D-4 x D-6)	81.82 (D-9 x D-6)	128.23 (D-12 x D-11)
3rd	-30.00 (D-2 x D-6)	50.00 (D-1 x D-5)	-8.29 (D-2 x D-4)	72.33 (D-2 x D-11)	94.64 (D-9 x D-6)
Mid parents					
1st	-54.95 (D-12 x D-11)	87.50 (D-5 x D-11)	-10.28 (D-2 x D-9)	120.51 (D-6 x D-11)	166.29 (D-6 x D-11)
2nd	-40.46 (D-6 x D-11)	62.50 (D-9 x D-6)	-10.27 (D-2 x D-1)	95.12 (D-9 x D-6)	141.88 (D-12 x D-11)
3rd	-34.50 (D-2 x D-6)	50.00 (D-2 x D-11)	-10.09 (D-2 x D-4)	74.88 (D-2 x D-11)	107.62 (D-1 x D-11)
Top parents					
1st	-8.71 (D-12 x D-11)	25.00 (D-5 x D-11)	-5.00 (D-2 x D-9)	19.63 (D-1 x D-5)	23.50 (D-1 x D-5)
2nd	-	8.25 (D-8 x D-13)	-3.11 (D-2 x D-4)	10.59 (D-4 x D-11)	14.00 (D-4 x D-11)
3rd	-	8.25 (D-4 x D-12)	-2.54 (D-2 x D-1)	1.27 (D-1 x D-6)	-
S.Ed	3.38	32.13	1.51	0.39	1.97

[1]Standard error (S.E.) was calculated based on the performance of 45 F_1 hybrids as per the heterosis (%) value.

DAG - Delhi ash gourd.

These hybrids can be commercially exploited for higher yield. Based on the combining ability (data not shown), DAG-4 was the top most parent for total number of female flowers per vine, number of fruits per vine, fruit weight and yield per vine; DAG-6 for number of male flowers; DAG-5 for first flowering node of female flower; and days from fruit set to maturity; DAG-2 for days to opening of male and female flowers; and DAG-13 for sex ratio. The highest negative heterosis for days to fruit maturity was obtained by the cross DAG-2 × DAG-9 (~ –10%) indicating that it can be successfully utilized in breeding for earliness in ash gourd. Two best hybrids namely, DAG-1 × DAG-5 (Fig. 1) and DAG-4 × DAG 11 (Fig. 2) displaying positive top parent heterosis of

23.50 and 14.00%, respectively for yield per vine can be taken up for commercial cultivation.

Fig. 1 : Fruits of F_1 cross DAG-1 × DAG-5 (oblong, 26.0 cm) long and 29.0 cm diameter, with average fruit weight of 8.5 kg).

Fig. 2 : Fruits of F_1 cross DAG-4 × DAG-11 (oblong, 24.50 cm) long and 32.0 cm diameter, with average fruit weight of 7.8 kg).

The mean performance and heterosis of the top three F_1 hybrids, measuring different traits, are given in Table 3. In order of their performance, ranking of the hybrids varied significantly for the different traits. For instance, hybrid $P_5 \times P_{10}$ ranked first for individual fruit weight, and fruit length. It gave the highest yield per vine (24.70 kg), 134.30% more than the better parent. Of all 20 traits measured, yield per vine exhibited the maximum positive heterosis over the better-parent values followed by individual fruit weight, and total number of female flowers per vine. Fruit length, fruit diameter, and flesh thickness displayed comparatively lower positive heterosis; whereas vine length, days to opening of first male and female flowers, and days to fruit maturity demonstrated low negative heterosis. The number of seeds per fruit, and seed weight per fruit showed intermediate levels of negative better-parent heterosis. Similar observations on the extent of heterosis for different traits were reported earlier in ash gourd by Gowda (1983) and Sureja *et al.* (2006).

Table 3 : Performance and heterosis of top three F_1 hybrids of ash gourd for different traits

Traits	Top three hybrids (performance, heterosis)			S.E.
First flowering node of male flower	$P_1 \times P_6$ ----	$P_2 \times P_7$ (20.70, −15.10)	$P_3 \times P_6$ (23.70, −13.40)	(1.12, 1.10)
First flowering node of female flower	$P_1 \times P_6$ (23.00, −20.70)	$P_2 \times P_5$ (24.30, −17.00)	$P_2 \times P_7$ (23.70, −16.50)	(1.31, 1.36)
Days to opening of first male flower	$P_3 \times P_8$ (63.70, −10.30)	$P_2 \times P_9$ (61.30, −7.10)	$P_5 \times P_{10}$ (64.70, −5.80)	(1.03, 1.07)
Days to opening of first female flower	$P_7 \times P_8$ (76.00, −12.30)	$P_2 \times P_6$ (78.70, −8.50)	$P_2 \times P_9$ (79.30, −7.75)	(0.90, 0.90)
Total number of male flower	$P_6 \times P_{10}$ (46.00, −33.70)	$P_6 \times P_7$ (38.30, −25.80)	$P_3 \times P_6$ (44.00, −20.00)	(2. 68, 2. 84)
Total number of female flower	$P_1 \times P_5$ (4.70, 100.00)	$P_9 \times P_{10}$ (7.00, 90.90)	$P_4 \times P_5$ (5.00, 87.50)	(0.67, 0.67)
Sex ratio (M/F)	$P_9 \times P_{10}$ (8.90, −51.60)	$P_5 \times P_{10}$ (10.60, −37.75)	$P_1 \times P_5$ (11.90, −30.00)	(3.34, 3.38)
Days to fruit maturity	$P_1 \times P_4$ (112.00, −10.15)	$P_2 \times P_5$ (122.30, −8.90)	$P_1 \times P_2$ (114.30, −8.30)	(1.40, 1.51)
Number of fruits per vine	$P_8 \times P_{10}$ (5.00, 87.50)	$P_6 \times P_8$ (4.30, 62.50)	$P_7 \times P_{10}$ (3.30, 42.85)	(0.44, 0.44)
Individual fruit weight (kg)	$P_5 \times P_{10}$ (7.17, 115.00)	$P_4 \times P_5$ (6.70, 81.80)	$P_1 \times P_{10}$ (5.90, 72.30)	(0.38, 0.39)
Fruit length (cm)	$P_5 \times P_{10}$ (24.30, 29.90)	$P_1 \times P_8$ (29.60, 24.00)	$P_7 \times P_{10}$ (22.70, 21.00)	(1.08, 1.14)
Fruit diameter (cm)	$P_4 \times P_5$ (23.00, 27.00)	$P_5 \times P_{10}$ (24.30, 26.90)	$P_1 \times P_6$ (22.80, 26.10)	(0.89, 0.91)
Flesh thickness (cm)	$P_4 \times P_5$ (6.80, 24.10)	$P_5 \times P_9$ (6.70, 23.00)	$P_4 \times P_9$ (6.40, 17.10)	(0.17, 0.18)
Rind thickness (mm)	$P_4 \times P_9$ (1.90, 19.60)	$P_5 \times P_9$ (1.90, 19.15)	$P_4 \times P_5$ (1.90, 15.50)	(0.09, 0.10)
Number of seeds per fruit	$P_4 \times P_7$ (230.30, −36.90)	$P_8 \times P_9$ (351.70, −31.30)	$P_2 \times P_9$ (303.30, −29.70)	(32.52, 32.13)
Seed weight per fruit (g)	$P_8 \times P_9$ (117.70, −34.60)	$P_1 \times P_9$ (118.33, −34.30)	$P_4 \times P_7$ (88.30, −24.30)	(10.92, 10.32)
Yield per vine (kg)	$P_5 \times P_{10}$ (24.70, 134.30)	$P_9 \times P_{10}$ (23.60, 128.20)	$P_4 \times P_5$ (18.20, 94.60)	(2.22, 1.97)
TSS (°B)	$P_8 \times P_{10}$ (3.10, 24.30)	$P_3 \times P_{10}$ (3.30, 23.75)	$P_7 \times P_{10}$ (3.10, 19.50)	(0.10, 0.10)
Dry matter (%)	$P_6 \times P_{10}$ (5.60, 45.30)	$P_2 \times P_4$ (5.70, 33.60)	$P_1 \times P_5$ (3.25, 27.00)	(0.23, 0.28)
Vine length (m)	$P_7 \times P_{10}$ (6.50, −9.40)	$P_2 \times P_8$ (4.40, −9.40)	$P_6 \times P_{10}$ (6.50, −8.20)	(0.28, 0.30)

Data in parentheses indicate mean values of three replications and better parent heterosis (%), respectively.

Hybrid vigour was estimated for twelve traits in the thirty cross combinations by Josephin (2008) and expressed over mid, better and standard parents in percentage. The results are presented in the Table 4.

Days to First Female Flowering

The negative significant relative heterosis per cent for days to first female flowering was recorded in the cross combinations IVAG-90 x Indu (-1.00*), DAG-4 x CO_2 (-1.97*) and HYBH-25 x CO_2 (-1.99*). The negative significant heterobeltiosis was recorded in the cross combination PAG 3 x Indu (-1.27*), Chal Kumar x Indu (-2.37 *), PAG 3 x Indu (-2.65*) and DAG 6 x Pusa Ujwal (-2.93*). The negative standard heterosis (diii) was recorded in NO 600 x Indu (-0.56*), Coimbatore local x Indu (-0.67*), Chal Kumar x CO_2 (-1.40*) BH 21 x Pusa Ujwal (-1.69*) and PAG 3 x CO_2 (-2.64*).

Node at First Female Flower Appearance

The significant negative relative heterosis for this character was recorded in 13 cross combinations. It was observed as the highest IVAG-3 x CO_2 (-3.46*) followed by DAG-6 x CO_2 (-3.99*) and DAG-4 x CO_2 (-5.31*).The significant and negative heterobeltiosis was recorded in IVAG-90 x CO_2 (-2.09*), IVAG-3 x Indu (-2.49*) HYBH-25 x Indu (-2.66*) and IVAG-3 x CO_2 (-3.96*). Twenty hybrids recorded significant and negative heterobeltiosis. The highest level of significant and negative standard heterosis was observed in PAG-3 x CO_2 (-3.39*), Coimbatore local x Pusa Ujwal (-5.54*), NO-600 x Indu (-16.79*) and Chal kumar x CO_2 (-43.06*).

Sex Ratio

The negative significant relative heterosis for sex ratio was measured in the hybrid combinations Chal kumar x Indu (-0.15*), IVAG 3 x Pusa Ujwal (-1.04*), IVAG 3 x Indu (-4.46*), Chal kumar x Pusa Ujwal (-6.71*) and PAG 3 x Indu (-6.75*). The heterobeltiosis was at the highest level in the cross combinations IVAG 90 x Indu (-7.58*) Chal kumar x Pusa Ujwal (-8.11*), Chal Kumar x Indu (-8.69*), PAG 3 x Indu (-11.83*) and IVAG 3 x Pusa Ujwal (-12.01*).The level of negative and significant standard heterosis was measured in BH 21 x Pusa Ujwal (-2.38*), Coimbatore Local x Pusa Ujwal (-4.33*), Chal Kumar x CO_2 (-9.84*) and PAG 3 x CO_2 (-10.55*).

Number of Fruits Per Vine

All the hybrids showed positive significant heterosis for number of fruits per vine. PAG 3 x Pusa Ujwal registered the highest relative heterosis (29.97), followed by PAG 3 x CO_2 (82.57), BH 21 x Pusa Ujwal (48.07) and PAG 30 x CO_2 (41.64). The heterobeltiosis was the highest in PAG 3 x CO 2 (51.57) followed by Chal Kumar x CO_2 (47.79), BH 21 x Pusa Ujwal (31.36) and Coimbatore Local x CO_2 (21.95). The hybrids PAG 3x CO_2 (43.50), Chal Kumar x CO_2 (39.93), BH 21x Pusa Ujwal (27.43) and Coimbatore Local x Indu (17.91) recorded higher values of standard heterosis.

Single Fruit Weight

The significant positive relative heterosis for single fruit weight was recorded in the hybrids IVAG 90 x CO_2 (33.42*), BH 21 x CO_2 (30.80*) and IVAG 3 x Indu (30.61*). The heterobeltiosis was high in the hybrid combinations NO 600 x Pusa Ujwal (13.58*), Chal Kumar x Pusa Ujwal.(4.62*), BH 21 x CO_2 (1.47*), IVAG 90 x CO_2 (1.45*) and NO 600 x CO_2 (1.31*). The standard heterosis was high in IVAG 3 x Indu (40.52*), DAG 4 x CO_2 (39.08*). IVAG 90 x CO_2. (39.05*), IVAG 3 x Pusa Ujwal (37.15*), NO 600 x Pusa Ujwal (31.44*) and in BH 21 x CO_2.and Chal Kumar x Pusa Ujwal (31.43*).

Yield Per Vine

The significant positive relative heterosis for yield per vine was recorded in PAG-3 x CO_2 (118.33), Chal kumar x CO_2 (109.32), HYBH-25 x CO_2 (101.64) IVAG-90 x CO_2 (98.49) and NO-600 x CO_2 (91.15). The significant positive heterobeltiosis was noticed in the hybrid combinations IVAG-90 x CO_2 (56.01*) and Chal kumar x CO_2 (55.99*), followed by PAG-3 x CO_2 (52.36*). The highest level of positive significant standard heterosis was measured in the hybrid PAG-3 x CO 2 (45.85), BH-21 x Pusa Ujwal (29.41*) and Coimbatore local x CO_2 (28.21*).

Table 4 : Heterosis (per cent) for flowering and yield traits of ash gourd

Hybrids	Days to first female flowering			Node at first female flower appearance			Sex ratio		
	di	dii	di	dii	di	dii	di	dii	diii
$L_1 \times T_1$	-2.38*	-2.57*	-2.38*	-10.61*	-12.53*	21.07*	-16.48*	-23.80*	0.38
$L_1 \times T_2$	-3.85*	-5.45*	-3.85*	-25.22*	-28.24*	-3.39*	-25.01*	-31.18*	-10.55*
$L_1 \times T_3$	2.09*	-0.12	2.09*	10.52*	7.48*	42.37*	11.86*	9.90*	23.62*
$L_2 \times T_1$	-5.77*	-6.70*	-5.77*	3.94*	3.02*	45.18*	-21.17*	-22.31*	-2.38*
$L_2 \times T_2$	4.85*	3.58*	4.85*	3.29*	2.15*	47.19*	6.81*	5.95*	37.80*
$L_2 \times T_3$	-3.84*	-5.19*	-3.84*	-9.39*	-14.46*	20.54*	-14.78*	-19.89*	2.36*
$L_3 \times T_1$	-2.32*	-2.57*	-2.32*	4.91*	3.19*	42.86*	5.58*	3.03*	42.72*
$L_3 \times T_2$	1.64*	0.17*	1.64*	1.49*	-2.09*	41.07*	-18.67*	-21.13*	9.04*
$L_3 \times T_3$	-1.00*	-3.08*	-1.00*	14.14*	10.41*	47.86*	1.93*	-7.58*	27.97*
$L_4 \times T_1$	-0.01	-0.08*	-0.01	33.86*	18.04*	63.39*	19.93*	15.93*	52.76*
$L_4 \times T_2$	4.88*	3.56*	4.88*	4.22*	-9.65*	30.18*	40.04*	36.22*	77.18*
$L_4 \times T_3$	-2.43*	-3.16*	-2.43*	1.16*	-6.69*	-16.79*	44.40*	38.31*	70.08*
$L_5 \times T_1$	-2.50*	-2.93*	-2.50*	-16.20*	-19.57*	21.07*	19.49*	12.35*	48.03*
$L_5 \times T_2$	-3.06*	-5.28*	-3.06*	-3.99*	-6.05*	41.43*	30.66*	23.61*	60.63*
$L_5 \times T_3$	-5.18*	-7.82*	-5.18*	-0.90	-9.25*	36.61*	59.12*	56.78*	81.89*
$L_6 \times T_1$	1.62*	0.87*	1.62*	-8.73*	-9.76*	37.62*	-6.71*	-8.11*	24.8*
$L_6 \times T_2$	-4.48*	-5.57*	-4.48*	0.20	-0.71*	-43.06*	-17.27*	-19.03*	-9.84*
$L_6 \times T_3$	-0.73	-2.37*	-0.73	-25.64*	-29.92*	-0.89	-0.15*	-8.69*	24.02*
$L_7 \times T_1$	-3.20*	-5.08*	-3.20*	2.67*	1.70*	43.49*	21.61*	21.31*	59.83*
$L_7 \times T_2$	-1.97*	-5.63*	-1.97*	-5.31*	-6.30*	35.0*	6.13*	5.70*	38.58*
$L_7 \times T_3$	-2.77*	-6.87*	-2.77*	10.68*	4.44*	47.32*	12.76*	4.80*	37.40*
$L_8 \times T_1$	2.93*	1.18*	2.93*	-20.78*	-22.72*	12.5*	-1.04*	-12.01*	48.82*
$L_8 \times T_2$	3.31*	3.14*	3.31*	-3.46*	-3.96*	39.82*	-7.67*	-18.37*	38.19*
$L_8 \times T_3$	4.12*	3.41*	4.12*	4.86*	-2.49*	41.96*	-4.46*	-20.46*	34.65*
$L_9 \times T_1$	-3.01*	-4.48*	-3.01*	-26.56*	-29.14*	-5.54*	-24.97*	-31.26*	8.66*
$L_9 \times T_2$	1.47*	1.11*	1.47*	-19.19*	-20.50*	18.39*	-33.63*	-39.55*	-4.33*
$L_9 \times T_3$	-3.90*	-4.73*	-3.90*	5.93*	-2.51*	45.18*	-20.93*	-32.33*	7.07*
$L_{10} \times T_1$	-6.18*	-7.32*	-6.18*	-18.15*	-22.85*	20.66*	-21.81*	-23.40*	0.79
$L_{10} \times T_2$	-1.99*	-2.65*	-1.99*	-19.40*	-22.58*	21.07*	-16.78*	-17.96*	6.69*
$L_{10} \times T_3$	-0.09	-1.27*	-0.09	8.14*	-2.66*	52.25*	-6.75*	-11.83*	11.43*
SEd	1.76	2.04	1.76	1.75	2.03	2.03	0.27	0.32	0.32

Hybrids	Number of fruits per vine			Single fruit weight			Yield per vine		
	di	dii	diii	di	dii	diii	di	dii	diii
$L_1 \times T_1$	29.97*	6.87*	3.60*	-9.42*	-26.17*	17.14*	51.63*	26.66*	24.27*
$L_1 \times T_2$	82.57*	51.57*	43.50*	-11.42*	-35.77*	1.90*	118.33*	52.36*	45.85*
$L_1 \times T_3$	6.07*	-18.37*	-5.32*	12.90*	-20.16*	26.67*	47.66*	25.39*	20.04*
$L_2 \times T_1$	48.07*	31.30*	27.43*	-11.20*	-21.32*	1.90*	54.35*	25.06*	29.41*
$L_2 \times T_2$	-26.31*	-33.96*	-37.48*	30.80*	1.47*	31.43*	15.52*	-21.09*	-18.33*
$L_2 \times T_3$	6.54*	-12.26*	1.82*	-8.68*	-31.17*	-10.88*	6.42*	-12.41*	9.36*
$L_3 \times T_1$	-4.87*	-28.79*	-30.92*	0.48	-13.12*	19.05*	26.03*	24.10*	-27.81*
$L_3 \times T_2$	4.16*	-21.38*	-25.57*	33.42*	1.45*	39.05*	98.49*	56.01*	3.34*
$L_3 \times T_3$	8.77*	-22.99*	-10.68*	7.04*	-20.83*	8.56*	46.59*	45.91*	-2.46*
$L_4 \times T_1$	-28.05*	-35.54*	-37.48*	21.85*	13.58*	31.44*	5.37*	-10.67*	-17.52*
$L_4 \times T_2$	-22.91*	11.32*	5.40*	25.29*	1.31*	17.14*	91.15*	34.79*	24.45*
$L_4 \times T_3$	1.23*	-15.82*	-2.41*	-11.81*	-30.86*	-20.00*	-3.02*	-16.39*	-22.82*
$L_5 \times T_1$	-34.19*	-37.38*	-39.26*	-5.53*	-17.49*	10.49*	-26.75*	-43.67*	-32.80*
$L_5 \times T_2$	-27.47*	-30.18*	-33.91*	11.31*	-14.65*	14.29*	-2.88*	-36.02*	-23.67*
$L_5 \times T_3$	3.48*	-9.18*	5.39*	11.34*	-16.78*	11.43*	24.78*	-2.64*	16.17*
$L_6 \times T_1$	-16.57*	-33.70*	-35.69*	16.50*	4.62*	31.43*	20.11*	9.98*	-15.05*
$L_6 \times T_2$	84.31*	47.79*	39.93*	-12.51*	-31.38*	-13.71*	109.32*	55.99*	20.50*
$L_6 \times T_3$	-19.53*	-39.93*	-30.33*	13.48*	-13.57*	8.58*	4.35*	-2.66*	-24.81*
$L_7 \times T_1$	14.47*	-7.98*	-10.68	8.34*	-9.15*	34.30*	61.56*	43.10*	19.10*
$L_7 \times T_2$	-23.25*	-37.73*	-41.04*	26.98*	-5.80*	39.08*	35.17*	-1.65*	-18.14*
$L_7 \times T_3$	18.51*	-10.62*	3.61*	14.50*	-17.27*	22.2*	68.42*	51.85*	26.40*
$L_8 \times T_1$	-2.59*	-17.12*	-19.61*	9.92*	-8.28*	37.15*	32.51*	8.10*	9.89*
$L_8 \times T_2$	17.08*	0.62*	-4.73*	16.46*	-13.94*	28.61*	75.36*	20.35*	22.35*
$L_8 \times T_3$	-20.34*	-36.86*	-26.76*	30.61*	-5.98*	40.52*	21.48*	0.68	2.37*
$L_9 \times T_1$	18.18*	3.74*	0.64	-4.23*	-21.06*	21.68*	42.38*	13.76*	22.14*
$L_9 \times T_2$	37.46*	21.95*	15.52*	-1.30*	-27.79*	11.43*	76.54*	19.41*	28.21*
$L_9 \times T_3$	24.56*	1.64*	17.91*	-10.43*	-36.13*	-1.56	33.12*	8.08*	16.06*
$L_{10} \times T_1$	39.53*	16.02*	12.54*	-9.94*	-21.77*	6.02*	59.04*	38.05*	19.26*
$L_{10} \times T_2$	41.64*	18.93*	12.54*	7.63*	-17.83*	11.43*	101.64*	45.65*	24.65*
$L_{10} \times T_3$	22.12*	-5.08*	10.16*	7.85*	-19.94*	8.58*	65.06*	38.85*	19.10*
SEd	0.40	0.46	0.46	0.25	0.29	0.29	1.95	2.26	2.26

Among the different types of molecular markers, random amplified polymorphic DNA (RAPD) markers have been used extensively in cucurbits to classify accessions (Horejsi and Staub, 1999) identify cultivars and hybrids (Meng *et al.*, 1996) and analyse genetic diversity (Lee *et al.*, 1996; Garcia *et al.*, 1998; Gwanama *et al.*, 2000 ; Levi *et al.*, 2001 ; Sureja *et al.*, 2006 ; Dey *et al.*, 2006). The inter-simple sequence repeat (ISSR), which is similar to RAPD has also been used in genetic diversity analysis of different cucurbits (Dje *et al.*, 2006 ; Levi *et al.*, 2004 ; Ritschel *et al.*, 2004 ; Chiba *et al.*, 2003) and phylogenetic studies in *Citrullus* spp., *Cucumis* spp. and *Praecitrullus fistulosus* (Levi *et al.*, 2005) and among different sub species of *Cucurbita pepo* (Paris *et al.*, 2003). Like RAPD, it does not require prior knowledge of the genome and cloning or specific primer design. It has higher reproducibility than RAPD because of high annealing temperatures and the cost of the analyses is lower than the cost of AFLPs to date there are no published reports on the analysis of genetic diversity based on ISSR markers and its relationship with hybrid performance in ash gourd. Verma *et al.* (2007) employed RAPD and a few ISSR markers to study the extent of genetic diversity among 10 inbred lines and to predict F_1 hybrid performance in ash gourd.

Ten inbred lines of ash gourd [*Benincasa hispida* (Thunb.) Cogn.] were crossed to produce 45 F_1 hybrids (without reciprocal), which were evaluated along with the parents for 20 growth and yield related traits, in a replicated field trial. High level of heterosis was observed among the hybrids for most of the traits including yield. These inbred lines were analysed by using 42 RAPD primers, which produced 282 DNA marker bands. A total of 130 RAPD markers were obtained with a mean of 3.1 per primer, which in combination discriminated all the inbreds from each other. Pair-wise genetic distance measurements ranged from 0.07 to 0.31, suggesting a wide genetic diversity for these inbreds. These inbreds were also analysed with five ISSR primers, of which four were informative. Twenty six ISSR marker bands were generated, of which 11 were polymorphic with an average of 2.80 per primer. The percentage of polymorphic bands produced was higher in ISSR markers (>80%) than generated through RAPD markers (46%).

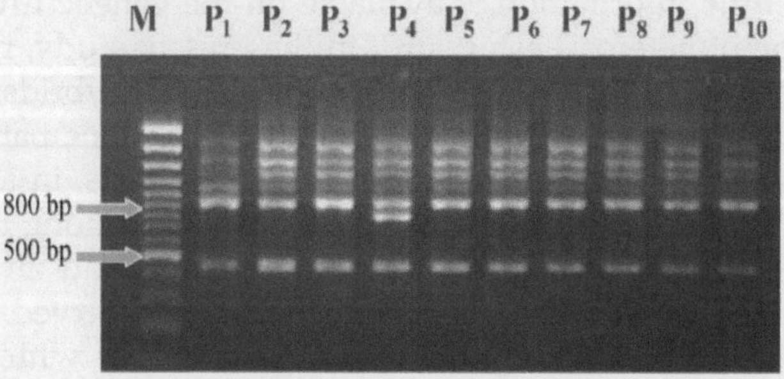

Fig. 3 : Gel showing polymorphic band with RAPD primer OPL-04 in inbred P_4 at 800 bp.

The degree of RAPD polymorphism expressed as percentage of polymorphic fragments in ash gourd (Fig. 3) was higher than that reported in melon (18%; Garcia - Mas *et al.*, 2000), watermelon (21%; Lee *et al.*, 1996) and pumpkin (23%; Gwanama *et al.*, 2000). In parallel to this experiment in ash gourd, Sureja *et al.* (2006) found that the number of bands per RAPD primer ranged from four to 12, with an average of 6 bands per primer. They also reported five primers (19.2% of primers used) with high percentage (>50%) of polymorphic bands (Fig. 4) and three primers (11.5% of the primers) with low percentage (<15%) polymorphic bands and the average number of polymorphic bands per primer was 1.8.

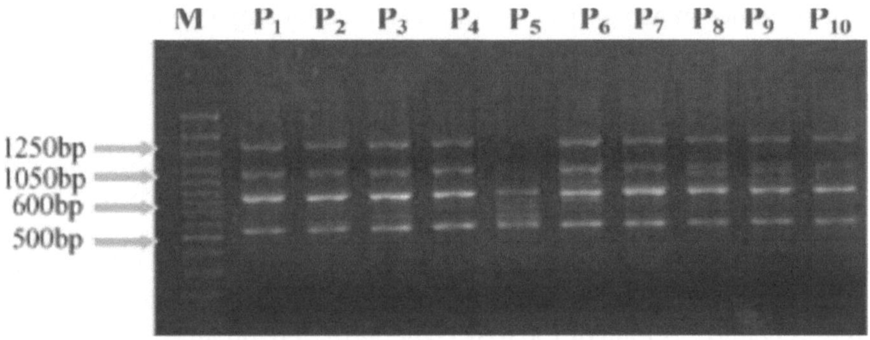

Fig. 4 : Gel showing no band with RAPD primer OPP-16 in inbred P_5 at 1050 and 1250 bp and presence of polymorphic bands in P_8, and P_9 at 900 bp.

The better parent heterosis (computed as the percentage deviation from the better parent value) was estimated by using SPAR 1 (statistical software) developed by Indian Agricultural Statistics Research Institute, New Delhi, India. Pair-wise genetic distances between the inbred lines were correlated with hybrid performance as well as with better parent heterosis estimated for each trait.

Very little information is available on the genetic diversity and exploitation of heterosis ash gourd. In the present study, nine inbred lines were crossed to produce all possible (72) F_1 hybrids, including reciprocals, which were evaluated along with the nine parents for 13 growth- and yield-related traits and 14 quality traits, in a replicated field trial. A high level of heterosis was observed among the hybrids for most of the traits examined, including iron and zinc content. Significant reciprocal differences in heterosis were observed. The inbred lines were analysed using 130 RAPD primers, of which 26 were informative and amplified 163 DNA marker bands. A total of 47 polymorphic bands were obtained with a mean of 1.8 per primer, which, in combination, discriminated all the inbreds from each other. Pair-wise genetic distance measurements ranged from 0.056 to 0.179, suggesting a narrow genetic base for the inbreds. Although the results indicated significant positive correlations of genetic distance with hybrid performance and heterosis, the magnitude of the correlations was not high enough to be of predictive value.

References

Arora, R.K. 2003. Genetic resources of underutilized vegetable crops and their region of diversity. *Veg. Sci.* 30: 1-9.

Chiba, N., K. Suwabe, T. Nunome and Hirai M. 2003. Development of microsatellite markers in melon (*Cucumis melo* L) and their application to major cucurbit crops. *Breed. Sci.* 53: 21–27.

Dey, S.S., T.K. Behera, A.D. Munshi, and Sirohi. P.S. 2006. Studies on genetic divergence in bitter gourd (*Momordica charantia* L.). *Indian J Hort.,* 64: 53-57.

Dje, Y., G.C. Tahi, B.I.A. Zoro, M. Malice, J.P. Baudoin and Bertin. P. 2006. Optimization of ISSR marker for African edible-seeded cucurbitaceae species genetic diversity analysis. *Afr. J. Biotech.,* 5 : 83–87.

Fonesca, S., and Patterson F.L. 1968. Hybrid vigour in a seven parent diallel cross in common winter wheat (*T. aestivum* L.). *Crop Sci.,* 8: 85-88.

Franco, J., J. Crossa, JM. Ribaut, J. Betran, ML. Warburton, and Khairallah M. 2001. A method for combining molecular markers and phenotypic attributes for classifying plant genotypes. *Theor Appl Genet* 103: 944-952.

Garcia, L., M. Jamilena, JI. Alvarez, T. Arnedo, JL. Oliver, and R. Lozano 1998. Genetic relationship among melon breeding lines revealed by RAPD markers and agronomic traits. *Theor. Appl. Genet.,* 96: 878-885

Gowda, R.V. 1983. Genetic analysis of yield and its components is ash gourd [*Benincasa hispida* (Thunb.) Cogn.]. M.Sc. Thesis. University of Agricultural Sciences, Bangalore, India.

Grover, JK., G. Adiga, V. Vats and Rathi. SS. 2001. Extracts of *Benincasa hispida* prevent development of experimental ulcer. *J Ethnopharmacol.,* 78:159–164

Gwanama, C., MT. Labuschange and Botha. AM. 2000. Analysis of genetic variation in *Cucurbita moschata* by random amplified polymorphic DNA (RAPD) markers. *Euphytica* 113: 19-24

Horejsi, T. and Staub. JE. 1999. Genetic variation in cucumber (*Cucumis sativus* L.) as assessed by random amplified polymorphic DNA. *Genet Res Crop Evol.,* 46: 337-350

Huang., Hui Yu, Huang, Ju Jen, Tso, K.Tim., Tsai, Ying Chieh and Chang C. K., 2004. Antioxidant and angiotension converting enzyme inhibition capacities of various parts of *Benincasa hispida* (wax gourd). *Nahrung.* 48: 230-33.

Josephin, A.L., 2008. Studies on development of F_1 hybrids in ash gourd [*Benincasa hispida* (Thunb.) Cogn.] for yield and quality. M.Sc(Hort) thesis TNAU, Coimbatore, India.

Kalloo, G. and Bergh. O. 1993. *Genetic improvement of vegetable crops.* Pergamon Press. pp.235-38.

Konarev, V.G. 1976. Physiological and biochemical aspects of heterosis. In: Heterosis in plant breeding. (Ed.) Janossy, A. and F.G.H. Lupton. Elsever, Sci. Pub. Co., New York. pp 81-82.

Lee, SJ., JS. Shin, KW. Park and Hong. YP. 1996. Detection of genetic diversity using RAPD-PCR and sugar analysis in watermelon [*Citrullus lanatus* (Thunb.)Mnsf.] germplasm. *Theor Appl Genet* 92: 719-725.

Levi, A., CE. Thomas, AP. Keinath and Wehner. TC. 2001. Genetic diversity among watermelon (*Citrullus lanatus* and *Citrullus colocynthis*) accessions. *Genet Res Crop Evol.* 48: 559-556

Levi, A., C.E. Thomas, M. Newman, O.U.K. Reddy, X. Zhang and Xu, Y. 2004 ISSR and AFLP markers differ among American Watermelon cultivars with limited genetic diversity, *J. Am. Soc. Horticult. Sci.* 129 : 553-558.

Levi,A., C.E. Thomas, A.M. Simmons and *J.A.* Thies, 2005 Analysis based on RAPD and ISSR markers reveals closer similarities among *Citrullus* and *Cucumis* species than with *Praecitrullus fistulosus* (Stocks) Pangalo, *Genet. Resour. Crop Evol.* 52 : 465–472

Mandal, J., P.S. Sirohi, and Behera T.K. 2002. Inheritance of fruit characters in ash gourd [*Benincasa hispida* (Thunb.) Cogn.]. *Veg. Sci,.* 29: 113-15.

Marr, K.L., Y.M. Xia, and Bhattarai N.K. 2007. Allozymic, morphological, phenological, linguistic plant use, and nutritional data of *Benincasa hispida (Cucurbitaceae). Econ. Bot.,* 61: 44–59.

Meng, XD., YY. Wei, H. Ma, WH. Zhang and Li. J.R. 1996 Identification of Chinese wax gourd and chieh-qua cultivars using RAPD markers. *Acta Agric. Shanghai.,* 12: 45-49

Mohanty, B.K., and Mishra R.S., 2000. Studies on combining ability for flowering traits in pumpkin. *Haryana J. Hort. Sci.,* 29: 220-22.

Nayar, NM. and More. TA. 1998 Cucurbits. Oxford & IBH Publishing Co. Pvt. Ltd., New Delhi, p 340

Pandey, S., S, Kumar, U. Mishra, A. Rai, M. Singh and Rai. M. 2008. Genetic diversity in Indian ash gourd *(Benincasa hispida)* accessions as revealed by quantitative traits and RAPD markers. *Sci. Hort.,* 118: 80–86.

Paris, H.S., N. Yonash, V. Portnoy, N. Mozes-Daube, G. Tzuri and Katzir. N. 2003. Assessment of genetic relationships in *Cucurbita pepo* (Cucurbitaceae) using DNA markers, *Theoret. Appl. Genet.,* 106 : 971–978.

Peter, KV., P. Indira, and Rajivan. K. 1991 Ash gourd variability in Southern India. *South Indian Hort.,* 46:91–94

Qian, W., S. Ge and Hong. D.Y. 2001. Genetic variation within and among populations of a wild rice *Oryza granulata* from China detected by RAPD and ISSR markers, *Theoret. Appl. Genet.,* 102 : 440–449.

Ramesh, M., V. Gayathri, AVNA. Rao, MC. Prabhakar and Rao. CS. 1989. Pharmacological action of fruit juice of *Benincasa hispida. Fitoterapia,* 60:241–247

Reddy, M.P., N. Sarla and Siddiq. E.A. 2002. Inter-simple sequence repeat (ISSR) polymorphism and its application in plant breeding. *Euphytica,* 128 : 9–17.

Ritschel, P.S., T.C.D.L. Lins, R.L. Tristan, G.S.C. Buso, J.A. Buso and Ferreira. M.E. 2004. Development of microsatellite markers from an enriched genomic library for genetic analysis of melon (*Cucumis melo* L), *BMC Plant Biol.,* 4 : 9–23.

Robinson, RW. and DS. Decker-Walters. 1999. Cucurbits. CAB International. Wallingford, Oxford, (UK)

Rubatzky, V.E. and Yamaguchi. M. 1997. *World Vegetables.* Chapman and Hall, New York, USA, 843 p.

Sirohi, PS., AD. Munshi, K. Gunjeet and Behra. TK. 2005. Cucurbits. (In) Dhillon BS, Tyagi RK, Saxena S. and Randhawa GJ (Eds.) Plant genetic resources: horticultural crops. Narosa Publishing House, New Delhi, pp 34–38 site, Western Highlands Province, Papua Guinea. *Archaeol. Ocea.* 38: 186-91.

Sureja, A.K. 2003. Studies on heterosis and its relationship with molecular diversity in ash gourd *[Benincasa hispida* (Thunb.) Cogn.]. Ph D (Hort.) Thesis, IARI, New Delhi, India.

Sureja, A.K., P.S.Sirohi, T.K. Behera and Mohapatra. T. 2006. Molecular diversity and its relationship with hybrid performance and heterosis in ash gourd [*Benincasa hispida* (Thumb.) Cogn.]. *J. Hort. Sci. Biotech.*, 81: 33-38.

Verma, VK., TK. Behra, AD. Munshi, SK. Parida and Mohapatra. T. 2007. Genetic diversity of ash gourd *[Benincasa hispida* (Thumb.) Cogn.] inbred lines based on RAPD and ISSR markers and their hybrid performance. *Sci Hort.*, 113:231–237.

Verma, Veerendra and Behera, T.K. 2007. Heterosis studies in ash gourd [*Benincasa hispida* (Thunb.) Cogn.] for yield and related traits. *J. Tropical Agric.*, 45: 52-55.

Walters, T. W. and Decker-Walters, D. S. 1989. Systematic re-evaluation of *Benincasa hispida* (Cucurbitaceae). *Econ. Bot.*, 43: 274–278.

Zietkiewicz, E., A. Rafalski and Labuda. D. 1994. Genome fingerprinting by simple sequence repeat (SSR)-anchored polymerase chain reaction amplification, *Genomics*, 20 : 176–183.

15

Safe Management of Borers in Vegetables

R. Srinivasan

Borer pests cause significant yield losses and are one of the most serious production constraints in tropical vegetable production. For instance, eggplant fruit and shoot borer (EFSB), *Leucinodes orbonalis* Guenée (Lepidoptera: Pyralidae) reduces yield as much as 70% (Dhandapani *et al.*, 2003). Tomato fruit borer, *Helicoverpa armigera* Hubner (Lepidoptera: Noctuidae) causes about 5 to 55% loss in almost all the tomato producing areas in the world (Kashyap and Batra, 1987). Up to 80% yield losses have been reported in various vegetable and grain legumes due to legume pod borer (LPB), *Maruca vitrata* F. (Lepidoptera: Pyralidae) damage (Singh *et al.*, 1990; Sharma, 1998; Gressel *et al.*, 2004; Hammig *et al.*, 2008; Zahid *et al.*, 2008). Due to higher yield losses, growers tend to apply extremely high quantities of chemical pesticides to control borer pests on vegetable crops. Because the borer larvae are exposed on external plant parts only for a brief time after hatching, the window for effective pesticide application is very brief; vegetable farmers spray throughout the growing season, but without satisfactory results. A survey of pesticide use in Bangladesh indicated that farmers spray up to 180 times with chemical insecticides during a year to protect their eggplant crop against EFSB (SUSVEGAsia, 2007). Pesticide application often exceeded 50 sprays per tomato crop season in south India (Nagaraju *et al.*, 2002). In Bangladesh, country bean (*Lablab purpureus*) is being sprayed at weekly or biweekly intervals—sometimes every day—to control LPB (Hoque *et al.,* 2001). On average, farmers used 10-24 kg/ha of pesticides in cabbage and cauliflower in India to manage the diamondback moth, *Plutella xylostella* (Lepidoptera: Plutellidae) and other pests (Weinberger and Srinivasan, 2009).

Excessive pesticide use adversely affects human health and increases the cost of production, making vegetables expensive for consumers. For instance, the share of pesticides to total material input costs was 55% for eggplant, followed by tomato (31%) and cabbage (49%) in the Philippines (Orden *et al.,* 1994). It was 40-50% on eggplant in Bangladesh (SUSVEGAsia, 2007). It is unlikely that any single method of pest management such as chemical pesticides can achieve a level of borer pest control acceptable to vegetable producers. Therefore, integrated pest management strategies have been developed and are being promoted among growers in South and Southeast Asia. This chapter aims to compile simple, affordable, safe and environmentally sound control technologies for the management of key borer pests on tropical vegetable crops.

1. Eggplant Fruit and Shoot Borer, *Leucinodes Orbonalis* Guenée (Lepidoptera: Pyralidae)

Eggplant fruit and shoot borer is mostly monophagous, although sometimes it feeds on tomato, potato, *Solanum indicum* L., *S. xanthocarpum* **Schrad.** & Wendl., *S. torvum* Swartz., and *S. nigrum* L. (David 2001; Alam *et al.,* 2003). During the early vegetative phase of the crop, the larva feeds inside the tender shoots, which results in wilting of young shoots, followed by drying and drop-off. The pest prefers to feed on the fruit during the fruiting stage of the crop; damaged fruit exhibits boreholes on the surface that often are sealed with excreta, rendering the fruit unfit for marketing and consumption.

Management

1.1. Cultural practices

As EFSB is practically a monophagous insect on eggplant, discontinuation of eggplant cultivation in a community for a few seasons will significantly reduce populations of this pest. Eggplant seedlings should not be grown near fields with previous or existing crops, or near dried eggplant heaps (Alam *et al.,* 2003). If seedlings must be grown in those areas, the seedling beds should be covered with 30-mesh nylon net to prevent the entry of EFSB moths and thus egg laying.

1.2. Host plant resistance

Planting resistant or moderately resistant cultivars can deter EFSB. For instance, accessions or varieties such as EG058, Pusa Purple Long, Pusa Purple Cluster, Pusa Purple Round, Banaras Long Purple, Arka Kesav, Arka Kusmakar, Punjab Barsati, Punjab Chamkila, and Kalyanpur-2 have been reported to be tolerant or resistant (Parker *et al.*, 1995; Alam *et al.*, 2003; Shivalingaswamy and Satpathy 2007). Except for EG058, which is an AVRDC accession, most of the varieties noted originated in India.

1.3. Mechanical control

Prompt removal and destruction of infested shoots and fruit at regular intervals until final harvest significantly reduces EFSB damage (Alam *et al.*, 2003). However, growers sometimes leave the culled shoots and fruit in or around the field, where the discarded plant material may serve as a source for future infestation. Pruning and destroying infested material is most effective when a whole community in a particular region follows the practice, rather than only an individual grower. Pruning does not adversely affect the plant growth or yield (Srinivasan and Huang, 2009).

1.4. Biological control

Parasitoids such as *Trathala flavoorbitalis* (Cameron), *Eriborus sinicus* Holmgren, and *Pristomerus testaceus* Morley are commonly found occurring on EFSB larvae (AVRDC, 1996a; Alam *et al.*, 2003). Reduced use of synthetic pesticides enhances the activities of these natural enemies. In addition, weekly releases of egg parasitoid *Trichogramma chilonis* Ishii @ 1g parasitized eggs/ha/week, and larval parasitoid *Bracon hebetor* Say @ 800-1000 adults/ha/week are effective in reducing EFSB (Alam *et al.*, 2006).

EFSB sex pheromone lures in traps can be installed at the rate of 100 traps per hectare (Cork *et al.*, 2003) to reduce fruit damage and increase yield. Traps should be placed either at canopy level or at slightly above the canopy level for effective attraction (Alam *et al.*, 2003).

2. Tomato Fruit Borer, *Helicoverpa Armigera* Hub. (Lepidotera: Noctuidae)

The tomato fruit borer is a polyphagous and highly mobile insect. It is a serious pest, causing significant yield losses in agricultural and horticultural crops. It has been recorded as a damaging pest on 181 cultivated and wild plant species in at least 45 families (Venette *et al.*, 2003; Srivastava *et al.*, 2010). The older larvae prefer to feed on flowers and young fruits. The larvae make holes and feed by thrusting their heads inside. The holes are circular and often surrounded by fecal pellets.

2.1. Cultural practices

Crop rotation with non-host crops is effective in managing the tomato fruit borer, because the insects emerging from pupae in the soil in the present season (especially in locations where *H. armigera* diapause during winter) which pupated during the previous crop cycle pose a serious threat. African marigold (*Tagetes erecta* L.) can be planted as a trap crop to reduce the incidence of *H. armigera* (Srinivasan *et al.*,1994). It is important to synchronize transplanting of both crops so that flowering coincides, which attracts *H. armigera* female adults. Tropical soda apple (*Solanum viarum* Dunal) also can be used as an effective trap crop to manage *H. armigera* (AVRDC, 2000; 2001).

2.2. Host plant resistance

Although resistant tomato cultivars can reduce pest damage, commercial cultivars with appreciable levels of resistance are not yet available. Germplasm screening at AVRDC – The World Vegetable Center revealed the presence of high levels of fruit borer resistance only in the wild *Solanum* species, particularly *S. habrochaites* and *S. pennellii*. Efforts to introgress resistance from the wild species into cultivated tomato resulted in small-fruited resistant accessions (Talekar *et al.*, 2006).

2.3. Biological control

H. armigera sex pheromone traps can be used to monitor, mass-trap, or disrupt the mating of male moths. Although sex pheromone traps baited with *H. armigera* pheromone lures can be used to trap more males, this method is less effective for polyphagous insects like *H. armigera*. Polyphagous populations are always higher due to the availability of multiple host plants in the tropics and two population parameters *viz.*, density and polymorphism, are important factors that affect the trap catches of *H. armigera* males (Kumar and Shivakumara, 2003). Placing a high concentration of sex pheromone (all or one component of the multi-component pheromone) in a slow-release formulation on a 5- and 10-m grid in the field results in a drastic reduction in male moths, which adversely affects mating in *H. armigera* (AVRDC 1988).

Egg parasitoids (*T. pretiosum* Riley and *T. chilonis*) and larval parasitoids (*Campoletis chlorideae* Uchida) can be conserved and/or released in tomato fields at regular intervals to check the build-up of *H. armigera* (Ballal and Singh, 2003; Gupta *et al.*, 2004). In addition, commercially available biopesticides based on *Bacillus thuringiensis*, *Helicoverpa armigera* nucleopolyhedrovirus (HaNPV) and neem (*Azadirachta indica* A. Juss.) can be used against *H. armigera*.

2.4. Chemical control

Chemical pesticides are widely used against this pest and they are effective against the early larval stages, before the larvae enter into the floral buds or fruit. Pesticide spraying should be scheduled soon after noticing the eggs or during early larval stages. It is advisable to follow a proper pesticide rotation. Before application, it is important to check the effectiveness of chemical pesticides in the region and their registration status for tomato.

3. Legume Pod Borer, *Maruca vitrata* F. (Lepidoptera: Pyralidae)

Legume pod borer is considered the most serious pest of vegetable and grain legumes in tropical Asia and Africa (Sharma, 1998). LPB larvae feed on floral buds, flowers, and pods by webbing. They occasionally feed on the peduncle and stems (Taylor, 1967). A total of 39 host plants of LPB, including two non-leguminous hosts (Rathore and Lal, 1998), have been reported. Although two other species, *M. amboinalis* and *M. nigroapicalis*, were observed in the Indo-Malaysian and Tonkin area, *M. vitrata* is believed to be the predominant species causing severe economic losses in food legumes.

3.1. Host plant resistance

Although *M. vitrata* is an economically important pest on several food legume crops, it is quite important in mungbean, yard-long bean and cowpea. Only 12 mungbean accessions were identified as resistant to *M. vitrata* in a germplasm screening of more than 6000 accessions

(AVRDC, 1993a; 1994; 1996a). V 1649, V 1857, V 3274, V 3276 and V 3279 are some of the highly resistant accessions. Similarly, intensive research on *M. vitrata*-resistant cowpea has been carried out at the International Institute of Tropical Agriculture in Nigeria, where more than 8000 accessions were screened (Brink and Belay, 2006). Although a few *Vigna vexillata* accessions and the wild cowpea cultivar TVu 946 have good resistance to *M. vitrata*, the crosses between *V. vexillata* accessions and cultivated lines were not successful (Fatokun 2002), and the crosses of cultivated lines with TVu 946 had produced progeny with unacceptable agronomic characters and an inadequate level of resistance (Singh, 1985). Thus, there is virtually no cultivar available with demonstrable resistance against *M. vitrata* (Machuka, 2002).

3.2. Biological control

Sex pheromone compounds already have been identified in *M. vitrata*, consisting of one major compound and two minor compounds. When the major and minor compounds were formulated in a ratio of 100:5:5 for lures, they attracted large numbers of male moths in Benin and Ghana; however, the major compound alone attracted a significantly higher number of moths in Burkina Faso, compared to the lures based on major and minor compounds (Downham *et al.*, 2004). The major compound either alone or combined with (E)-10-hexadecenol in 90:10 ratio [(E)-10-hexadecenol was not reported from *M. vitrata*] attracted a significantly higher number of male moths in Andhra Pradesh, southern India (Hassan, 2007). Pheromone traps could be used against *M. vitrata*, but the traps need some refinement to be suitable to a specific local environment.

Although a substantial number of parasitoid species have been reported to attack *M. vitrata* in tropical Asia and Africa, a few selected candidates are quite promising. A strain of *Apanteles taragamae*, a solitary braconid endoparasitoid, parasitized as high as 63% of *M. vitrata* in Taiwan (Huang *et al.*, 2003). Another braconid wasp, *Bassus asper*, was the most prevalent parasitoid with parasitism rates up to 17.1%; it was found in all cropping periods in the Philippines (Ulrichs *et al.*, 2001). *Phanerotoma leucobasis*, the predominant braconid wasp, inflicted about 30% parasitism in Benin (Arodokoun *et al.*, 2006). Recently, *P. philippinensis* was found to be the most effective egg-larval parasitoid of *M. vitrata* in Thailand (Yule and Srinivasan, 2012). In addition, *Therophilus marucae* (Braconidae) was found to occur widely in Vietnam, Lao PDR and Taiwan (Srinivasan *et al.*, 2012). Hence, conserving and/ or releasing these candidates in food legume fields at regular intervals could contain the build-up of *M. vitrata*.

Biopesticide formulations could effectively reduce the incidences of *M. vitrata*. Several isolates of *Beauveria bassiana* and *Metarhizium anisopliae* were highly pathogenic to the eggs (achieving 89–100% mortality) and larvae (94–100% mortality) of *M. vitrata* (Ekesi *et al.*, 2002). *Maruca vitrata* multiple nucleopolyhedrovirus (MaviMNPV), discovered and characterized at AVRDC (Lee *et al.*, 2007) is effective in managing *M. vitrata* either alone or in combination with other biopesticides and natural enemies (Srinivasan *et al.*, 2009; Tamo *et al.*, 2010). Formulations based on *B. thuringiensis* subsp. *kurstaki* and *B. thuringiensis* subsp. *aizawai* were highly effective in reducing damage of *M. vitrata* on vegetable legumes (AVRDC, 1996a; 1997), as the larvae are highly susceptible to different *B. thuringiensis* toxins such as Cry1Ab and Cry1Ca (Srinivasan, 2008). Thus, different biopesticides could be used to manage *M. vitrata* on food legumes.

3.3. Chemical control

Chemical pesticides are widely used against *M. vitrata*. Although pesticides are effective against early larval stages especially before the larvae enter the floral buds or pods, indiscriminate use leads to the development of pest resistance (Ekesi *et al.*, 1999). A proper pesticide rotation must be followed. Before application, it is important to check the effectiveness of chemical pesticides in the region and their registration status for vegetable legumes.

4. Bean Flies, *Ophiomyia* spp. and *Melanagromyza sojae* (Diptera: Agromyzidae)

Three species of bean flies *viz.*, *Ophiomyia phaseoli*, *O. centrosematis* and *Melanagromyza sojae* are important pests of food legumes including soybean and mungbean in tropical and subtropical Asia. The larvae attack young plants, especially in the first three to four weeks after germination, and complete destruction of the crop is common. Swollen and cracked stems, wilted or dead seedlings are common symptoms of bean fly attacks; the leaves of older plants under attack may be yellow and stunted. Although they occur throughout the year, bean flies only cause severe damage in dry seasons, especially during the fall (Talekar and Chen, 1983).

4.1. Host plant resistance

AVRDC has identified sources of resistance to bean flies in selected vegetable legumes. About 9000 soybean accessions were tested against bean flies; however, only ten *Glycine soja* and two *Neonotonia wightii* accessions proved to be resistant to *M. sojae* (Chiang and Talekar, 1980; AVRDC, 1987). Two *G. max* accessions (PI 227687 and PI 171444) were found to be moderately resistant to *O. phaseoli* (Talekar and Tangkano, 1993). More than 3700 mungbean accessions were screened against bean flies; however, only three accessions (V 2396, V 3495 and V 4281) were consistently resistant to bean flies in Taiwan (Chiang and Talekar, 1980). Since V 4281 was the most resistant accession, it was crossed with a susceptible, but high yielding accession, V 2184. Five breeding lines from this cross, *viz.*, VC 2839-56, VC 2839-70, VC 2839-71, VC 2839-75 and VC 2839-89 in the F_8 generation were selected because of their higher yield potential and level of resistance comparable to V 4281 (Talekar, 1987). Sources of resistance to *O. phaseoli* exist in common bean and cowpea (Talekar, 1990). However, concerted efforts are required to develop agromyzid resistant vegetable legumes.

4.2. Cultural control

Various cultural practices such as ridging of young plants, planting of vegetable legumes after a green manure crop, crop rotation, and mulching with rice straw enhance plant growth and induce tolerance to bean fly damage. Late plantings should be avoided, since infestations of bean fly are heavier then (Parker *et al.*, 1995).

4.3. Biological control

About 26 eulophid and four braconid parasitoids were reported to attack the agromyzid bean flies (Konishi, 2004). However, these parasitoids may not play any significant role in checking the bean fly population because the concealed immature stages such as larvae and pupae of bean flies reduce the effectiveness of most natural enemies.

4.4. Chemical control

Treating seed with systemic chemical pesticides before sowing, or applying pesticides to the soil when seed is sown, can provide adequate control of bean flies for two to three weeks (Parker *et al.*, 1995; Rahaman and Prodhan, 2007). One or two additional foliar sprayings may be necessary to protect the crop.

5. Diamondback Moth, *Plutella xylostella* (Lepidoptera: Plutellidae)

Diamondback moth (DBM) is a cosmopolitan and destructive insect pest of vegetable brassicas. The larvae feed and create holes in the leaves; however, severe damage is caused when the larvae tunnel into the heads of crops like cabbage, sometimes causing almost 100% crop loss. Hence, pesticides still occupy the predominant role in managing DBM. Due to extensive and inappropriate pesticide use, DBM has developed resistance to almost every class of insecticide that has been used against it, including novel insecticides such as chlorantraniliprole and flubendiamide (Sukonthabhirom *et al.*, 2011) and biopesticides such as *B. thuringiensis* (Tabashnik *et al.*, 1990). A recent conservative estimate of total costs associated with DBM management is about US$4-5 billion (Zalucki *et al.*, 2012).

5.1. Cultural control

Trap cropping with Indian mustard (*Brassica juncea*) is an effective way of reducing DBM damage. Growing of two rows of mustard between every 25 rows of cabbage as a trap crop is suggested. One row of mustard should be sown 15 days before cabbage planting, so that mustard attains a height of about 6-8 cm and provides thick and bushy foliage to help attract early DBM arrivals. Mustard starts flowering about 40 days after sowing and stops producing new foliage. This results in onset of senescence and leaves become unsuitable for larval feeding. To maintain continuous foliage, a second sowing is recommended in the adjacent ridge on the 25th day after planting cabbage (Srinivasan and Krishnamoorthy 1992). This method is commonly known as sequential trap cropping (Shelton and Badenes-Pérez, 2006). The first and last row of plots also should be planted with the trap crop. The DBM colonize on the trap crop, averting damage on the main cabbage crop. The trap crop should be sprayed with chemical pesticide.

Yellow rocket, *Barbarea vulgaris* var. *arcuata* can be used as a dead-end trap crop for the DBM. Yellow rocket is highly attractive to DBM, but DBM offspring cannot survive on it. Dead-end trap crops do not require any pesticide treatment to prevent pest populations from moving onto the main crop. The dead-end trap crop should be planted in field borders where it can intercept DBM adults and thus reduce damage on the main crop (Shelton and Badenes-Pérez, 2006).

5.2. Mechanical control

Brassica seedling production can be carried out under 32-mesh nylon net to reduce DBM infestation early in the season. Growers can thus reduce the amount of chemical pesticides used at the beginning of the crop cycle, which will help to protect natural enemies. In peri-urban areas where the brassicas are grown on smaller farms, plots can be confined with 32-mesh nylon net barriers on all four sides using a 2-m high net barrier (AVRDC, 2000; 2001).

5.3. Biological control

Traps baited with female sex pheromone lures can be placed at the rate of 50 per hectare randomly in the field, and replaced twice during the season (AVRDC, 1991). A recent study has shown that sex pheromone traps installed at the rate of 30-45 traps/ha were effective for controlling DBM in the field, and reduced insecticide use by 30% per planting season in southwestern China (Zhao *et al.*, 2011). It must be noted that different DBM populations may react differently to the same sex pheromone lure (Feng *et al.*, 2011). Hence, the pheromone lures should be validated in a new region before using them as a pest management tool.

Releasing the egg parasitoid, *Trichogrammatoidea bactrae,* once a week for six weeks starting within a week after transplanting brassicas is a recommended method to manage DBM (AVRDC, 1991). Release of a larval parasitoid (*Diadegma semiclausum*) and the pupal parasitoid (*Diadromus collaris*) significantly reduced damage from DBM in highland areas. The larval parasitoid *Cotesia plutellae* is well-adapted to manage DBM in lowlands. Unlike the egg parasitoid, augmentative releases for the larval and pupal parasitoids at regular intervals are not necessary. These larval and pupal parasitoids already have been introduced into several countries in Asia (AVRDC, 1993b; 1996b).

Although DBM was the first insect to develop resistance to Bt toxins in open field conditions (Tabashnik *et al.*, 1998), diligent deployment of *B. thuringiensis* biopesticides still offers significant protection of brassica crops from DBM damage. For example, the toxins Cry1Ac, Cry1Aa and Cry1Ca were equally toxic to the DBM population in southern Taiwan (Srinivasan and Hsu, 2008). Cry1Ba2 was more toxic to larvae of *DBM* than Cry1Ca4 (Shelton *et al.*, 2009). Hence, for effective treatment, it is important to choose *B. thuringiensis* formulations based on the most sensitive Bt toxin. Because *B. thuringiensis* biopesticides are not harmful to DBM parasitoids, both these components synergistically reduce DBM damage. Strains of entomopathogenic fungi such as *Metarhizium anisopliae* were also effective against DBM (AVRDC, 1999). Improved formulations of *Beauveria bassiana* showed promising prolonged impact in suppressing DBM populations in field conditions (Ghosh *et al.*, 2011). Biological control is an integral component in the sustainable management of DBM.

5.4. Chemical control

Pesticide misuse can accelerate the development of resistance in pests such as DBM. Utmost care should be taken when selecting chemical pesticides to use against DBM. For instance, use of broad-spectrum chemical pesticides early in the season often disrupts the control exerted by natural enemies (Talekar and Shelton, 1993). For effective integration of chemical pesticides with other DBM pest management components, countries should have a national DBM integrated pest management program that involves various stakeholders in different states or provinces. For example, based on pesticide use patterns and pest population pressure, the year was divided into two relatively equal periods (windows) in Australia; different pesticides registered for DBM control on vegetable brassicas were allocated to one of the two windows (Baker, 2011). However, because of the differences in the crucifer pest complex, timing of the peak periods of DBM pressure, and consequent pesticide use patterns in different regions, three different regional versions of the strategy, each with different calendar dates for the two window periods were devised (Baker, 2011; Ridland and Endersby, 2011). The rotation and windows approach for pesticide application can effectively manage DBM on vegetable brassicas.

6. Cabbage Head Caterpillar, *Crocidolomia binotalis* (Lepidoptera: Pyralidae)

It is interesting to note that cabbage head caterpillar (*C. binotalis*, CHC) was the most destructive insect pest on vegetable brassicas in the Old World before the use of chemical pesticides (Smyth *et al.*, 2003a). However, CHC may resurge as a serious pest due to the lack of effective bio-control agents and pesticide resistance, especially when DBM is brought under reasonable control by a guild of parasitoids. The early instar larvae migrate toward the growing center of vegetable brassicas such as cabbage, where they conceal themselves in webbing. Subsequent feeding can damage the plant's apical meristem and thus destroy the entire crop (Smyth *et al.*, 2003b).

6.1. Cultural control

Trap cropping with Indian mustard is effective against CHC (Srinivasan and Krishnamoorthy 1992). Clean cultivation is recommended, because CHC can feed and develop on certain indigenous vegetables such as African cabbage (*Cleome gynandra*), or other *Cleome* spp. that often grow as weeds.

6.2. Mechanical control

Erecting nylon net barriers around vegetable brassica plots prevents the entry of CHC adults and thus reduces larval damage. A 2.3-m high nylon net should be attached to a 2-m high framework with a 30-cm lip extending outward and downward at an angle of 80-85° to the walls (AVRDC, 1999). However, a barrier net alone is not adequate to achieve satisfactory control and it should be integrated with other compatible pest management components.

6.3. Biological control

Although male CHC moths responded to sex pheromone lures in the laboratory, the lures did not attract the pests in the field in Taiwan (AVRDC, 1996a). However, traps baited with a 10:1 mixture of (Z)-11-hexadecenyl acetate and (Z)-9-tetradecenyl acetate attracted significantly high numbers of male moths in vegetable brassica fields in Indonesia (Adati *et al.*, 2007). It is possible that pheromone lures can be developed as a monitoring and/or mass-trapping tool in CHC management.

A generalist egg parasitoid, *T. chilonis*, was found to efficiently parasitize egg masses of CHC in Samoa (Uelese *et al.*, 2011). Although further detailed studies of CHC biology and ecology and its potential host range are required before further recommendations can be made, there exists evidence for the development of *T. chilonis* as a biological control agent of CHC. Although the polyphagous pentatomid bug, *Eocanthecona furcellata*, predates more than 70% of the third instar CHC larvae, the searching ability in the field decreases (AVRDC, 1996a). Because there is no species-specific efficient parasitoid available for CHC, the potential of *T. chilonis* and *E. furcellata* can be studied in detail in further experiments.

B. thuringiensis-based biopesticides are an effective tool against CHC. This pest is quite susceptible to most of the Cry1A toxins such as Cry1Aa, Cry1Ab, and Cry1Ac (Srinivasan and Hsu, 2008) and to *B. thuringiensis* subsp. *kurstaki*- based formulations (Ooi, 1980; Sastrosiswojo and Setiawati, 1992; Malathi and Sriramulu, 2000), in which the Cry1A toxins are the major components. A recent study has shown that CHC is susceptible to Cry1Ba and Cry1Ca (Shelton *et al.*, 2009). *B. thuringiensis* subsp. *aizawai*- based formulations, in which the Cry1C toxin is the major component, could also be used.

6.4. Chemical Control

Proper selection and judicious use of chemical pesticides targeting DBM should also control CHC on vegetable brassicas, as CHC also was found to be susceptible to newer pesticides such as emamectin benzoate, spinosad and indoxacarb (Kannan *et al.*, 2011). Chemical controls should be carefully monitored at the regional or national level to effectively contain DBM and secondary lepidopterans such as CHC.

7. Cabbage Webworm, *Hellula undalis* (Lepidoptera: Pyralidae)

The cabbage webworm (CWW) is a serious pest of vegetable brassicas because of its preferential feeding on the young terminal bud and the unopened leaves (Sivapragasam and Chua, 1997). Outbreaks with yield losses up to 100% have been reported in Asia, Africa and the Pacific (Kalbfleisch, 2006). CWW has been reported as a major pest in lowland production systems (Sivapragasam and Chua, 1997) and in summer (AVRDC, 1978).

7.1. Cultural control

Clean cultivation is recommended for CWW because it feeds and develops on certain weeds such as *Cleome rutidosperma* and *C. viscose*, on which infestations may reach up to 60%. Both species are widespread and play a possible key role as alternate hosts for CWW larvae during the off-season in South and Southeast Asia (Sivapragasam and Chua, 1997; Kalbfleisch, 2006).

7.2. Biological control

Wing traps baited with female sex pheromone (E,E-11,13-hexadecadienal @ 10 µg per trap) can be placed at a height of 0.5 m above ground at a distance of 15 m, and replaced every two to six weeks depending on the trap catches (Kalbfleisch, 2006).

Release of an egg parasitoid, *T. bactrae,* once in a week for five weeks starting within a week after transplanting the brassicas is recommended to manage CWW (Sivapragasam, 1996). In addition, a few larval parasitoids such as *Bassus* sp., *T. flavoorbitalis, Chelonus* sp. and *Phanerotoma* sp. have been reported to infest CWW in Malaysia. Although they were reported to play an insignificant role in regulating the CWW larvae (Sivapragasam and Chua, 1997), it is important to confirm the identity of the various species and to check whether they are species-specific parasitoids of CWW.

CWW was highly sensitive to Cry1Ca, but less susceptible to Cry1A toxins (Srinivasan and Hsu, 2008). Shelton *et al.*(2009) also confirmed that CWW was susceptible to Cry1Ba and Cry1Ca. This was further confirmed by the high susceptibility of CWW to *B. thuringiensis* subsp. *aizawai* based formulations (Srinivasan and Hsu, 2008), but not for the *B. thuringiensis* subsp. *kurstaki* based formulations in Taiwan (AVRDC, 1987). However, the latter was shown to be effective in the Philippines (Ulrichs and Mewis 2003) and India (Singh *et al.,* 2000). Hence, after local validation, *B. thuringiensis* formulations can be used to manage CWW. In addition, CWW can be controlled by deploying biopesticides that focus specifically on the shoot tips. Shoot-tip treatment using *B. thuringiensis* once in a week until head formation was highly effective (Sivapragasam and Aziz, 1992).

7.3. Chemical control

The chemical control strategy suggested for DBM and CHC should also contain CWW. However, an improved mode of pesticide application via brassica shoot treatment is suggested (Sivapragasam, 1996) as this method minimizes the adverse effects of chemical pesticides on the natural enemies present in brassica production systems.

8. Melon Fly, *Bactrocera eucurbitae* (Diptera: Tephritidae)

The melon fly is widely distributed in temperate, tropical and subtropical regions of the world. It has been recorded as a damaging pest on 81 host plants in several plant families, but predominantly in Cucurbitaceae. The extent of losses varies between 30 to 100%, depending on the plant species and the season (Dhillon *et al.,* 2005). Maggots feed inside the fruit, leading to distortion or rotting.

8.1. Host plant resistance

Although sources of resistance to melon fly have been identified in pumpkin, bottle gourd, sponge gourd, ridge gourd and round melon (Nath, 1966; Mahajan *et al.*, 1997), intensive screening was carried out in *Momordica* sp. (Pal *et al.*, 1984: Srinivasan, 1991; Thakur *et al.*, 1992, 1994 & 1996; Tewatia *et al.*, 1997; Dhillon *et al.*, 2005a; Gogi *et al.*, 2010). Resistance was not stable across locations. For instance, varieties such as Faisalabad-long and Col-II, reported as the most resistant in Pakistan (Gogi *et al.*, 2010) were highly susceptible to melon fly in Taiwan.

8.2. Cultural control

The collection and destruction of melon fly-infested fruit will reduce subsequent population build-up. If field sanitation is adopted at the community level, it will significantly reduce the melon fly population. Burying infested fruit 15-45 cm deep in the soil reduces the survival rate of melon fly maggots and prevents adult fly emergence (Klungness *et al.*, 2005; Kumar *et al.*, 2011).

8.3. Mechanical control

Bagging fruit soon after formation prevents adult female melon flies from laying eggs (Kumar *et al.*, 2011) and reduces subsequent damage. Akhtaruzzaman *et al.*(1999) suggested cucumber fruits should be bagged 3 days after anthesis, and the bags should be retained for 5 days for effective control.

8.4. Biological control

Kairomonal attractant cue-lure [4-(p-acetoxyphenyl)-2-butanone] traps are highly effective for monitoring and mass-trapping of *B. cucurbitae* in bitter gourd and other crops (Pawar *et al.*, 1991; Vargas *et al.*, 2000). Similarly, protein baits are highly attractive to female melon flies (Kumar *et al.*, 2011). Protein baits mainly consist of proteins and sugars, which are essential food for the flies to survive and reproductively mature. Protein bait lures and protein bait sprays are commercially available.

Although the larval – pupal parasitoid, *Psyttalia* (=*Opius*) *fletcheri* was reported to be the predominant parasitoid of melon fly in India (Srinivasan, 1994), the parasitism of melon fly by this parasitoid was less than 5% (Wong *et al.*, 1989; Nishida, 1963). An egg – pupal parasitoid, *Fopius arisanus,* is believed to have high potential in reducing the melon fly and it has been introduced into Hawaii and Mexico (Bautis,ta *et al.*, 2004; Zenil *et al.*, 2004).

8.5. Chemical control

Spraying chemical pesticides against melon fly is not advisable because of the ineffectiveness of pesticides as well as resistance issues. However, attractants can be combined with small quantities of pesticides to develop an 'attract and kill' system. For example, protein bait sprays mixed with chemical pesticides can be spot-sprayed on the roosting host plants of melon fly. Crops like maize, cassava, sorghum and castor should be planted around the main crop (bitter gourd, water melon, etc). These roosting host plants can be spot-sprayed with protein bait sprays once a week (or more often during the wet season) to kill the newly emerged flies before they mature (Vargas *et al.*, 2008; Mcquate, 2011). The combined application of cue lure and a toxicant is the most effective way to contain male flies (Vargas *et al.*, 2008).

Conclusions

Cultural practices such as crop rotation, clean cultivation and trap cropping are effective in reducing overall populations of borer pests. Although resistant sources have been identified, varieties with appreciable levels of pest resistance as well as acceptable horticultural traits are scarce. With the advent of biotechnological tools, breeding barriers may be overcome to develop borer resistant vegetable varieties. Although protected cultivation such as net-houses is a high-input technology, it could be adopted in peri-urban vegetable production systems where vegetable production is a highly profitable business due to readily available markets. Egg parasitoids such as *T. chilonis*, *T. pretiosum* or *T. bactrae* can be released against their target pests at regular intervals. Inoculative releases of parasitoids such as *D. semiclausum* and *T. marucae* can be made in newer regions where they are not present. Withholding of broad-spectrum chemical pesticides could augment the performance of generalist parasitoids such as *T. flavo-orbitalis* or *A. taragamae*. Commercially available biopesticides based on *B. thuringiensis*, nucleopolyhedrovirus, entomopathogenic fungi, and neem could be applied either alone or in combination to manage borer pests on vegetable crops. However, biopesticides should be applied when pests are in early larval stages, which often occur on the plant surface. Sex pheromone traps are highly useful as monitoring or mass-trapping tools in managing lepidopteran borer pests. Chemical pesticides can be integrated with other components of pest management in vegetable crops; however, it is important to choose selective pesticides, and follow a mode of action rotation with a windows approach to effectively curtail development of resistance. The effectiveness of chemical pesticides in the region and their registration status for the target crop should be checked before using.

References

[AVRDC] AVRDC – The World Vegetable Center. 1978. Cabbage webworm. In: Progress Report of the Asian Vegetable Research and Development Center, Shanhua, Taiwan. pp. 34-35.

[AVRDC] AVRDC – The World Vegetable Center. 1987. 1984 Progress Report. Asian Vegetable Research and Development Center, Shanhua, Taiwan. 480 p.

[AVRDC] AVRDC – The World Vegetable Center. 1988. 1986 progress report. Asian Vegetable Research and Development Center, Shanhua, Taiwan

[AVRDC] AVRDC – The World Vegetable Center. 1991. 1990 Progress Report. Asian Vegetable Research and Development Center, Shanhua, Taiwan

[AVRDC] AVRDC – The World Vegetable Center. 1993a. AVRDC 1992 Progress Report. Asian Vegetable Research and Development Center, Shanhua, Tainan, Taiwan (ROC).

[AVRDC] AVRDC – The World Vegetable Center. 1993b. Vegetable Research and Development in Southeast Asia: The AVNET Final Report. AVRDC, P.O. Box 205, Taipei 10099. 50 P.

[AVRDC] AVRDC – The World Vegetable Center. 1994. AVRDC 1993 Progress Report. Asian Vegetable Research and Development Center, Shanhua, Tainan, Taiwan (ROC).

[AVRDC] AVRDC – The World Vegetable Center. 1996a. AVRDC 1995 Report. Asian Vegetable Research and Development Center, Shanhua, Tainan, Taiwan (ROC).

[AVRDC] AVRDC – The World Vegetable Center. 1996b. Vegetable research networking in South Asia: SAVERNET Phase I final report; AVRDC, P.O. Box 42, Shanhua, Tainan, Taiwan 741, ROC. 76 p.

[AVRDC] AVRDC – The World Vegetable Center. 1997. AVRDC 1996 Report. Asian Vegetable Research and Development Center, Shanhua, Tainan, Taiwan. 172 p.

[AVRDC] AVRDC – The World Vegetable Center. 1999. AVRDC Report 1998. Asian Vegetable Research and Development Center, Shanhua, Tainan, Taiwan. vii+148 p.

[AVRDC] AVRDC – The World Vegetable Center, 2000. AVRDC Progress Report, 1999. Asian Vegetable Research and Development Center, Shanhua, Taiwan. pp.152

[AVRDC] AVRDC – The World Vegetable Center. 2001. AVRDC Report 2000. Asian Vegetable Research and Development Center, Shanhua, Tainan, Taiwan. vii + 152 pp.

Adati J, Toriumi W, Sudiarta P, Susila W, Sumiartha K, Suastika G and Purwanto H. 2007. Effect of synthetic sex pheromone traps on the cabbage cluster caterpillar, *Crocidolomia pavonana* and its host preference in the vegetable yield of Indonesia. Journal of ISSAAS [International Society for Southeast Asian Agricultural Sciences], 13(2): 105

Akhtaruzzaman M, Alam MZ, and Ali-Sardar MM. 1999. Suppressing fruit fly infestation by bagging cucumber at different days after anthesis. *Bangladesh Journal of Entomology* 9: 103-112.

Alam SN, Dutta NK, Ziaur Rahman AKM and Sarker MA. 2006a. Annual Report 2005-2006. Division of Entomology, BARI, Joydebpur, Gazipur, 86 pp.

Alam SN, Rashid MA, Rouf FMA, Jhala RC, Patel JR, Satpathy S, Shivalingaswamy TM, Rai S, Wahundeniya I, Cork A, Ammaranan C and Talekar NS. 2003. Development of an integrated pest management strategy for eggplant fruit and shoot borer in South Asia, Technical Bulletin TB28, AVRDC - The World Vegetable Center, Shanhua, Taiwan. 66p.

Arodokoun DY, Tamò M, Cloutier C and Brodeur J. 2006. Larval parasitoids occurring on *Maruca vitrata* Fabricius (Lepidoptera: Pyralidae) in Benin, West Africa. *Agriculture, Ecosystems and Environment* 113: 320–325.

Baker, GJ. 2011. Crucifer vegetable insecticide resistance management strategies and issues in Australia. pp. 229-235. In: Srinivasan R, Shelton AM, Collins HL, eds. 2011. Proceedings of the Sixth International Workshop on Management of the Diamondback Moth and Other Crucifer Insect Pests, 21 25 March 2011, Kasetsart University, Nakhon Pathom, Thailand. AVRDC - The World Vegetable Center, Publication No. 11 755. AVRDC - The World Vegetable Center, Taiwan. 321 p.

Ballal CR and Singh SP. 2003. The Effectiveness of *Trichogramma chilonis, Trichogramma pretiosum* and *Trichogramma brasiliense* (Hymenoptera: Trichogrammatidae) as Parasitoids of *Helicoverpa armigera* (Lepidoptera: Noctuidae) on Sunflower (*Helianthus annuus*) and Redgram (*Cajanus cajan*). Biocontrol Science and Technology, 13: 231-240

Bautista R, Harris E, Vargas R and Jang E. (2004). Parasitization of melon fly (Diptera: Tephritidae) by *Fopius arisanus* and *Psyttalia fletcheri* (Hymenoptera: Braconidae) and the effect of fruit substrates on host preference by parasitoids. *ARS-Research.* http://www.ars.usda.gov/research/publications/Publications.htm?seq_no_115=155470 (26 July 2004).

Brink M and Belay G (Editors). 2006. Plant Resources of Tropical Africa 1. Cereals and pulses. PROTA Foundation, Wageningen, Netherlands / Backhuys Publishers, Leiden, Netherlands / CTA, Wageningen, Netherlands. 298 pp.

Chiang HS and Talekar NS. 1980. Identification of sources of resistance to bean fly and two other agromyzid flies in soybean and mungbean. Journal of Economic Entomology, 73:197-199.

Cork A, Alam SN, Rouf FMA, and Talekar NS. 2003. Female pheromone of brinjal fruit and shoot borer, *Leucinodes orbonalis*: trap optimization and preliminary mass trapping trials. *Bulletin of Entomological Research,* 93: 107–113.

David BV. 2001. Elements of Economic Entomology (Revised and Enlarged Edition). Popular Book Depot, Chennai, India. 590 p.

Dhandapani N, Shelkar UR and Murugan M. 2003. Bio-intensive pest management in major vegetable crops: An Indian perspective. *Journal of Food, Agriculture and Environment,* 1(2): 330-339.

Dhillon MK, Singh R, Naresh JS and Sharma HC. 2005. The melon fruit fly, *Bactrocera cucurbitae*: A review of its biology and management. 16pp. *Journal of Insect Science* 5:40, available online: insectscience.org/5.40

Dhillon MK, Naresh JS, Ram Singh and Sharma NK. 2005a. Evaluation of bitter gourd (*Momordica charantia* L.) genotypes to melon fruit fly, *Bactrocera cucurbitae* (Coquillett). *Indian Journal of Plant Protection* 33: 55-59.

Downham MCA, Tamo M, Hall DR, Datinon B, Adetonah S and Farman, DI. 2004. Developing pheromone traps and lures for *Maruca vitrata* in Benin, West Africa. *Entomologia Experimentalis et Applicata.* 110: 151–158.

Ekesi S, Adamu RS and Maniania NK. 2002. Ovicidal activity of entomopathogenic hyphomycetes to the legume pod borer, *Maruca vitrata* and the pod sucking bug, *Clavigralla tomentosicollis. Crop Protection.* 21(7): 589–595.

Ekesi S. 1999. Insecticide resistance in field populations of the legume pod borer, *Maruca vitrata* Fabricius in Nigeria. *Int. J. Pest Manag.* 45: 57–59.

Fatokun CA. 2002. Breeding cowpea for resistance to insect pests: attempted 52 crosses between cowpea and *Vigna vexillata.* Pp. 52-61. In: Fatokun, C.A., S.A. Tarawali, B.B. Singh, P.M. Kormawa, and M. Tamò (editors). 2002. Challenges and opportunities for enhancing sustainable cowpea pro-duction. Proceedings of the World Cowpea Conference III held at the International Institute of Tropical Agriculture (IITA), Ibadan, Nigeria, 4–8 September 2000. IITA, Ibadan, Nigeria.

Feng X, Li ZY, Wu QJ, Chen AD, Wu YD, Hou YM, He YR, Li JH, Xie SH, Zhang JM, Wie F and Ma CS. 2011. Rising to the challenge: a national project on sustainable control of DBM in China. pp. 13-16. In: Srinivasan R, Shelton AM, Collins HL, eds. 2011. Proceedings of the Sixth International Workshop on Management of the Diamondback Moth and Other Crucifer Insect Pests, 21 25 March 2011, Kasetsart University, Nakhon Pathom, Thailand. AVRDC – The World Vegetable Center, Publication No. 11 755. AVRDC – The World Vegetable Center, Taiwan. 321 p.

Ghosh SK, Chaudhary M and Kumar P. 2011. Myco Jaal® – A novel formulation of *Beauveria bassiana* for managing diamondback moth (*Plutella xylostella*) in sub tropical and tropical crucifer production system. pp. 143-148. In: Srinivasan R, Shelton AM, Collins HL, eds. 2011. Proceedings of the Sixth International Workshop on Management of the Diamondback Moth and Other Crucifer Insect Pests, 21 25 March 2011, Kasetsart University, Nakhon Pathom, Thailand. AVRDC – The World Vegetable Center, Publication No. 11 755. AVRDC – The World Vegetable Center, Taiwan. 321 p.

Gogi MD, Ashfaq M, Arif MJ, Sarfraz RM and Nawab NN. 2010. Investigating phenotypic structures and allelochemical compounds of the fruits of *Momordica charantia* L. genotypes as sources of resistance against *Bactrocera cucurbitae* (Coquillett) (Diptera: Tephritidae). Crop Protection, 29: 884-890

Gressel J, Hanafi A, Head G, Marasas W, Obilana B, Ochanda J, Souissi T and Tzotzos G. 2004. Major heretofore intractable biotic constraints to African food security that may be amenable to novel biotechnological solutions. *Crop Protection.* 23(8): 661–689.

Gupta RK, Desh Raj and Nirmala Devi. 2004. Biological and Impact Assessment Studies on *Campoletis chlorideae* Uchida: A Promising Solitary Larval Endoparasitiod of *Helicoverpa armigera* (Hübner). *Journal of Asia Pacific Entomology.* 7: 239 – 247

Hammig MD, Shepard BM, Carner GR, Dilts R and Rauf A. 2008. Areawide pest management for non-rice food crops in Southeast Asia. p. 326-350. In: Koul, O., Cuperus, G. and Norman Elliott (Eds.), Areawide Pest Management: Theory and Implementation, CABI, UK. 480 p.

Hassan MN. 2007. Re-investigation of the female sex pheromone of the legume pod-borer, *Maruca vitrata* (Lepidoptera: Crambidae). PhD Dissertation submitted to University of Greenwich, UK. 244 p.

Hoque, M., Mollik, S.R., Nazimuddin, M., Alam, S. N., Khorsheduzzaman, M., Jasmine, H. S., Sultana, N.A., Rahman, M., Ahmed, B., Rajotte, E. and Luther, G. 2002. Survey of Bean Pod Borer (*Maruca testulalis*), Whitefly and Aphids and Their Natural Enemies on Country Bean (*Lablab purpureus*). Tenth Annual Report of the Integrated Pest Management Collaborative Research Support Program (IPM CRSP). http://www.oired.vt.edu/ipmcrsp/Publications/AnnualReports/2003/Bangladesh/8BangladeshY_10_ANNUAL_REPORT1.pdf. p. 123–124.

Huang CC, Peng WK and Talekar NS. 2003. Parasitoids and other natural enemies of *Maruca vitrata* feeding on *Sesbania cannabina* in Taiwan. *Biol. Control.* 48: 407–416.

Kalbfleisch S. 2006. Integrated Pest Management of *Hellula undalis* on Crucifers in Central Luzon, Philippines, with E,E-11,13-Hexadecadienal as Synthetic Sex Pheromone', unpublished dissertation, Technische Universitat Munchen.

Kannan M, Vijayaraghavan C, Jayaprakash SA and Uthamsamy S. 2011. Studies on the biology and toxicity of newer insecticide molecules on cabbage head caterpillar, *Crocidolomia binotalis* (Zeller) (Lepidoptera: Pyralidae) in India. pp. 26-32. In: Srinivasan R, Shelton AM, Collins HL, eds. 2011. Proceedings of the Sixth International Workshop on Management of the Diamondback Moth and Other Crucifer Insect Pests, 21 25 March 2011, Kasetsart University, Nakhon Pathom, Thailand. AVRDC - The World Vegetable Center, Publication No. 11 755. AVRDC - The World Vegetable Center, Taiwan. 321 p.

Kashyap RK and Batra BR. 1987. Influence of some crop management practices on the incidence of *Heliothis armigera* (Hubner) and yield of tomato (*Lycopersicon esculentum* Mill) in India. *Tropical Pest Management*, 33: 166-169

Klungness LM, Jang EB, Mau RFL, Vargas RI, Sugano JS and Fujitani E. 2005. New approaches to sanitation in a cropping system susceptible to tephritid fruit flies (Diptera: Tephritidae) in Hawaii. *Journal of Applied Science and Environmental Management* 9: 5-15.

Konishi K. 2004. An illustrated key to the species of hymenopterous parasitoids of leaf mining agromyzid pests. pp 40-56. *In* Biological Control of Agricultural Pests in Asia. Theory and Practice, 2004 Seminar JASS'04 Fukuoka, Japan 12-22 October 2004.

Kumar ARV and Shivakumara B. 2003. Variable response of male *Helicoverpa armigera* moth to sex pheromone blends: A case of behavioural polymorphism? *Current Science* 84: 705-709.

Kumar P, Alma Linda A, Ketelaar JW and Shanmugam V. 2011. Field Exercise Guides on Fruit Flies Integrated Pest Management. Pp. 63. http://ipm.ait.asia/wp-content/uploads/2010/11/Field_Excercise_Guide_on_Fruit_Fly_IPM-N.pdf

Lee ST, Srinivasan R, Wu YJ and Talekar NS. 2007. Occurrence and characterization of a nucleopolyhedrovirus from *Maruca vitrata* (Lepidoptera, Pyralidae) isolated in Taiwan. *BioControl.* 52: 801–819.

Machuka J. 2002. Potential role of transgenic approaches in the control of cowpea insect pests. In: *Challenges and Opportunities for Enhancing Sustainable Cowpea Production.* (Eds. CA Fatokun, SA Tarawali, BB Singh, PM Kormawa and M Tamò). International Institute of Tropical Agriculture. IITA, Ibadan, pp. 213-222

Mahajan V, Mukherjee SC and Shaw SS. 1997. Use resistant vegetable varieties: A best alternative to tackle diseases and insect pests. *Farmer and Parliament* 33: (7-8) : 29-30.

Malathi S and Sriramulu M. 2000. Laboratory Efficacy of Biotic Insecticides against Lepidopterous Pests Fed on Treated Cabbage Leaves. Shashpa, 7: 63-66.

Mcquate, G.T. 2011. Assessment of attractiveness of cassava as a roosting plant for melon fly, *Bactrocera cucurbitae*, and oriental fruit fly, *B. dorsalis*. *Journal of Insect Science*. 11:1536-2442.

Nagaraju N, Venkatesh HM, Warburton H, Muniyappa V, Chancellor TCB and Colvin J. 2002. Farmers' perceptions and practices for managing tomato leaf curl virus disease in southern India. *International Journal of Pest Management*, 48:333-338.

Nath P. 1966. Varietal resistance of gourds to the fruit fly. *Indian Journal of Horticulture* 23: 69-78.

Nishida T. 1963. *Zoogeographical and ecological studies of Dacus cucurbitae* (Diptera: Tephritidae) in India. Technical Bulletin No. 54. Hawaiian Agricultural Station, University of Hawaii, Hawaii.

Ooi PAC. 1980. The Pathogenicity of *Bacillus thuringiensis* for *Crocidolomia binotalis*. MAPPS Newsletter, 3: 4

Orden MEM, Patricio MG and Canoy VV. 1994. Extent of pesticide use in vegetable production in Nueva Ecija: Empirical evidence and policy implications. *Research and Development Highlights 1994*, Central Luzon State University, Republic of the Philippines. 196-213 PP.

Pal AB, Srinivasan K and Doijode SD. 1984. Sources of resistance to melon fruit fly in bitter gourd and possible mechanisms of resistance. *SABRAO Journal* 16: 57-69.

Parker BL, Talekar NS and Skinner M. 1995. Field guide: Insect pests of selected vegetables in tropical and subtropical Asia. Asian Vegetable Research and Development Center, Shanhua, Tainan, Taiwan, ROC. Publication no. 94-427. 170 p.

Pawar DB, Mote UN and Lawande KE. 1991. Monitoring of fruit fly population in bitter gourd crop with the help of lure trap. *Journal of Research, Maharashtra Agricultural Universities* 16: 281.

Rahaman MA and Prodhan MDH. 2007. Management of borboti bean fly (*Ophiomyia sp*) using different synthetic pesticides. *Int. J. Sustain. Crop Prod.* 2(3):5-8

Rathore YS and Lal SS. 1998. Phylogenetic relationship of host plants of *Maruca vitrata*. *Indian Journal of Pulses Research*, 11(2):152–155.

Ridland PM and Endersby NM. 2011. Some Australian populations of diamondback moth, *Plutella xylostella* (L.) show reduced susceptibility to fipronil. pp. 195-203. In: Srinivasan R, Shelton AM, Collins HL, eds. 2011. Proceedings of the Sixth International Workshop on Management of the Diamondback Moth and Other Crucifer Insect Pests, 21 25 March 2011, Kasetsart University, Nakhon Pathom, Thailand. AVRDC – The World Vegetable Center, Publication No. 11 755. AVRDC – The World Vegetable Center, Taiwan. 321 p.

Sastrosiswojo S and Setiawati W. 1992. Biology and Control of *Crocidolomia binotalis* in Indonesia. In: Management of Diamondback Moth and Other Crucifer Pests: Proceedings of the Second International Workshop, ed. N.S. Talekar, Shanhua, Taiwan: Asian Vegetable Research and Development Center, pp. 81-87.

Sharma HC. 1998. Bionomics, host plant resistance, and management of the legume pod borer, Maruca vitrata – a review. *Crop Protection*. 17: 373–386.

Shelton AM and Badenes-Pérez FR. 2006. Concepts and applications of trap cropping in pest management. *Annual Review of Entomology* 51: 285-309.

Shelton AM, Gujar GT, Chen M, Rauf A, Srinivasan R, Kalia V, Mittal A, Kumari A, Ramesh K, Borkakatti A, Zhao JZ, Endersby N, Russell D, Wu YD and Uijtewaal B. 2009. Assessing the susceptibility of cruciferous Lepidoptera to Cry1Ba2 and Cry1Ca4 for future transgenic cruciferous vegetables. *Journal of Economic Entomology*, 102(6): 2217-2223.

Shivalingaswamy TM and Satpathy, S. 2007. Integrated pest management in vegetable crops. In: Jain PC, Bhargava MC (eds.), Entomology: Novel Approaches, New India Publishing Agency, New Delhi, India. p. 353-375.

Singh IP, Seema K and Singh R. 2000. Bioefficacy of Insecticides against Cabbage Head Borer (*Hellula undalis* Fab.) and Dissipation of Chlorpyriphos in Cabbage. *Journal of Applied Biology*, 10: 194-198.

Singh SR, Jackai LEN, Dos Santos JHR and Adalla CB. 1990. Insect pests of cowpeas, pp. 43–90, *in* S.R. Singh (ed.). Insect Pests of Tropical Legumes. John Wiley & Sons, Chichester, United Kingdom.

Singh SR. 1985. Insects damaging cowpeas in Asia. In: Cowpea research, production and utilization, eds. Singh SR, Rachie KO, John Willey and Sons, Chichester, U.K. pp. 245-248.

Sivapragasam A and Aziz AM. 1992. Cabbage webworm on crucifers in Malaysia. In: Talekar NS. (ed.). Diamondback moth and other Crucifer pests: Proceedings of the Second International Workshop, Tainan, Taiwan. Asian Vegetable Research and Development Center, AVRDC Publ. No. 92-368. pp. 75-80.

Sivapragasam A and Chua TH. 1997. Preference for sites within plant by larvae of the cabbage webworm, *hellula undalis* (Fabr.) (Lep., Pyralidae). *Journal of Applied Entomology*, 121: 361-365.

Sivapragasam A. 1996. Integrated management of cabbage with special reference to the control of *Hellula undalis* (Fabr.). In: Hokyo N, Norton G. (eds.), Proc. Int. Workshop on the Pest Management Strategies in Asian Monsoon *Agroecosystems*. Pp. 223-234.

Smyth RR, Hoffmann MP and Shelton AM. 2003a. Effects of host plant phenology on oviposition preference of *Crocidolomia pavonana* (Lepidoptera: Pyralidae), *Environmental Entomology*, 32: 756-764.

Smyth RR, Hoffmann MP and Shelton AM. 2003b. Larval performance in relation to labile oviposition preference of *Crocidolomia pavonana* [F.] (Lepidoptera: Pyralidae) among phenological stages of cabbage. *Environmental Entomology*, 32: 765-770.

Srinivasan K, Krishna Moorthy PN and Raviprasad TN. 1994. African marigold as a trap crop for the management of the fruit borer, *Helicoverpa armigera* on tomato. *International Journal of Pest Management*, 40: 56-63.

Srinivasan K and Krishnamoorthy P. 1992. Development and adoption of integrated pest management for major pests of cabbage using Indian Mustard as trap crop. In: Management of Diamondback moth and other crucifer pests. Proceedings of the second International Workshop. (Ed.) Talekar, NS. Asian Vegetable Research and Development Center, Shanhua, Taiwan. 603p.

Srinivasan K. 1991. Pest management in cucurbits - An overview of work done under AICVIP. In: *Group Discussion of Entomologists Working in the Coordinated Projects of Horticultural Crops*, 28-29 January 1991, pp. 44-52. Central Institute of Horticulture for Northern Plains,Lucknow, Uttar Pradesh, India.

Srinivasan K. 1994. Recent trends in insect pest management in vegetable crops. In: Dhaliwal GS, Arora R editors. Trends in Agricultural Insect Pest Management, pp. 345-372. Commonwealth Publishers, New Delhi, India.

Srinivasan R and Hsu YC. 2008. Susceptibility of major lepidopterans to ä-endotoxins and a formulation of *Bacillus thuringiensis* (*B.t.*) on vegetable brassicas in Taiwan. *Biocontrol Science and Technology*. 18(9): 935 – 939

Srinivasan R and Huang CC. 2009. The effect of simulated borer infested shoot pruning on yield parameters of eggplant. *Journal of Asia Pacific Entomology*. 12: 41 – 43

Srinivasan R, Lin MY, Sopana Yule, Hsu YC, Hai VM, Khanh LD, Phommachan C and Kodsimouang S. 2012. Exploration and identification of parasitoids of legume pod borer, *Maruca vitrata* in Southeast Asia. In: Abstracts of International Conference on Entomology, February 17-19, 2012, Patiala, India.

Srinivasan R, Tamo M, Lee ST, Lin MY, Huang CC, and Hsu YC. 2009. Towards developing a biological control program for legume pod borer, *Maruca vitrata*. In: Sanjeev Gupta, M. Ali and B.B. Singh (eds.), Grain Legumes: Genetic Improvement, Management and Trade, Indian Society of Pulses Research and Development, Kanpur, India. pp.183-196.

Srinivasan R. 2008. Susceptibility of legume pod borer (LPB), *Maruca vitrata* to ä-endotoxins of *Bacillus thuringiensis* (Bt) in Taiwan. *Journal of Invertebrate Pathology*, 97: 79-81

Srivastava CP, Nitin Joshi and Trivedi TP. 2010. Forecasting of *Helicoverpa armigera* populations and impact of climate change. *The Indian Journal of Agricultural Sciences* 80(1): 3-10.

Sukonthabhirom S, Dumrongsak D, Jumroon S, Saroch T, Chaweng A and Tanaka T. 2011. Update on DBM diamide resistance from Thailand: causal factors and learnings. Pp. 190-194. In: Srinivasan R, Shelton AM and Collins HL, eds. 2011. Proceedings of the Sixth International Workshop on Management of the Diamondback Moth and Other Crucifer Insect Pests, 21 25 March 2011, Kasetsart University, Nakhon Pathom, Thailand. AVRDC – The World Vegetable Center, Publication No. 11 755. AVRDC – The World Vegetable Center, Taiwan. 321 p.

SUSVEG-Asia. 2007. SUSVEG-Asia Brinjal integrated pest management (IPM). http://susveg-asia.nri.org/susvegasiabrinjalipm4.html (Accessed on 19 September 2012)

Tabashnik BE, Cushing NL, Finson N and Johnson MW. 1990. Field development of resistance to *Bacillus thuringiensis* in diamondback moth (Lepidoptera: Plutellidae). *J. Econ. Entomol.*, 83: 1671-1676.

Tabashnik, BE, Liu YB, Malvar T, Heckel DG, Masson L and Ferre´J. 1998. Insect resistance to *Bacillus thuringiensis:* uniform or diverse? Philos. *Trans. R. Soc. Lond.* B 353: 1751-1756.

Talekar NS, Opena RT and Hanson P. 2006. *Helicoverpa armigera* management: a review of AVRDC's research on host plant resistance in tomato. *Crop Protection*, 25(5): 461-467.

Talekar NS and Shelton AM. 1993. Biology, ecology, and management of the diamondback moth. *Annu. Rev. Entomol.* 38: 275-301.

Talekar NS and Tengkano W. 1993. Mechanism of resistance to bean fly (Diptera: Agromyzidae) in soybean. *Journal of Economic Entomology*, 86: 981-985.

Talekar NS and Chen BS. 1983. Seasonality of Insect Pests of Soybean and Mungbean in Taiwan. *Journal of Economic Entomology*, 76: 34-37

Talekar NS. 1987. Host plant resistance to insects attacking soybean and mungbean in the tropics. *Insect Science and Its Application*, 8: 777-782.

Talekar NS. 1990. Agromyzid flies of food legumes in the tropics. Wiley Eastern Limited, New Delhi.

Tamò M, Srinivasan R, Dannon E, Agboton C, Datinon B, Dabire C, Baoua I, Ba M, Haruna B and Pittendrigh BR. 2010. Biological control: a major component for the long-term cowpea pest Management strategy. In: Boukar O, Coulibaly O, Fatokun C, Lopez K, Tamò M. (eds.). Enhancing cowpea value chains through research advances. Proceedings of the 5th World Cowpea Research Conference, 26 September - 1 October 2010 Saly, Senegal (in press).

Taylor TA. 1967. The bionomics of *Maruca testulalis* Gey. (Lepidoptera: Pyralidae), a major pest of cowpeas in Nigeria. *J. West Afr. Sci. Assoc.* 12: 111–129.

Tewatia AS, Dhankhar BS and Dhankhar SK. 1997. Growth and yield characteristics of melon fruit fly resistant and highly susceptible genotypes of bitter gourd- A note. *Haryana Journal of Horticultural Science* 25: 253-255.

Thakur JC, Khattra AS and Brar KS. 1992. Comparative resistance to fruit fly in bitter gourd. *Haryana Journal of Horticultural Science* 21: 285-288.

Thakur JC, Khattra AS and Brar KS. 1994. Stability analysis for economic traits and infestation of melon fruit fly (*Dacus cucurbitae*) in bitter gourd (*Momordica charantia*). *Indian Journal of Agricultural Sciences* 64: 378-381.

Thakur JC, Khattra AS and Brar KS. 1996. Correlation studies between economic traits, fruit fly infestation and yield in bitter gourd. *Punjab Vegetable Growers* 31: 37-40.

Uelese A. 2011. The potential of *Trichogramma chilonis* Ishii (Hymenoptera: Trichogrammatidae) as a biological control agent for *Crocidolomia pavonana* F. (Lepidoptera: Crambidae). pp. 50-51. In: AVRDC - The World Vegetable Center. 2011. Abstracts from the Sixth International Workshop on Management of the Diamondback Moth and Other Crucifer Insect Pests, 21-25 March 2011, Kasetsart Unversity, Nakhon Pathom, Thailand. AVRDC Publication No. 11-746. AVRDC - The World Vegetable Center, Taiwan. 101 p.

Ulrichs C, Mewis I, Schnitzler WH and Burleigh JR. 2001. Parasitoids of the bean podborer, *Maruca vitrata* F. (Lepidoptera: Pyraustinae), a pest of *Vigna sesquipedalis* in the Philippine lowlands. Mitteilungen der Deutschen Gesellschaft fur allgemeine und angewandte *Entomologie.* 13(1-6): 283–288.

Ulrichs C and Mewis I. 2003. Evaluation and Risk-Assessment of Bio-Rational Pesticides for Controlling Crucifer Pests in Urban Horticulture in the Philippines. In: Balder H, Strauch KH, Backhaus GF. (eds.), Proceedings of the Second International Symposium on Plant Health in Urban Horticulture. pp. 211-221.

Vargas RI, Stark JD, Kido MH, Ketter HM and Whitehand LC. 2000. Methyl eugenol and cue-lure traps for suppression of male oriental fruit flies and melon flies (Diptera: Tephritidae) in Hawaii: effects of lure mixtures and weathering. *J Econ Entomol.* 93(1):81-87.

Vargas RI. Mau RFL, Jang EB, Faust RM and Wong L. 2008. The Hawaii Fruit Fly Area-wide Pest Management Programme. *Publications from USDA-ARS / UNL Faculty.* Paper 656. http://digitalcommons.unl.edu/usdaarsfacpub/656

Venette RC, Davis EE, Zaspel J, Heisler H and Larson M. 2003. Mini Risk Assessment: Old World bollworm, *Helicoverpa armigera* Hübner [Lepidoptera: Noctuidae]. University of Minnesota, St. Paul, MN 55108, USA. http://www.aphis.usda.gov/plant_health/plant_pest_info/pest_detection/downloads/pra/harmigerapra.pdf

Weinberger K and Srinivasan R. 2009. Farmers' management of cabbage and cauliflower pests in India and their approaches to crop protection. *Journal of Asia Pacific Entomology.* 12: 253 - 259

Wong TTY, Cunningham RT, Mcinnis DO and Gilmore JE. 1989. Seasonal distribution and abundance of *Dacus cucurbitae* (Diptera: Tephritidae) in Rota, Commonwealth of the Mariana Islands. *Environmental Entomology* 18: 1079-1082.

Yule S and Srinivasan R. 2012. *Phanerotoma philippinensis*, a potential parasitoid for managing legume pod borer, *Maruca vitrata*. In: Abstracts of XXIV International Congress of Entomology, August 19-25, 2012, Daegu, South Korea.

Zahid MA, Islam MM and Begum MR. 2008. Determination of economic injury levels of *Maruca vitrata* in mungbean. *J. Agric. Rural Dev.* 6(1&2): 91–97.

Zalucki MP, Shabbir A, Silva R, Adamson D, Liu SS and Furlong MJ. 2012. Estimating the economic cost of one of the World's major insect pests, *Plutella xylostella* (Lepidoptera: Plutellidae): Just how long is a piece of string? *Journal of Economic Entomology*, 105(4):1115-1129.

Zenil M, Liedo P, Williams T, Valle J, Cancino J and Montoya P. 2004. Reproductive biology of *Fopius arisanus* (Hymenoptera: Braconidae) on *Ceratitis capitata* and *Anastrepha* spp. (Diptera: Tephritidae). *Biological Control* 29 (2004) 169–178.

Zhao XQ, Li XY, Yin YQ and Chen AD. 2011. Occurrence and control of Plutella xylostella (Lepidoptera: Plutellidae) in Yunnan, China. pp. 273-277. In: Srinivasan R, Shelton AM and Collins HL, eds. 2011. Proceedings of the Sixth International Workshop on Management of the Diamondback Moth and Other Crucifer Insect Pests, 21 25 March 2011, Kasetsart University, Nakhon Pathom, Thailand. AVRDC – The World Vegetable Center, Publication No. 11 755. AVRDC – The World Vegetable Center, Taiwan. 321 p.

16

Mode and Mechanism of Pollination in Horticultural Crops

Vanaja, T.

Horticulture includes fruits, vegetables, floriculture, spices, tubers, plantation crops, medicinal and aromatic plants and mushrooms. The sector has established its importance in improving land use, promoting crop diversification, generating employment and above all providing nutritional security to the people. Even though India is the second largest producer of horticulture produces in the world, our per capita consumption of fruits and vegetables are very low.

Pollination

Flowers need to be pollinated in order to fertilize the ovule(s). The fertilized ovules become the seeds; they produce hormones which ensure that the flower sets fruit and that the fruit grows. In a few cases, fruits grow even without the stimulus from the seeds, so-called parthenocarpic fruits. Examples include pineapple, banana, and some cultivars of other fruits, e.g. seedless guava and mandarin. If the pollen come from the same flower or from a flower on the same tree this is called self-pollination. If the pollen come from a tree with a different genetic make-up, this is called cross-pollination. Nowadays many fruit crops are cloned. Since all the plants in a clone have the same genetic make-up, pollination within a clone is self pollination. Continued self-pollinations in successive generations lead to inbreeding, meaning a narrowing of genetic variation. As a rule this leads to reduced vigour and fitness. This may be the reason why in many plants, form or function of the flowers makes self-pollination difficult or impossible, thus favouring cross-pollination. One arrangement is that the pistil is not receptive at the time that the stamens of that flower shed their pollen, another that the stamens are too short for the anthers to shed the pollen on the stigma. Self-pollination is impossible if the genetic make-up of the pollen is not compatible with that of the pistil, so that the pollen tube cannot grow down the style towards the ovules (e.g. pineapple).Apart from these adaptations to limit self-pollination, plants generally produce more fruit following cross-pollination. If you grow a single clone of a fruit crop, nearby seedling trees of that species may provide the pollen to cross-pollinate the flowers on trees. It is generally safer to grow a few clones together in your orchard.

A good pollinator has viable pollen that is genetically compatible with the cultivar to be pollinated and flowers at the same time. Unfortunately this information is not available for most tropical fruit crops. Therefore it may be wise to plant at least three cultivars together and to try to make sure that the flowering periods of these cultivars overlap. Pollen may be transported by wind, but to carry the pollen from tree to tree most tropical fruit crops rely on insects (honey bees), a few also on birds (humming birds for pineapple in South America), or bats (baobab, durian). The flowering habit and pollen transfer in some minor fruit crops have not yet been described. Flowering and pollination of several important fruit crops – such as avocado, rambutan, oriental persimmon – are too complex to be explained.

Dioecy presents a serious problem to the grower, especially where the crop is grown from seed, as in papaya, salak, rambai,Amazon tree grape, butter fruit tree and oyster nut. Half the seedlings will be male, but we cannot identify them until the trees flower, at which time male flowers can be removed, leaving only 1 or 2 out of 20 to pollinate the female flowers. The only solution is to adopt clonal propagation: plant a female clone and interplant with a compatible male clone in 5 – 10 % of the tree positions. It is clear that cross-pollination is an important issue in fruit growing. The grower should plant alternate rows of different cultivars or, if the pollen come from male trees, interplant these trees in the rows. A windscreen may be needed to create shelter for the insects during flowering season. A beekeeper can assist by placing hives in the orchard. In some cases the grower has to lend a hand;hand pollination is often practiced on custard apple, passion fruit and salak.

Significance of Pollination

The mode of reproduction, which in turn is the system of pollination and fertilization, becomes important in classical or traditional breeding, because it determines a crop's genetic constitution. Genetic composition is the deciding factor to develop suitable breeding and selection methods, and it is also essential for artificial manipulation of crop to breed improved types (Dabholkar, 2006).

Mode and Mechanism of Pollination in Major Horticulture Crops

Flowers borne singly or in small inflorescences (as in soursop, sugarapple,guava) are usually perfect, but inflorescences with many flowers (like those in mango and rambutan) often contain imperfect flowers as well. Strict segregation of the sexes is found in papaya for instance: male and female inflorescences are on separate plants. This segregation of the sexes on separate trees, which prevents self pollination, is called dioecy. Jackfruit and breadfruit are examples of male and female inflorescences borne on the same tree (monoecy).In banana and coconut, flowers of both sexes are found in the same inflorescence, the staminate flowers at the tip and the pistillate flowers at the base. The large mango inflorescence bears both staminate and perfect flowers. Flowering habit not only varies depending on the cultivar, it also varies from year to year. The percentage pistillate flowers in coconut or that of perfect flowers in mango may be much higher in one year than in another, generally it is higher when growing conditions are favourable. Growing conditions also influence flower quality. A healthy tree of moderate vigour bears strong flowers, the stigma being receptive for pollination over a longer period of time to improve fruit set. Mode and mechanism of pollination of major horticultural crops ,representative of various regions are explained in detail.

Aloe

The flowers of aloe show protandry. Major floral visitors are birds, honey bees and ants observed in plant population and sunbird (*Necterenia asiatica* var. *asiatica*) plays a major role in pollen-transfer. Pollen-grains remained viable even after 76 hours of storage in humidity chamber at both room temperature and refrigerated condition. Pollination experiments showed that selfing, crossing and open-pollination occur in *A. barbadensis*.

Anthurium

Cultivated Anthurium species are protogynous with gynoecium maturing first(Bhattacharya, 2006) from the base to the top, hence out breeding is the nature. Protogyny prevents self fertilization as the stigmatic surface becomes receptive 7-10 days before the pollen is shed. Anthesis period is between 8 to 10 am as they are favoured by moderately low temperature and high relative humidity. Even though a large number of pollen grains are produced per anther, pollen fertility is low.

Apple

Apples grow from the flowers which bloom on an apple tree in the spring. For a flower to develop into an apple, the pollen that is produced by one flower must be transferred to another flower. Cross-pollination occurs when pollen is moved from the flowers on one apple tree to the flowers on another. This most often happens when bees come to the apple flowers to collect nectar and pollen and fly from one tree to another. Though an apple tree may produce some fruit without cross-pollination, it will yield a lot more fruit if cross-pollination occur. This is why an orchard can't be made up of only one apple variety. At least two different kinds of apple trees are needed for cross pollination. Some apple trees bloom much earlier than others. Usually, apples which produce ripe fruit early in the season are also early bloomers while those that fruit later in the season also bloom later. It is helpful for cross-pollination to plant apple varieties which bloom at the same time near each other in an orchard. Also, pollination is helped by not planting two trees of the same variety next to each other, instead, alternate varieties in rows.

Apricot, Peach, Nectarine and Sour Cherry

All are self pollinated. But the J.H.Hale peach is not self fruitful but can be pollinated by almost any other peach variety with a synchronous bloom period except Elberta. Self unfruitful varieties of Apricot include Perfection, Riland, and Rival. These self unfruitful varieties can be pollinated by any other apricot variety.

Banana

In banana, several flower bracts replace leaves at the transition of vegetative to the floral stage, and an inflorescence comprising a rachis and associated inflorescence bud comprised of tightly overlapping sets of bracts separating whorls of flowers (hands) emerge. On the date of rachis and inflorescence emergence, pollination barrier procedures were performed involving bagging of the inflorescence buds, isolating them from natural pollinators, while anthesis of successive flowers occurred along with intermodal growth of the rachis, with whorls of female flowers emerging between overlapping bracts derived from the inflorescence bud.

Bhindi

The flower structure in bhindi is homogamous in which both anthers and stigma mature simultaneously resulting in self pollination. Stigma is the most receptive on the day of flower opening, anther dehiscence, and dusting of the stigma with pollen being almost simultaneous. Protogyny has also been reported which is mainly responsible for allogamy in bhindi. Cross pollination by insects is also possible due to it's showy corolla and protogyny. The extent of cross pollination depends upon cultivars, competitive flora, insect population and season(Peter, 1998).

Black Pepper

In the cultivated black pepper, the plants are mostly gynomonoecious (*i.e.,* bearing female and bisexual flowers) or trimonoecious (*i.e.,* bearing female, male and bisexual flowers) (Krishnamurthi, 1969). Bisexual flowers are protogynous and stigmas are exerted 3-8 days before anther dehiscence (Martin and Gregory, 1962). Flowers open between 17.30 and 02.30 hrs with peak opening during 19.30-20.30 hrs (Kanakamany, 1982). Stigmas remain receptive for ten days with peak receptivity 3-5 days after exertion (Martin and Gregory, 1962). Nambiar *et al.,* (1978) reported that distal flowers in inflorescence lost their receptivity in a fewer days ,indicated by a viscous condition. Since pepper is mostly dioecious in wild state, cross pollination is essential for sexual reproduction in such types. Cultivated pepper, however, has mostly bisexual flowers fertilized by self pollination. Cultivated plants have been selected for genetic constancy and reliable fertility since the beginning of domestication(Frankel and Galun, 1977). Kinds of pollination, which take place in pepper are, within the same flower, between flowers in an inflorescence, and between flowers in separate inflorescence (Ibrahim *et al.,* 1985).

Brinjal

Flowers are solitary or borne in clusters of 2-5 flowered cyme. Flowers are usually bisexual and perfect. Anthesis starts in the morning by 7 and continues till noon. Stigma becomes receptive after two hours of flower opening. Anther dehiscence is not uniform. Normally it is after two hours of flower opening. Under field conditions stigma receptivity and pollen viability last for 2-3 days. Self pollination is the mode of reproduction, but natural cross pollination up to 15% may occur. The adaptations for self pollination are bisexual nature of flowers, nearness of anthers and stigma, simultaneous maturity of androecium and gynoecium, long styled nature and flower held downwards. Adaptations for cross pollination are; delay in anthesis, closed anther condition, prolonged viability of pollen grains and receptivity of stigma.

Cassava

Inflorescences are borne as axillary recemes near the tip of branches, with unisexual flowers. Male flowers are borne towards tip of the inflorescence. The actual blooming of flowers is from 10 to 12 noon depending on the environmental conditions. Female flowers open first (protogynous) and this phase continues for 3-5 days. After that there is an interphase of 1-2 weeks. These male phase continues for about a week. Mode of pollination is cross pollination, predominantly entomophilous. Adaptations for cross pollination are unisexual nature of flowers, protogynous condition, distinct interphase, presence of secretary glands and pollen grains heavy and sticky. The female flowers can be opened fully, highly branched sticky stigma with one week viable pollen grains.

Citrus

Citrus combines an unusual mixture of reproductive characteristics including non-climacteric development. In contrast to climacteric fruits (Giovannoni,2004), the mechanism of ripening in non-climacteric fruits is basically unknown. Citrus also exhibits a long juvenility phase and nearly all important citrus species including sweet oranges, mandarins, lemons and grapefruits show some degree of apomixis. Furthermore, many of them are parthenocarpic, sterile or self-incompatible and/or develop defective pollen (Baldwin, 1993; Davies and Albrigo, 1994). In seeded citrus cultivars, fruit development is linked to presence of seeds and, therefore, it depends upon pollination and fertilization. Self-pollination usually takes place in the unopened or opening flower, often allowing pollination before anthesis. Cross-pollination occurs between plants of different genetic background by insect transport of pollen. However, many current citrus cultivars are mostly seedless varieties showing high parthenocarpy, in many instances due to gametic sterility. Generative sterility can be relative or absolute. Relative gametic sterility may be due to self-incompatibility as in Clementine and to cross incompatibility. On the other hand, absolute gametic sterility is associated with pollen and/or embryo-sac sterility.

Citrus species show a relatively long juvenility period of two to five years before the trees reach the mature stage to produce flowers. The inflorescence developed in citrus may be either leafless or leafy and these may carry a single flower or several of them (Goldschmidt and Huberman, 1974). The ratio of each kind of inflorescence in the tree varies with flowering intensity and cultivar. Citrus species usually produce a large number of flowers over the years. In subtropical regions, citrus bloom occurs during the spring flush along with the vegetative sprouting. Under these environmental conditions, flowering takes place after a period of bud quiescence and the exposure to the low temperatures and short days of winter. Generally, summer and fall flushes are less intense and produce almost exclusively vegetative shoots. The importance of temperature as a major factor of flower induction is well established and recognized for a long time (Moss, 1969., Nebauer *et al.*, 2006). Temperatures under 20°C are demonstrated to contribute to flower bud induction in a time dependent manner (Moss, 1969; Southwik and Davenport, 1986; García-Luis *et al.*, 1992).In tropical climates, however, bud sprouting and flowering come about without interruption throughout the year although the main bloom still occurs during the spring (Monselise, 1985; Spiegel-Roy and Goldschmidt, 1996). In contrast to the sub-tropical stimuli, in tropical conditions, citrus apparently flowers in response to drought periods. In addition to low temperature, water deficit has also been recognized for a long time as another strong inductor of flowering in citrus (Cassin *et al.*, 1969). Moreover, water deficit has been proved to increase the ratio of floral shoots and the total number of flowers (Southwick and Davenport, 1986). There are many instances suggesting that the flowering response to inductive conditions is also influenced by endogenous factors. For example, citrus buds on previous summer shoots and buds at apical positions produce more flowers than older or lateral buds (Valiente and Albrigo, 2004). It is also well known that the fruit load has a strong negative effect on spring sprouting of both vegetative and generative buds and therefore constitutes a major inhibitor of flowering.

Conversely, the absence of fruit or a scarce fruiting induces huge flowering intensities in the next season. Thus, crop load is likely the main cause of the "alternate bearing" behavior of many citrus species and varieties including many seeded mandarins with alternate reduced flowering and fruiting ("off year") with increased flower induction and fruit production ("on year").

In general, fruit set also depends on successful pollination and fertilization since the presence of fertilized ovules normally triggers fruit development. In seeded citrus, the decision to initiate fruit development certainly requires pollination and fertilization. If the flower is not pollinated, the development of the gynoecium arrests, the whole flower senesces and eventually abscises. In the seeded sweet orange cultivars of 'Pineapple' for example, lack of fertilization will inevitably lead to abscission of the ovary since all emasculated and non-pollinated flowers arrested growth and fell shortly after anthesis (Ben-Cheikh *et al.*, 1997).

Coconut

Plant is monoecious with male and female flowers in the same inflorescence. The female flower is much larger than male flower. Flowering occurs continuously. Coconut palms are believed to be largely cross-pollinated, although some dwarf varieties are self-pollinating. The male phase lasts 5-7 days and usually begins 22 days after the opening of spathe(inflorescence). Usually the phase does not overlap so that the chance of intra spadix pollination is negligible.

Cow pea

Inflorescence is an axillary receme with a few flowers crowded near the tip of the peduncle in alternate pairs on nodes. The flowers bloom early in the morning, 7-9 am, close before noon and fall by evening. The anther dehiscence takes place between 10 pm to 12 midnight. It takes 11- 14 days for a flower to develop and bloom. The flower cycle in cowpea is for 3 weeks. Mode of pollination is self pollination. Adaptations for self pollination are bisexual nature of flowers, close proximity of anthers and stigma, simultaneous maturity of anthers and stigma, and cliestogamous condition.

Grape

Flowers are small (0.4cm), indiscrete, and green, borne in racemose panicles, opposite leaves at the base of current season's growth. The calyptra, or cap is the corolla, in which the petals are fused at the apex;

it abscises at the base of the flower and pops off at anthesis. Species in *Euvitis* may have 100+ flowers per cluster, whereas muscadine grapes have only 10-30. Vinifera and Concord grapes are perfect-flowered and self-fruitful, whereas some muscadine cultivars have only pistillate flowers. Flowering in grapes occur at the basal nodes of current season's growth in all species (left). Perfect flowered muscadines have longer stamens, while the pistillate flowers have short, reflexed stamens. Most grapes are self-fruitful and do not require pollinizers; however, pistillate muscadines (e.g., 'Fry', 'Higgins', 'Jumbo') must be interplanted with perfect-flowered cultivars for pollination. Pollination is accomplished by wind, and to a lesser extent insects.

Mango

Inflorescence is a terminal panicle with 1000-6000 male and bisexual flowers depending up on the variety and climatic factors. Flowering duration is usually of short(2 to 3 weeks). Flowers open in the early morning from 4 - 7am (maximum from 9.30 to 10.30 am). Dehiscence of anthers takes place at 11.30 am and it continues up to 3.45pm with sticky pollen. Stigma becomes receptive even 18 hours before flower opening. Due to self incompatibility, the mode of pollination is cross pollination (entomophily).Nectar is present to attract the insects.

Melon

Flowers are numerous and borne in leaf axils, male and female flowers are borne on the same plant. Male flowers appear first and female flowers appear 3-7 days after. Female flowers are borne singly and male flowers may be in clusters of 3-5. The ratio of female to male flowers varies with cultivars. The number of male flowers always exceeds that of female flowers. The flower opening takes place in the morning hours usually 6 to 10 am. This varies with species and climate. In a plant, male flowers start opening 3-7 days earlier than female flowers. Cross pollination driven by insects is the common method of pollination. Unisexual flowers, attractive colour of flowers, presence of nectar, sticky anthers and appearance of male and female flowers on different leaf axils are the adaptations for cross pollination.

Papaya

Papaya is a polygamous plant and has many sex forms. There are 3 basic sex types - staminate or male, hermaphrodite and pistillate or female. Of these, only female is stable whereas flowers of hermaphrodite and male vary in sex expression under different environmental conditions. Some papaya bear only short stalked female flower or bisexual flowers, while others may bear only male flowers, clustered on panicles 1.5-1.8 m long. Some papaya may have both male and female flowers. The flowers of papaya are usually white and small. Pollination of papaya is done by wind.

Pear

Most pear varieties are self incompatible. But all pears are suitable pollinizers for other varieties which bloom on the same day. One exception is Seckel variety which is not a good pollinizer. Eventhough Anjou, Bartlett and Kieffer are partially self fruitful, they should be cross pollinated to produce heavy and regular crops. As Pear flowers produce only a small amount of nectar which is low in sugar, more pollinizers and bees are needed than any other tree fruit.

Plum

Most European plums are cross pollinated, but European plums like Stanley and Damson are self pollinated. Japanese plums such as Santa Rosa, Burbank, Redheart, Shiro, Mithley, and Osark premier require pollination from another Japanese or an American -Japanese hybrid.

Pomegranate

Normal flowering of pomegranate varieties occurs, in general, between March-April and July-August. It continues for up to 10-12 weeks or more depending on variety and geographical situation.The period of full bloom lasts about one month, and it was observed that flowering and fruit set occur in about 3 or 4 distinct waves (Ben Arie *et al.*,1984; Hussein *et al* 1994; El Sese, 1988).Pomegranate is considered as a monoecious species and develops (on the same tree) two kinds of flowers: (i) male flowers (sterile) with short styles and atrophied ovaries containing a few eggs, which are generally, "bell-shaped"; and (ii) hermaphrodite flowers (fertile = perfect) with normal ovary developing to fruit, which are, in general, "vase-shaped" (Shulman *et al.*, 1984; Chaudhari and Desai, 1993). The percentage of male flowers is, in general, very important (more than 60-70%) depending on varieties and seasons. The male types drop and rarely set fruits leaving the hermaphrodite type to produce majority of the crop. A positive correlation was found between the bearing capacity and the percentage of perfect flowers (El Sese, 1988; Chaudhari and Desai, 1993).In some situations, perfect flowers are predominant at the beginning of the season, followed by the development of unfruitful flowers. Almost all of the normal flowers are produced between the onset of flowering and the end of full bloom (Shulman *et al.*, 1984; El Sese, 1988; Husseine *et al.*, 1994; Abou- Aziz *et al.*, 1995). For other cultivars and in other situations, early in the season, male flowers are predominant, while later, flowers are mostly perfect. Male flower production peaks earlier than for normal flowers (El Kassas *et al.*, 1992). Fruit set varies with flowering period (waves). It is, in general, high during the early flowering, gradually decreasing towards full bloom, then increasing again. Fruit retention (at harvest) increased from about 30% during early flowering to about 80% during full bloom and/or late flowering (El Sese, 1988; Hussein *et al.*, 1994). Flowers produced 4 to 5 weeks after the onset of blooming give the highest fruit set (90%) with the lowest fruit cracking and the best fruit quality. The pomegranate is described variously as self-pollinated, self and cross-pollinated, highly cross-pollinated or often cross-pollinated. Recent investigations demonstrated that it is capable of both open and self pollination. The use of a marker gene, governing the pigmentation of bud flower and petiole base, confirmed the fact that pomegranate (cvs 'Ganesh' and 'Kabul Yellow') is self- and, in low proportion (13%), cross-pollinated (Jalikop and Sampath Kumar, 1990). Trials conducted with several cultivars showed that fruit set was 79% and 43.3% for, respectively, intact open and self pollinated flowers or 26.4% and 66.2% for the same after emasculation (Karale *et al.*, 1993).

Red gram(Pigeon pea)

The flowers are borne in short recemes. The ovary is superior, subsessile, densely pubescent. Out of large number of flowers only 10% set fruits. Flowers are cleistogamous favouring self pollination. However unlike other legumes, considerable degree of out crossing takes place varying from 0.15 -7%.

Rubber

Male flowers are small in size but more in number than female flowers. Pollens are sticky. Pollination is by insects. When the inflorescence was enclosed in insect proof bags, no fruit set was observed.

Sweet cherry

Varieties Stella, Lapins, and Starkrimson are self fruitful. Most other varieties of Sweet cherry require cross pollination. Varieties like Bing, Lambert, and Royal Ann will not pollinate each other. In commercial orchards, beehives should be placed in the orchard on the first day of bloom.

Tomato

Tomato is typically a day neutral crop. It is mainly self pollinated as style grows up through the anther tube, and in some cases where style is short and self pollination is the rule. But a certain percentage of cross pollination also occurs depending upon variety and environment. Tomato is highly self pollinated due to homomorphism and chasmogamy.

References

Abou-Aziz, A.B., El Kassas, S.E., Boutros, B.N., El Sese, A.M. and Soliman, S.S. 1995. Flowering and fruit setting of Manfalouti pomegranate in response to soil moisture and irrigation regime. *Assiut Journal of Agricultural Sciences*, 26(1): 129-147.

Baldwin EA .1993. Citrus fruit. In: Seymour GB, Taylor JE,Tucker GA (eds), Biochemistry of Fruit Ripening,pp.107-149. Chapman & Hall, London.

Ben Arie, R.N., Segal, N. and Guelfat-Reich, S. 1984. The maturation and ripening of the Wonderful pomegranate.*J. Amer.Soc.Hort.Sci.*, 109(6):898 - 902.

Ben-Cheikh *et al.*, 1997. The pollinated fruitlet. *Braz. J. Plant Physiol.*, 19(4):333-362.

Bhattacharya,S.K. 2006. Advances in Ornamental Horticulture. Vol 2.

Cassin J, Bourdeaut J, Fougue A, Furon V, Gaillard JP,LeBourdelles J, Montagut G and Moreuil C 1969).The influence of climate upon blooming of citrus in tropical areas. Proc. 1st Int. Citrus Symp. 1:315-323.

Chaudhari, S.M. and Desai, U.T. 1993. Effects of plant growth regulators on flower sex in Pomegranate(*Punica granatum* L.) *Indian Journal of Agricultural Sciences* 63(1): 34-35.

Dabholkar, A. R. 2006. Reproductive systems in crop plants. *In* General Plant Breeding. Published by Ashok Kumar Mittal, Concept Publishing Company.New Delhi, India.

Davis FS and Albrigo LG .1994. Citrus. (In) Crop Production Science in Horticulture Series, pp.134-135. CAB International, Wallington.

El Kassas, S.E., Amen, K.I.A., Hussein, A.A. and Osman, S.M. 1992. Effects of certain methods of weed control and nitrogen fertilization on the yield, fruit quality and some nutrient contents of Manfalouty pomegranate trees. I. Flowering and fruit setting. *Assiut Journal of Agricultural Sciences,* 23(3): 183-198.

El Sese, A.M. 1988. Physiological studies on flowering and fruiting habits of some pomegranate cultivars under Assiut conditions. *Assiut Journal of Agricultural Sciences,*19(4): 320-336.

Frankel, R. and Galun, E. 1977. *Pollination Mechanism, Reproduction and Plant Breeding.* Monographs on TAG.Springer Verlag, Berlin.

Garcia-Luis A, Kanduser M, Santamarina P and Guardiola JL .1992. Low temperature influence on flowering in Citrus. The separation of inductive and bud dormancy releasing effects. *Physiol. Plant.* 86:648-682.

Giovannoni J .2004.Genetic regulation of fruit development and ripening. Plant Cell 16:170-180.

Goldschmidt EE and Huberman M .1974. The coordination of organ growth in developing Citrus flowers: a possibility for sink type regulation. *J. Exp. Bot.* 25:534-541.

Hussein, M.A, El Sese, A.M., El Mahdy, T.K. and Abd-El-Sabour, B. 1994. Physiological studies on thinning effects on the yield and fruit quality of Manfalouty pomegranate. A. Flowering behavior and sevin, NAA and hand thinning effects on fruit thinning, yield and certain fruit characteristics. *Assiut Journal of Agricultural Sciences,* 25(3):27-40.

Ibrahim, K.K. *et al.* 1985. *Indian Spices.* 22:21-25.

Jalikop, S.H. and Sampath Kumar, P. 1990. Use of a gene marker to study the mode of pollination in Pomegranate.*J.Hort.Sci.* 65(2): 221-223.

Kanakamany, N.T. 1982. *M.Sc.(Ag) Thesisi* Kerala Agricultural University, Trichur.

Karale, A.R., Supe, V.S., Kaulgud, S.N. and Kale, P.N. 1993. Pollination and fruit set studies in pomegranate. *Journal of Maharashtra Agricultural Universities.*18(3): 364-366.

Krishnamurthi, A. 1969. *The Wealth of India*: Raw Materials, 8: 350.

Martin, F.W. and Gregory, L.E. 1962. *Crop Sci.,* 3: 295-299..

Monselise SP .1985. Citrus and related species. (In) Halevy AH (ed), CRC Handbook of Flowering, Vol. 2,pp.275-294. CRC Press, Boca Raton.

Moss GI .1969. Influence of temperature and photoperiod on flower induction and inflorescence development in sweet orange (*Citrus sinensis* L. Osbeck). *J. Hort. Sci.*44:311-320.

Nambiar, P.K.V. *et al.* 1978. *J Plantn. Crops,* 6: 4 – 11.

Nebauer SG, Avila C, García-Luis A. and Guardiola JL .2006). Seasonal variation in the competence of the buds of three cultivars from different Citrus species to flower. *Trees* 20:507-514.

Peter, K.V. 1998. Genetics and breeding of vegetables. Indian Council of Agricultural Research, New Delhi.

Shulman, Y., Fainberstein, L. and Lavee, S. 1984. Pomegranate fruit development and maturation. *Journal of Horticultural Sciences* 59(2): 265-274.

Spiegel-Roy P, Goldschmidt EE .1996. Biology of Citrus.Cambridge University Press, Cambridge.

Southwick SM and Davenport TL .1986. Characterization of water stress and low temperature effects on flower induction in Citrus. *Plant Physiol.* 81:26-29.

Valiente JI, Albrigo LG .2004. Flower bud induction of sweet orange trees [*Citrus sinensis* (L.) Osbeck]: effect of low temperatures, crop load, and bud age. *J.Am. Soc. Hort. Sci.* 129:158-164.

Colour Plates

Chapter 1: Transgenic Ornamentals with Reduced Ethylene Production and Perception

Fig. 5 : Senescence of carnation flowers of the non-transformed 'Nora' (top) and the sACO-1 transgenic line (bottom). Flowers of both lines were harvested at the full-opening stage (day 0), and then left in a glass vessel with their cut stem end in distilled water. Photographs were taken 0, 5, 6, 7, 8, 9 and 10 days after the start of the experiment.

'Sei-Marine' No. 8 No. 18 No. 41

Fig. 12 : Leaf yellowing profiles of the control and putative transgenic lines of chrysanthemum after treatment with ethylene. In vitro plantlets of the control non-transformed 'Sei-Marine' and the lines of the mDG-ERS1(etr1-4) transformant were enclosed in a tight plastic container with ethylene at 10 μl l^{-1} and left for 6 days. According to the magnitude of leaf yellowing, the lines were classified into three groups of ethylene sensitivity; the lines with ethylene sensitivity similar to that of 'Sei-Marine' (ES), the lines with reduced ethylene sensitivity (RES1 lines showing yellowing in one or two of the lowermost leaves) and the lines with more reduced ethylene sensitivity (RES2 lines showing no leaf yellowing). In the photographs are shown the control 'Sei-Marine' and representatives of the ES, RES1 and RES2 lines, i.e., No. 8, No. 18 and No. 41 in this order.

Sei-Marine No. 33

Fig. 13 : Leaf yellowing profiles of the non-transformed 'Sei-Marine' control and a transgenic line No. 33 after treatment with ethylene. Shoots of 20-cm long were detached from soil-grown plants, enclosed in a glass container with 10 μl l^{-1} ethylene, and left for 9 days at 25°C under continuous white fluorescent light.

Sei-Marine No. 17 No. 33

Fig. 15 : Leaf senescence profiles of the non-transformed 'Sei-Marine' control and transgenic lines No.17 and No. 33 of chrysanthemum after natural senescence. Shoots of 20-cm long were detached from soil-grown plants, enclosed in a plastic bag and left for 1 week in the dark at room temperature.

Chapter 4: Climate Change Research Amenities with Special Emphasis on CO_2 Fertilization Effect

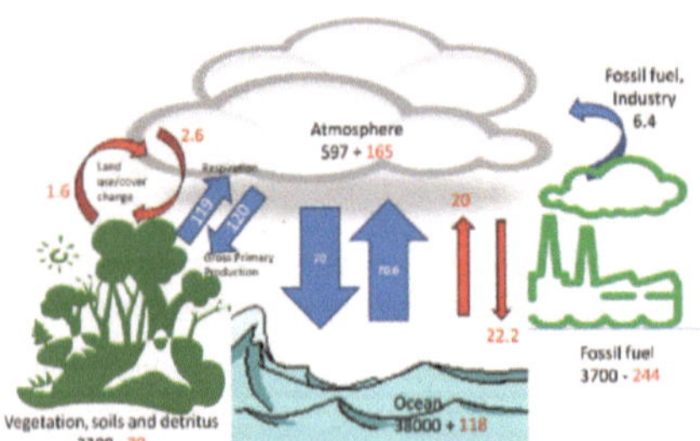

Fig. 1 : The carbon flux on Earth showing the reserve in GtC and fluxes in GtC yr^{-1}. The pre-industrial natural and the anthropogenic fluxes are represented by blue and red arrows, respectively (Adapted from Fig. 7.3, IPCC Fourth Assessment Report, 2007: Contribution of Working Group I)

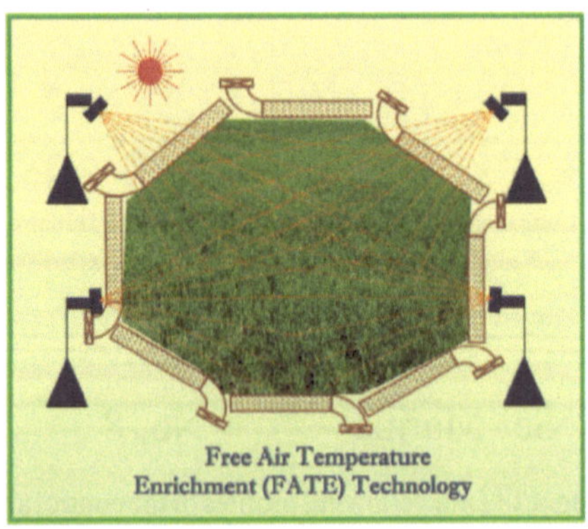

Fig. 8 : The Free-Air-Temperature-Enrichment technology (*Source :* Sheel Biotech limited)

Chapter 5: Advances in Molecular Breeding of Perennial Spice Crops

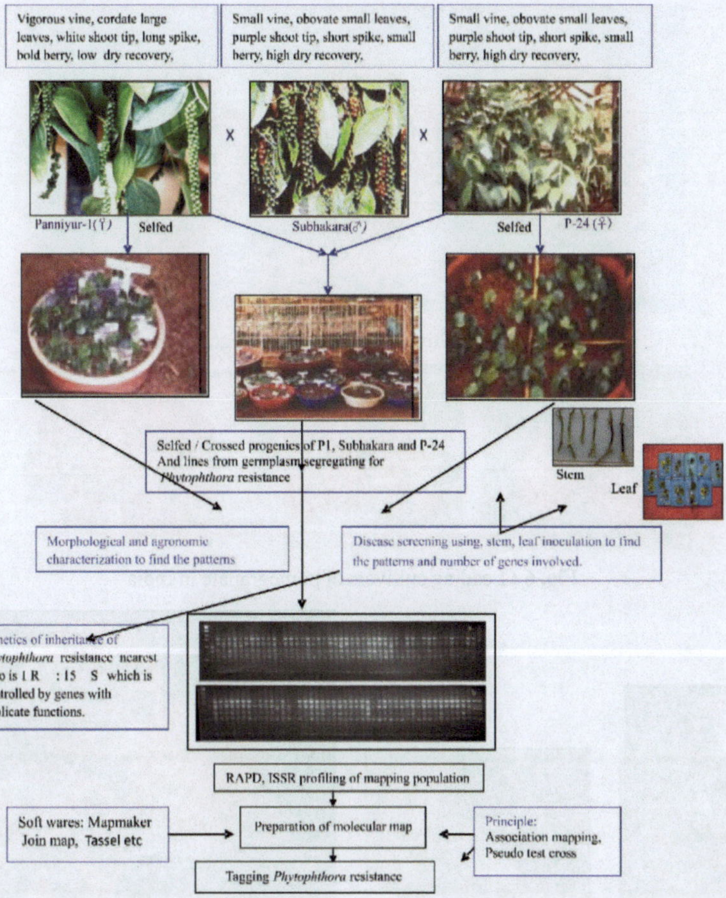

Fig 3 : Strategy for preparation of molecular map and tagging of *Phytophthora* resistance in Black pepper

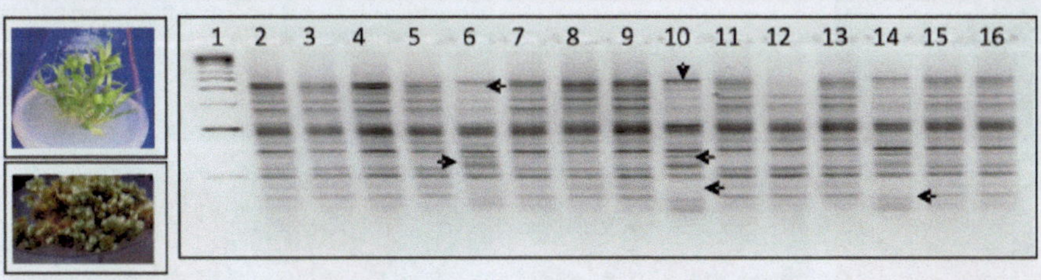

Fig. 6 : RAPD polymorphism in micropropagated plants of ginger. (Nirmal Babu *et al.* 2003)

Chapter 10: Postharvest Physiology and Technology of Pomegranate Fruit

Fig. 5 : Intercultivar variation in pomegranate arils

Fig. 6 : Leading cultivars of pomegranate in India

Fig. 8 : Harvesting operations in pomegranate

Fig. 9 : Surface abrasions and bruising during postharvest handling

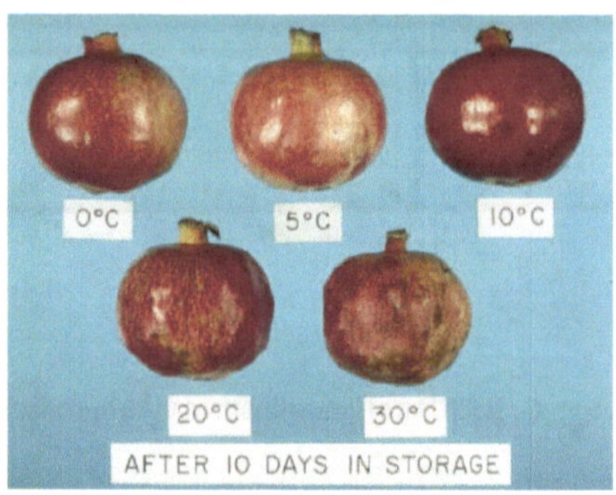

Fig. 14 : Water loss in pomegranate fruits at different temperatures

Fig. 16 : External symptoms of chilling injury

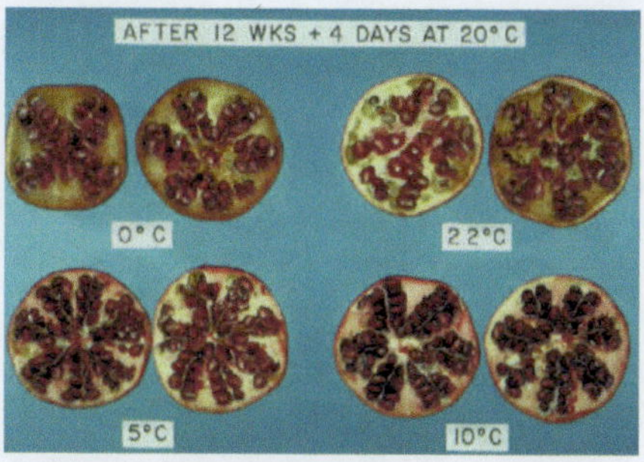

Fig. 17 : Effect of temperatures on chilling injury of pomegranates

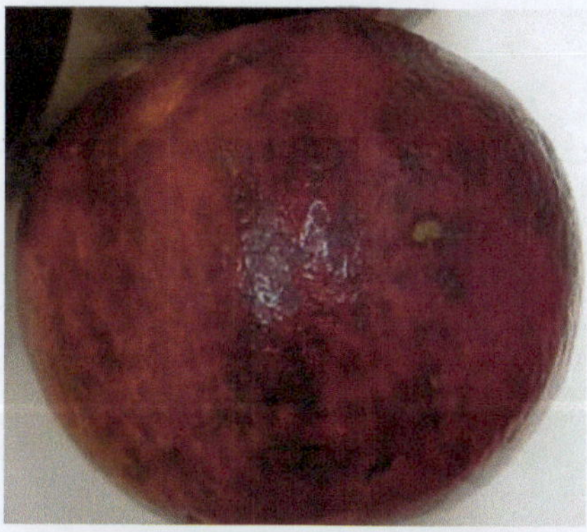

Fig. 18 : Superficial scald symptoms

Fig. 19 : Fruit cracking

Fig. 20 : Sunburn symptoms in pomegranate

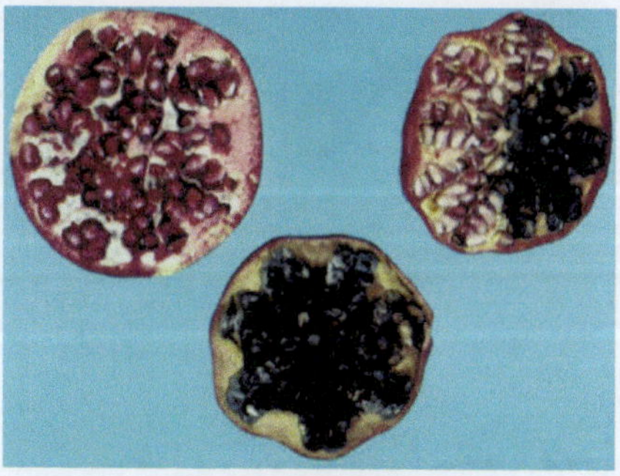

Fig. 21 : Alternaria rot in pomegranate

Index

A

B

D

E

M

N

O

P

Q

R

U

V

W

X